IET MATERIALS, CIRCUITS AND DEVICES SERIES 34

RF and Microwave Module Level Design and Integration

Other volumes in this series:

Volume 2	**Analogue IC Design: The current-mode approach** C. Toumazou, F.J. Lidgey and D.G. Haigh (Editors)
Volume 3	**Analogue–Digital ASICs: Circuit techniques, design tools and applications** R.S. Soin, F. Maloberti and J. France (Editors)
Volume 4	**Algorithmic and Knowledge-based CAD for VLSI** G.E. Taylor and G. Russell (Editors)
Volume 5	**Switched Currents: An analogue technique for digital technology** C. Toumazou, J.B.C. Hughes and N.C. Battersby (Editors)
Volume 6	**High-frequency Circuit Engineering** F. Nibler *et al.*
Volume 8	**Low-power High-frequency Microelectronics: A unified approach** G. Machado (Editor)
Volume 9	**VLSI Testing: Digital and mixed analogue/digital techniques** S.L. Hurst
Volume 10	**Distributed Feedback Semiconductor Lasers** J.E. Carroll, J.E.A. Whiteaway and R.G.S. Plumb
Volume 11	**Selected Topics in Advanced Solid State and Fibre Optic Sensors** S.M. Vaezi-Nejad (Editor)
Volume 12	**Strained Silicon Heterostructures: Materials and devices** C.K. Maiti, N.B. Chakrabarti and S.K. Ray
Volume 13	**RFIC and MMIC Design and Technology** I.D. Robertson and S. Lucyzyn (Editors)
Volume 14	**Design of High Frequency Integrated Analogue Filters** Y. Sun (Editor)
Volume 15	**Foundations of Digital Signal Processing: Theory, algorithms and hardware design** P. Gaydecki
Volume 16	**Wireless Communications Circuits and Systems** Y. Sun (Editor)
Volume 17	**The Switching Function: Analysis of power electronic circuits** C. Marouchos
Volume 18	**System on Chip: Next generation electronics** B. Al-Hashimi (Editor)
Volume 19	**Test and Diagnosis of Analogue, Mixed-signal and RF Integrated Circuits: The system on chip approach** Y. Sun (Editor)
Volume 20	**Low Power and Low Voltage Circuit Design with the FGMOS Transistor** E. Rodriguez-Villegas
Volume 21	**Technology Computer Aided Design for Si, SiGe and GaAs Integrated Circuits** C.K. Maiti and G.A. Armstrong
Volume 22	**Nanotechnologies** M. Wautelet *et al.*
Volume 23	**Understandable Electric Circuits** M. Wang
Volume 24	**Fundamentals of Electromagnetic Levitation: Engineering sustainability through efficiency** A.J. Sangster
Volume 25	**Optical MEMS for Chemical Analysis and Biomedicine** H. Jiang (Editor)
Volume 26	**High Speed Data Converters** A.M.A. Ali
Volume 27	**Nano-Scaled Semiconductor Devices** E.A. Gutiérrez-D (Editor)
Volume 29	**Nano-CMOS and Post-CMOS Electronics: Devices and modelling** Saraju P. Mohanty and Ashok Srivastava
Volume 30	**Nano-CMOS and Post-CMOS Electronics: Circuits and design** Saraju P. Mohanty and Ashok Srivastava
Volume 32	**Oscillator Circuits: Frontiers in design, analysis and applications** Y. Nishio (Editor)
Volume 33	**High Frequency MOSFET Gate Drivers** Z. Zhang and Y. Liu
Volume 35	**Design of Terahertz CMOS Integrated Circuits for High-Speed Wireless Communication** M. Fujishima and S. Amakawa
Volume 38	**System Design with Memristor Technologies** L. Guckert and E.E. Swartzlander Jr.
Volume 39	**Functionality-Enhanced Devices: An alternative to Moore's Law** P.-E. Gaillardon (Editor)
Volume 40	**Digitally Enhanced Mixed Signal Systems** C. Jabbour, P. Desgreys and D. Dallett (Editors)
Volume 42	**Radio Frequency and Microwave Power Amplifiers: Theory, design and applications** A. Grebennikov (Editor)
Volume 43	**Negative Group Delay Devices: From concepts to applications** B. Ravelo (Editor)
Volume 47	**Understandable Electric Circuits: Key concepts. 2nd Edition** M. Wang
Volume 58	**Magnetorheological Materials and Their Applications** S. Choi and W. Li (Editors)
Volume 60	**IP Core Protection and Hardware-Assisted Security for Consumer Electronics** A. Sengupta and S. Mohanty
Volume 68	**High Quality Liquid Crystal Displays and Smart Devices Vol. 1 and Vol. 2** S. Ishihara, S. Kobayashi and Y. Ukai (Editors)
Volume 69	**Fibre Bragg Gratings in Harsh and Space Environments: Principles and applications** B. Aïssa, E.I. Haddad, R.V. Kruzelecky, W.R. Jamroz

RF and Microwave Module Level Design and Integration

Mohammad Almalkawi

The Institution of Engineering and Technology

Published by The Institution of Engineering and Technology, London, United Kingdom

The Institution of Engineering and Technology is registered as a Charity in England & Wales (no. 211014) and Scotland (no. SC038698).

First published 2019

The Institution of Engineering and Technology
Michael Faraday House
Six Hills Way, Stevenage
Herts, SG1 2AY, United Kingdom

www.theiet.org

British Library Cataloguing in Publication Data
A catalogue record for this product is available from the British Library

ISBN 978-1-78561-359-3 (hardback)
ISBN 978-1-78561-360-9 (PDF)

Typeset in India by MPS Limited
Printed in the UK by CPI Group (UK) Ltd, Croydon

To my parents, my wife Reema, and
my daughters Dareen and Aleen

Contents

List of figures

List of tables

Preface

This book is intended to bring together fundamental concepts and elements needed in the design, integration, and characterization phases of modern radio frequency (RF) and microwave modules and to familiarize the interested reader with the common challenges in this rapidly changing field.

While design for manufacturability (DFM) and design for testability (DFT) are two important practices that stress the need for careful attention at the early design stages of a product, module integration to achieve optimal RF performance is an art form that takes years to master. In this context, the objective of this book is to provide the necessary practical and theoretical background needed to tackle a multitude of issues accompanied by the design of contemporary RF modules. The uniqueness of this book among others resides in its inclusion of diverse aspects of RF module design and analysis concepts under one cover. The topics covered herein can serve a broad range of individuals: those just entering the field, graduate-level students, and researchers, as well as practicing engineers seeking new domains to expand their knowledge, understand design tradeoffs, and looking for a quick guide in digesting the several aspects of the RF module design processes and their common challenges.

The book comprises eight chapters that are to some extent independent but consequential. Each chapter is self-contained with a list of references for more insight into the literature. The material in this book is organized as follows.

Chapter 1 preludes with an overview on the existing packaging techniques regarding their developments, challenges, and tradeoffs which are essential factors in the decision-making process when selecting the proper packaging technology for a given device or a system. Different types of substrate technologies and module assemblies are discussed. Also, important steps in the microelectronics fabrication and characterization processes are presented.

Chapter 2 deals with the distributed and lumped elements used in the design, modeling, and implementation of RF and microwave circuits. This includes resistors, capacitors, and inductors in different configurations and implementation forms. Other derivative elements and subjects such as electrostatic discharge (ESD), mutual inductance, and via holes are treated in this chapter. Usage and applicability of lumped versus distributed reactive elements and several of their design considerations are also discussed. Common miniaturization techniques applied to distributed elements and resonators are described. Several Q-factor calculation methods and studies on Q-factor improvement of integrated inductors are elaborated through the course of this chapter.

Chapter 3 is devoted to reviewing several port analysis techniques and their essential properties. Some of the analysis methods are not applicable for solving high-frequency microwave circuits while others can accurately cover both low- and high-frequency circuits. Therefore, the port analysis methods in this chapter are categorized based on the analysis frequency of interest into voltage–current-based parameters and traveling-wave-based parameters. Linear equivalent circuit models of a two-port network are discussed. Then, canonical forms for the realization of one-port lossless networks and the Darlington synthesis of two-port networks using the pole-zero extraction method are explained.

Chapter 4 covers impedance matching techniques and circuit transformations applicable for use in RF microelectronics and distributed monolithic microwave integrated circuits (MMICs). This chapter provides an outlook on the resulting limitations due to selecting a specific matching configuration over another.

Chapter 5 begins with different aspects of electromagnetic (EM) coupling mechanisms. The emphasis is on explaining the different EM coupling types, their properties, and when such coupling is considered to be a desired or an undesired phenomenon. Part of this chapter is allocated to discuss symmetrical and asymmetrical coupled transmission line concepts which are essential to the realization of a large number of MMICs and multichip modules (MCMs). Laminate design considerations and EM shielding methods to mitigate unwanted EM coupling are also addressed.

Chapter 6 discusses a number of computer-aided design (CAD) tools and their methodologies that are widely used in the design, analysis, and modeling domains of electronic circuits and modules. This chapter presents basic and advanced principles of system-level modeling and simulation that allow analysis of complex RF systems. This encompasses computational electromagnetic methods (CEMs), modified nodal analysis (MNA), surrogate modeling methods based on machine learning algorithms, and co-simulation methods between linear and nonlinear, thermal and electrical, and acoustical and electrical domains. Topics in power handling, thermal management, electromagnetic compatibility (EMC) analysis, and the use of Six Sigma and Monte Carlo tolerance analysis in the context of analyzing production yield data and circuit sensitivity are also covered.

Chapter 7 summarizes the main components used in the realization of modern RF and microwave front-end modules and discusses the usage and applicability of each component based on their intrinsic performance capabilities. Also, a portion of this chapter highlights the market growth potential for some of the key components over the next few years. The discussed components include transistor power amplifiers (PAs) and their different classes of operation based on their biasing conditions; low noise amplifiers (LNAs); filters and multiplexers with focus on the high performance acoustic wave (AW) filters and their configurations; switches; phase shifters; oscillators, mixers; complementary metal–oxide semiconductor (CMOS) controllers; baluns; circulators and isolators; power limiters; and the concept of antenna diversity and its benefits.

Chapter 8 covers different measurement setups and definitions that are commonly used in the characterization of RF and microwave modules including small- and large-signal measures. These are only representative examples of common setups and do not represent a comprehensive list of possible configurations. Important measures that characterize the performance of modern wireless transceiver modules such as noise figure, sensitivity, de-sense, transmitter linearity, intermodulation distortion, adjacent channel leakage ratio, error vector magnitude, and electromagnetic compatibility, and how other submodules in a system could impact them are discussed. Advantages of on-wafer RF measurements are also illustrated, especially, in the case of mass produced products.

The book is expected to serve the immediate future needs, and several aspects of the book will continue to be of interest. It is written in the simplest manner possible while retaining the necessary technical complexity. I hope this book satisfies readers' interest and provide the knowledge they seek in the area of RF module design and integration.

Mohammad J. Almalkawi
Skyworks Solutions, Inc.
Greensboro, NC, USA
January 2019

Acknowledgments

I would like to express my sincere appreciation to those whom I have had the opportunity to work with during my professional career and the teachers whom I learned from throughout my educational journey.

As with any reasoning-based approach, this book has been built upon the works and efforts of a large number of researchers and authors, and I have been fortunate to benefit from their work. I would like also to express my appreciation for the knowledge management ecosystem between the authors and the publishing houses that facilitate access to specialized knowledge.

The completion of this book has depended on a number of very supportive individuals in a variety of ways in my workplace at Skyworks Solutions, Inc. I would like to thank the management, and my colleagues at Skyworks Solutions Josh Caron and CS Lam for their support and encouragement. Also, I am grateful to Tami Stegmaier and Jennifer Bayot from the Corporate Communications at Skyworks Solutions who carefully revised my manuscript.

I must also thank the book Editorial Board and the staff of the publisher, the Institute of Engineering and Technology (IET), for their strong support, cooperation, and meticulous editing and formatting services. I am especially indebted to the professionals at IET, Paul Deards, Editor for the IET Academic Books, and Nicki Dennis, Consulting Editor, in getting this project started, the reviewers at IET for their effort in outlining the book material, and Olivia Wilkins, Assistant Editor, for keeping track of the book progress.

Finally, I would like to thank my wife, Reema, and my daughters, Dareen and Aleen for their patience, love, and support during the course of developing the book material. This book would not have been possible without them.

List of abbreviations

2D	two-dimensional
3D	three-dimensional
3G	third generation
3GPP	third generation partnership project
4G	fourth generation
5G	fifth generation
ABC	absorbing boundary condition
ADS	advanced design system
AM–AM	amplitude modulation–amplitude modulation
AM–PM	amplitude modulation–phase modulation
ANN	artificial neural network
AC	alternating current
ACLR	adjacent channel leakage ratio
ACPR	adjacent channel power ratio
APT	average power tracking
ASIC	application-specific integrated circuit
AW	acoustic-wave
BAW	bulk acoustic wave
BEM	boundary element method
BER	bit error rate
BGA	ball grid array
BJT	bipolar junction transistor
BPF	bandpass filter
C4	controlled collapse chip connection
CA	carrier aggregation
CAD	computer-aided design
CDF	cumulative distribution function
CEM	computational electromagnetic
CLT	central limit theorem

CMOS	complementary metal oxide semiconductor
CMRR	common mode rejection ratio
CPW	coplanar waveguide
CQFP	ceramic quad flat pack
CSP	chip scale packaging
CTE	Coefficient of thermal expansion
CVD	chemical vapor deposition
CW	continuous wave
dB	decibel
DC	direct current
ET	envelope tracking
DFM	design for manufacturability
DFT	design for testability
DFN	dual flat no lead
DI	deionized
DIP	dual inline packages
DPD	digital predistortion
DPMO	defects per million opportunities
DUT	device under test
EM	electromagnetic
EMI/EMC	electromagnetic interference/electromagnetic compatibility
ESD	electrostatic discharge
ET	envelope tracking
EVM	error vector magnitude
FCBGA	flip-chip ball grid array
FDD	frequency division duplex
FDTD	finite difference time domain
FEM	finite element method
FET	field-effect transistor
FIT	finite integration technique
FOM	figures of merit
FPGA	field-programmable gate array
GaAs	gallium arsenide
GaN	gallium nitride
GEMS	general electromagnetic solver
GHz	gigahertz

GPOF	generalized pencil of function
GUI	graphical user interface
HB	harmonic balance
HBM	high power microwave
HDI	high density interconnect
HFSS	high-frequency structure simulator
HPUE	high power user equipment
HTCC	high temperature co-fired ceramic
IC	integrated circuit
IE	integral equation
IF	intermediate frequency
IL	insertion loss
IMD	intermodulation distortion
IIP3	input third-order intercept point
KCL	Kirchhoff's current law
KVL	Kirchhoff's voltage law
LNA	low noise amplifier
LO	local oscillator
LTE	long-term evolution
LPF	low-pass filter
LTCC	low temperature co-fired ceramic
MBVD	modified Butterworth-Van-Dyke
MHz	megahertz
MIM	metal-insulator-metal
MIMO	multiple-input multiple-output
MIPI	mobile industry processor interface
MLFMM	multilevel fast multipole method
MNA	modified nodal analysis
MoM	method of moments
MQFP	metal quad flat pack
NF	noise figure
OEM	original equipment manufacturer
PA	power amplifier
PCB	printed circuit board
PDF	probability density function
PDK	product development kit

PEC	perfect electric conductor
PGA	pin grid array
PML	perfectly matched layer
PVD	physical vapor deposition
PVS	polymer voltage suppression
PPM	parts per million
PQFP	plastic quad flat pack
PWM	pulse width modulation
QFN	quad flat no lead
QFP	quad flat package
R&D	research and development
RBC	radiation boundary condition
RBF	radial basis function
RF	radio-frequency
RFIC	radio-frequency integrated circuit
RFFE	RF front-end
RL	return loss
RMS	root mean square
SAM	system assembly modeling
SAW	surface acoustic wave
SDC	silicon daisy chain
SEM	scanning electron microscope
Si	silicon
SI	signal integrity
SiP	system in package
SISO	single-input and single-output
SIW	substrate integrated waveguide
SMD	surface mount device
SMR	solidly mounted resonator
SNR	signal-to-noise ratio
SOS	silicon on sapphire
SOI	silicon on insulator
SoC	system on chip
SoP	system on package
SPDT	single pole, double throw
SPICE	simulation program with integrated circuit emphasis

STC	space-time coding
SUMO	surrogate modeling
SVM	support vector machine
TAM	total available market
TBC	transparent boundary condition
TCV	temperature coefficient of velocity
TDD	time-division duplexing
TEM	transmission electron microscopy
TFBAR	thin film bulk acoustic resonator
THD	total harmonic distortion
TL	transmission line
TOI	third-order intercept
TSV	through-silicon via
TVS	transient voltage suppressor
TQFP	thin quad flat pack
UPML	uniaxial perfectly matched layer
VLSI	very-large-scale integration
VNA	vector network analyzer
VSG	vector signal generator
WLP	wafer level package
WPD	Wilkinson power divider

Chapter 1

RF and microwave device and module packaging

1.1 Introduction

The proliferation of portable devices and the rapid evolution of wireless communication technologies has led communication system integrators to continually call for reduced size, light-weight, high speed, and complex multifunctional products. As a consequence, giant chipmakers are encountering an ever-growing demand on developing processes and packaging technologies that allow a high level of component integration at reduced manufacturing cost in addition to the stringent requirements for competitive performance radio frequency (RF) modules. Yet, ensuring reliable operation of electronic systems is becoming increasingly important owing to the rising dependence of our daily life and economic structure of our society on modern electronics. All of these requirements necessitate that design engineers have to perform detailed analysis and testing in advance before committing to mass production. The fact that several RF subsystems are eventually going to operate in conjunction with other diverse modules and the lack of insight regarding the resultant system performance entails performing a thorough evaluation to ensure overall system compatibility, a factor that adds up to the manufacturing cost. In some application-specific integrated circuits (ASICs) and modules that require unavoidable several redesign cycles, engineering cost can be justified only for large production quantities. Realizing the significance of the subject (i.e., efficient RF and microwave module design and compliance with other integrated systems) semiconductor device manufacturers continue to make huge investments in research and development (R&D) and despite the advances in industry design flows and in-house developed tools, ensuring reliability and high yield of integrated RF modules remains an outstanding challenge.

1.2 Packaging technologies

The recent advances in microelectronics process and packaging technologies have enabled the integration of complex RF electronic systems on single chips as well as on compact size modules. Each application has different requirements, and there are many considerations to be taken into account when selecting the best packaging technology for a particular application. Knowing that the smallest possible package will always be the size of the chip itself, chipmakers are in a race to stay ahead of the competition by trying to achieve a small form factor packages close to the chip size as it directly impacts the business revenue. However, shrinking the chip size contributes significantly to its electrostatic discharge (ESD) sensitivity, limits its power handling capability, and exacerbates parasitic effects especially in passive devices. Moreover, the higher the chip pin count and the smaller pad pitch, the higher the possibility of unwanted EM coupling between them. In high precision performance devices, a small amount of unwanted coupling through the device package can lead to serious isolation problems between the input and output peripherals of the device. This is commonly feasible in the case of selective high precision bandpass filter designs [1] where coupling can deteriorate the attenuation level at the out-of-band rejections resulting in a violation of design specifications and eventually a product yield loss. Therefore, it can be concluded that performance, cost, and size are the main market drivers in the radio-frequency integrated circuits (RFICs) domain.

Several packaging solutions have been developed for the semiconductor microelectronic devices either for chip or module level. The role of these packages is to provide enclosure or mechanical support necessary for transporting, handling, and assembling the device. Packaging is also needed to provide a heat transfer mechanism for better thermal performance, increase the lifetime of the product by protecting the semiconductor device from ionic contamination due to moisture paths and sources, and create electrical connections from a die component to the outside circuits. In general, packaging allows a higher level of component integration and depending on the technology in use it can lead to significant cost reduction in the case of mass production.

Another use of packages which may not appear of interest to the design engineers is packaging samples of each particular technology that are used to construct various mechanical samples referred to as silicon daisy chain (SDC) samples which are mainly used for mechanical and process equipment set-up and evaluation. This type of analysis allows packaging engineers to improve assembly processes and build good knowledge regarding possible causes of failures which in some way can be important information to the design engineers.

An RF module package enclosing multichips implemented using hybrid technologies can contain diverse material types such as semiconductors, high conductivity alloys, insulators, and organics. The selection of the packaging materials depends on multiple factors, for example, the package geometry, thermal expansion coefficient, thermal conductivity, and whether the required package is hermetic or nonhermetic. Packaging solutions for RF devices are available in two main categories in terms of the medium encapsulating a die component or a module. These two solutions include molded and air cavity packages. Depending on the medium in use and the design tolerance, it's possible that the selection of encapsulating material may impact the device response. Materials with high moisture absorption rate can lead to what is called a "popcorn" effect when moisture collects and explodes in the assembly process. Additionally, lightweight and low-cost packaging solutions are usually made of polymeric materials rather than ceramic and/or glass. Such mold material comes into direct contact with the die whether a standalone die or serval dies in a module, and hence these polymeric materials exhibit a higher dielectric constant than air. Therefore, a molded package will experience higher parasitics that can limit the device performance and impact the ruggedness of the design. On the other side, a mold material can increase the thermal conductivity and aid as an additional path for heat transfer. From cost and process complexity perspective, a molded packaging solution is a preferred low cost and simple packaging option than air.

Sections 1.2.1–1.2.5 will discuss commonly used packaging technologies in the semiconductor industry and highlight their trade-offs when it comes to choosing the suitable packaging solution for a particular application.

1.2.1 System on chip

The introduction of the system on chip (SoC) packaging in the mid-1990s was mainly to address issues related to the increasing delay of the off-chip interconnects. SoC is a single chip that integrates multiple functionalities under its package. Modern semiconductor fabrication technologies are capable of integrating devices on a single chip with little or no additional fabrication steps. However, SoC carries with it some challenges in achieving high yield for the entire system since a SoC may be subject to EM interference between its integrated components and requires relatively longer design cycles to obtain optimal performance. Additionally, RF front-end (RFFE) components such as switches and filters are difficult to implement using standard silicon technologies.

1.2.2 System on package

System on package (SoP) comprises of a single package that provides an RF solution by integrating RF passives (RLC elements), antenna, complementary metal oxide semiconductor (CMOS) chips, and other analog chips that might as well have its own packaging technology. Integration of SoC is well handled in SoP since RF-passive components such as capacitors and high-Q inductors can be fabricated on the substrate of the package. Another advantage of using SoP is that it allows low latency RF solutions since shorter interconnect paths are viable and enables more functionality to the package. Under a single SoP the added functionality can comprise embedded mixed-signal (i.e., digital and analog) devices such as switches, couplers, filters, field-programmable gate arrays (FPGAs), antennas, and optoelectronic sensing components such as lasers and optical-detectors as depicted in Figure 1.1. The most important factor is that a SoP provides a low-cost solution when compared to its counterpart SoC [2] by decreasing the need for discrete components and thereby reducing the assembly time and total package size.

Challenges associated with this type of packaging technology can include the electromagnetic interference and electromagnetic compatibility (EMI/EMC) between the added chipsets and components, difficulty in having clear insight into the possible design trade-offs between its integrated components in order to obtain an optimal system-level performance, and potential thermal management issues due to the increased number of operating devices within the confined space of the package.

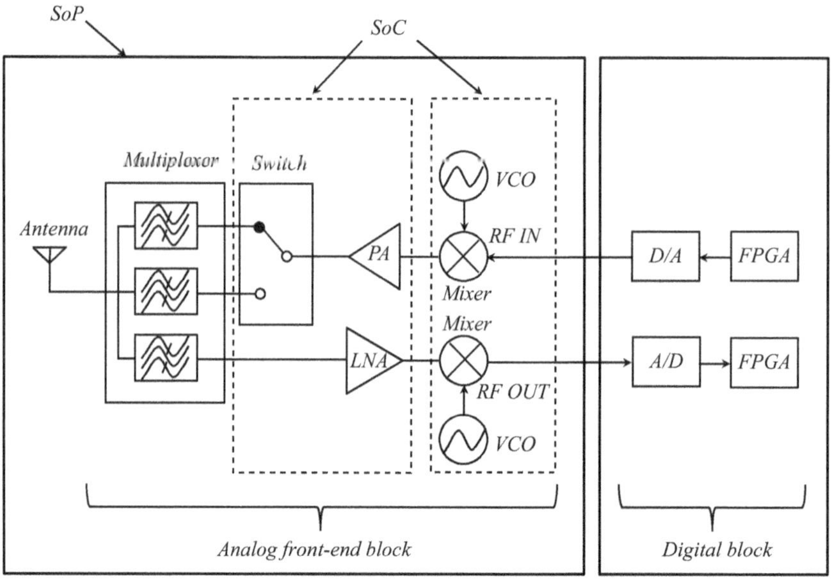

Figure 1.1 An RF transceiver module illustrating an SoC in SoP

1.2.3 System in package

Stacked ICs, stacked packages, and/or stacked wafers enclosed under a module package form what is known as a system in package (SiP). SiP is advantageous over SoC and SoP by virtue of its smaller form factor. This is arising from the fact that an SiP requires relatively less number of components and routings between the devices to achieve the intended functionality, which results in low cost, low latency solution, and allows a high level of circuit integration. However, the complexity of SiP imposes a set of design challenges embodied in the multitude possibilities that an SiP layout can take, the several device stacking approaches that can be employed, and the difficulty in analyzing the module-level performance using multidomain design and optimization methods. 3D-IC integration also faces challenges related to the thermal management of the die stacks as the power density per unit volume increases significantly if compared to 2D populated packages [3]. Moreover, laboratory evaluation and performance troubleshooting may be challenging due to difficulties in accessing and testing the components or RF paths of interest.

Figure 1.2 illustrates a SiP solution that encompasses stacked-die chip scale packages (CSPs) using wire bond connections and flip-chips instead of placing the individual dies side by side. The three dies in Figure 1.2(a) are placed on top of each other to minimize the horizontal space on a printed circuit board (PCB). Figure 1.2(b) shows three wafers stacked together before dicing them and represents a different form of an SiP.

According to Moore's Law, the number of transistors in very-large-scale integration (VLSI) technology is doubling every 1.8 years, leading to a growing demand for methods and techniques to resolve bottlenecks in the domain of

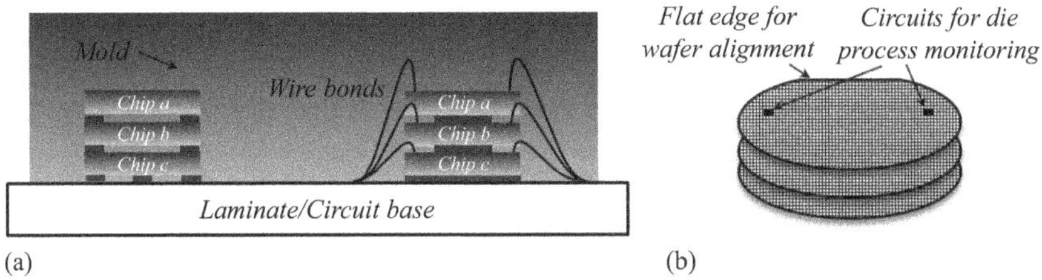

Figure 1.2 An SiP illustrating (a) vertically stacked dies using chip-level and external wire bond interconnects, and (b) vertically stacked wafers

interconnect modeling, layout construction, analysis, and testability. In a highly dense, sophisticated system, off-chip interconnects may potentially impact the system performance through increased insertion loss, signal delay, noise crosstalk, and signal integrity problems. As a result, vertical chip integration has been introduced as a potential solution to extend the capabilities of modern SoC by enabling multiple wafers bonding or bare dies as an alternative solution for next-generation designs in order to achieve improved electrical performance, reduced power consumption, and smaller form-factor realized by shortening the average lengths of the chip-to-chip and the intrachip interconnections. This can be feasible by using one of the key interconnect solutions, a through-silicon via (TSV) technology, which enables vertical electrical connectivity.

1.2.4 Wafer level packaging

Traditionally, the process of packaging individual units after dicing them from a wafer has been for several decades the norm for packaging semiconductor integrated circuits. Such an approach, however, is not currently adopted by major semiconductor manufacturers due to the high manufacturing cost associated with the increasing RF content in today's modules. For this reason, the advent of wafer level packaging (WLP) has brought a paradigm shift in the development of low-cost packages. WLP is a technology of packaging devices at the wafer level before dicing them. WLP is performed as an extension of the wafer fabrication process using standard tools and processes. Eventually, the manufactured WLP die will have metalized pads on the surface of the chip and solder dots deposited on each of the pads before dicing the wafer. This, in turn, makes WLP compatible with traditional PCB assembly processes and allows device testing to be performed on the wafer itself. Hence, it is a relatively low cost and efficient process, especially when the wafer size increases and as the die shrinks. The wafer size has been increasing over the past few decades progressing from 4, 6, 8 to 12 inches in diameter. This results in an increase in the number of dies per wafer and thus a lower manufacturing cost. In terms of electrical performance, WLP is advantageous over other packaging technologies, in a sense that it could lead to significantly reduced EM parasitic couplings once WLP devices are integrated in a dense RF module because of the relatively shorter interconnects between the devices and the PCB, unlike using wire bonds in some types of the CSP technology.

Sections 1.2.4.1 and 1.2.4.2 discuss two types of WLP technology commonly used in today's semiconductor device manufacturing.

1.2.4.1 WLP flip-chip technology

Flip-chip technology is also known as controlled collapse chip connection (C4) and is one of the chip assembly technologies developed by IBM in the 1960s [4]. Although packaging technologies based on wire bonding are more flexible in terms of providing free inductors for laboratory tuning after the hardware is built and also provides good thermal conductance, the use of solder bumps to make electrical connection between the substrate and the chip using a flip-chip packaging technology provides relative size reduction, reduced latency, and better isolation between its input and output pins. Figure 1.3 illustrates the basic structure of a flip-chip die on a substrate with solder balls on top of Cu pillars grown on the chip surface. The solder joints are usually encapsulated by an under-fill mold compound to provide mechanical support to the solder joints.

1.2.4.2 WLP chip scale packaging

CSP has been one of the most adopted packaging methods across the microelectronic and semiconductor industries. Although there are several types of CSP technologies that are already available and in use by the microchip manufacturers, new types continue to appear in response to the demand for products that support new functionalities and new specific applications. These packaging requirements may vary depending on the desired reliability level, cost, added functionality, and overall size. CSP, as the name implies, has a package size that is approximately the same as the die size which is one of its main advantages. CSP is continually evolving to achieve the smallest possible package-to-die size ratio by employing the WLP manufacturing process. The ball grid array (BGA) style of the CSP interconnects under the package as illustrated in Figure 1.4 allows having several interconnects while simplifying PCB routing, increasing PCB assembly yield, and reducing manufacturing costs.

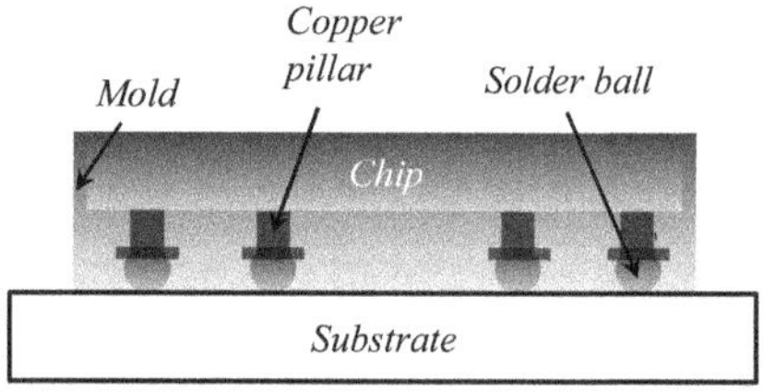

Figure 1.3 Example of basic structure of a flip-chip

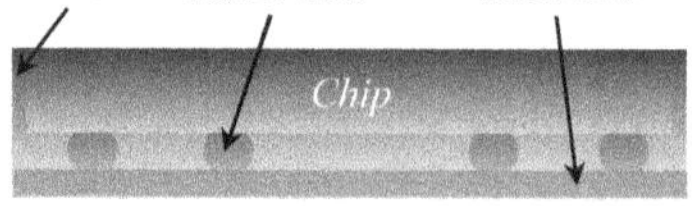

Figure 1.4 Example of basic structure of a CSP

1.2.5 Other packaging technologies

Other IC packages are also available in several forms, allowing seamless integration into customized application and module packages. Quad flat package (QFP) is one of the earliest surface mount IC packaging technologies where the structure of the package is formed of four sides with extended leads for interconnections as depicted in Figure 1.5(a). The protruded leads are attached to the package frame forming a metal-insulator-metal (MIM) type capacitor between the lead and the die metal that can be exploited as a matching element. This technology is suitable for millimeter size ICs where the number of periphery pins of the package can reach to more than 100 pins. Several derivatives do exist for this type of packaging depending on the material being used such as ceramic quad flat pack (CQFP), thin quad flat pack (TQFP), plastic quad flat pack (PQFP), and metal quad flat pack (MQFP).

Quad flat no lead (QFN) shown in Figure 1.5(b) is one of several surface mount packaging technologies that are formed of plastic packages with planar copper lead frame and heat spreader pad used as a heat sink. Wire bonding can be used for interconnection, and since wire bonds are not just conductors but also inductors, they usually impact the performance of the packaged device under this technology unless it is accounted for as part of the design. While QFN is formed of four sides for interconnections, dual flat no lead (DFN) is also available and formed of two sides of interconnecting planar leads.

All packaging technologies discussed in this subsection, however, are not applicable for use in the contemporary microelectronic SoP and SiP. Their use has boomed during the revolution of VLSI in the 1990s and decayed with the advent of more compact and dense wafer-level packaging technologies. The evolution of chip and module packaging technologies over the past few decades was driven mainly by the increasing requirements in pin counts and has witnessed a tremendous increase in the functionality, component density, and integration levels. Figure 1.6 shows the IC packaging evolution since 1970 where dual inline packages (DIP)

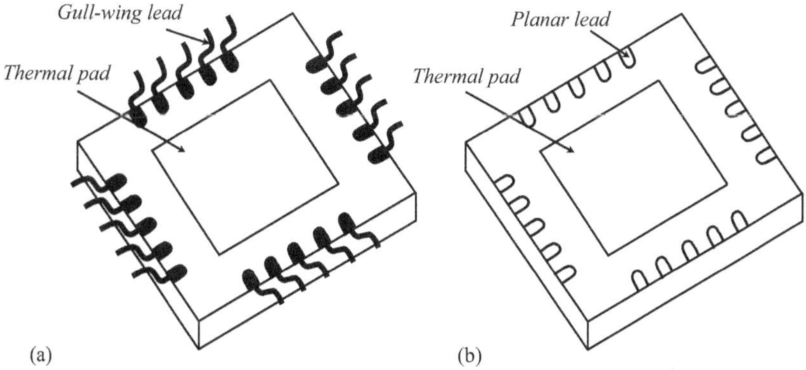

Figure 1.5 Integrated circuit packaging technologies: (a) quad flat package and (b) quad flat no lead

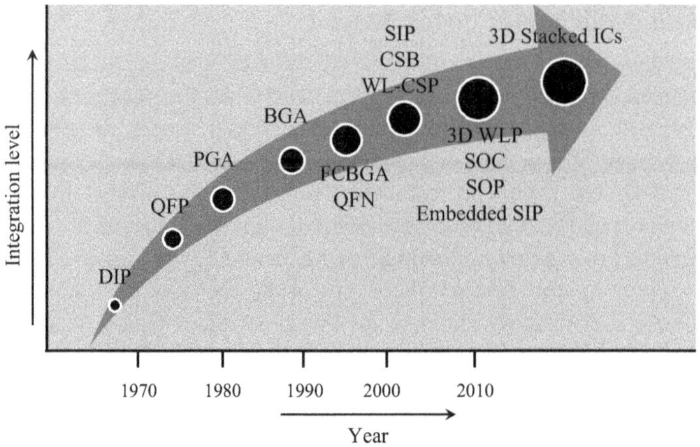

Figure 1.6 Evolution of IC packaging technologies

started to play a role in the electronics IC industry, then ICs with more pins started to appear such as QFP and even higher pin count technologies such as pin grid array (PGA) and flip-chip ball grid array (FCBGA).

1.3 Microelectronics fabrication

Microelectronics fabrication and component assembly are critical steps in the production chain of any RF product since it determines the product time-to-market and the number of redesign cycles that can be carried out over a limited product development time. Microelectronics fabrication is challenging due to the small size of the fabricated components and the features it can include which makes microelectronics fabrication vulnerable to environmental, thermal, and electrical conditions. Understanding these challenges and many others has led to the emergence of microelectronics process engineering as an independent discipline to cope with the challenges and rapid advances in this area.

Starting from a CAD design and eventually ending with a working device requires many fabrication steps and procedures to be followed. For high volume production and targeting high yield, these steps need to be thoroughly validated in order to provide consistent results whereupon the process can be locked for this particular device. Process experiments can include several trials of materials selection, methods to grow material layers, layer-to-layer alignment methods, and material patterning methods. Each of which can be different from one manufacturer to another making plenty of scope for anyone to invent new ones. To cope with the complexity of microelectronics fabrication, individuals with diverse backgrounds (e.g., physics, chemical, electrical, mechanical engineering, and material science) and training skills on latest technologies and equipment are typically involved to work in synergy to make such process successful.

1.3.1 Photolithography fabrication

Photolithography is the process of transferring patterns from a layout or a mask onto a surface of a semiconductor wafer by means of optical waves. The transfer mechanism of the patterns starts by using a light-sensitive polymer material called photoresist patterned with shapes defined on a mask. Once the pattern is transferred from the mask to the photoresist, dry or wet etching method is then used to transfer the pattern from the photoresist onto the surface of the wafer. The process itself consists of several steps and requires several tools to be established [5]. Basic photolithography steps are summarized in the chart illustrated in Figure 1.7 with a brief explanation of each step.

Wafer cleaning is usually conducted through a standard set of steps and starts by running the wafer into a solvent to clean particles and organic residues that may exist on

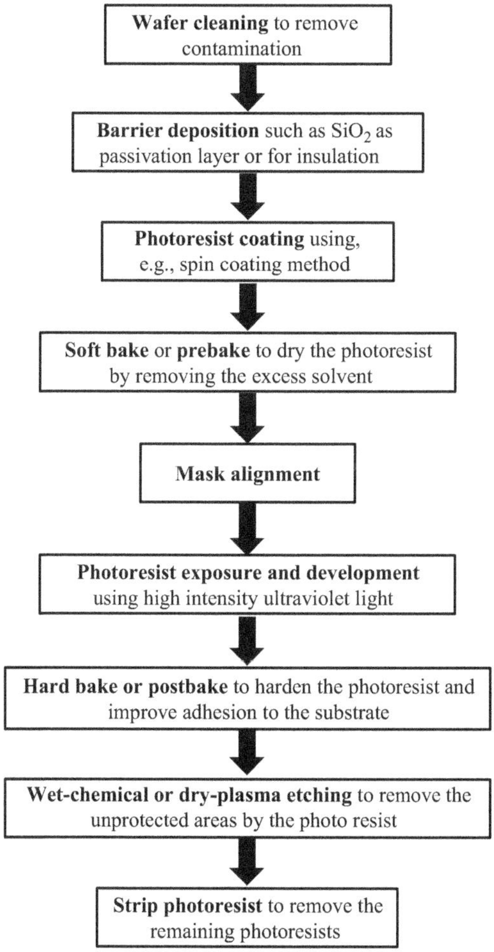

Figure 1.7 Basic steps involved in the photolithographic fabrication process flow

the wafer surface. Next, wafers are thoroughly soaked in deionized water (DI) to remove any remaining solvent, and then followed by RCA clean. The RCA clean oxidizes the wafer providing a thin protective layer of oxide to the surface of the wafer. The wafers are then dipped in hydrofluoric acid (HF) to remove the oxide layer from the wafer surface. Finally, wafers are DI rinsed and then blown dry with nitrogen.

Deposition of thin film materials is usually conducted using either chemical vapor deposition (CVD) or physical vapor deposition (PVD). As a concept, both coating processes are very broad. More specific and derivative techniques are usually used by the semiconductor manufacturers. PVD uses a pure source material that is gasified via evaporation. The PVD heat supply is accomplished by the application of high power electricity or laser ablation. The vaporized material will then diffuse through a vacuum to the substrate surface to create the desired film. However, CVD doesn't require pure source material, but the material is mixed with a volatile precursor as a material carrier. The coating material is then injected through a gas delivery system to a reaction chamber at a certain temperature that encloses the substrate. The vapored material then reacts with the substrate and eventually deposits on it. At this stage, the precursor material decomposes and leaves the material deposited in the substrate. CVD processes require much higher temperatures than PVD processes. Depending on the application, one method may be preferred over the other due to several manufacturing aspects such as process complexity, desired film thickness, availability of source material, and cost reasons. To this end, after cleaning the wafer, barrier or passivation layer such as silicon dioxide (SiO_2) is usually deposited on the surface of the wafer.

A photoresist layer is then applied to the surface of the wafer utilizing a standard spin-coating technique whereby high-speed spinning of the wafer substrate with several thousand rounds per minute (rpm) generates a centrifugal force capable of uniformly distributing the photoresist layer on the wafer surface. Depending on the reaction to the applied light, the photoresist can be classified into positive or negative resist. In a positive resist, the resist exposed to the applied light is the portion that becomes soluble thus can be removed by the photoresist developer solution whereas the unexposed portion of the resist remains insoluble. Negative photoresists act in an opposite manner to positive resists. Between the two, positive resist became more popular as they offer relatively better process controllability for small geometric features.

Next, soft baking is used to remove all solvents after the photoresist coating process at which photoresist becomes photosensitive and hence soft-baking is a critical step determining the photo-imaging quality.

After the first photomask is applied, each of the consequent masks must be aligned to the previous one so that the pattern can be transferred accurately onto the wafer surface. Mask alignment is carried out using alignment marks on the mask and the wafer to register the patterns before light exposure. Once the mask is aligned, the photoresist is then exposed through the patterned mask with a high-intensity light source and thereby the intended pattern is developed.

Next in the photolithography process is the hard-baking step which is necessary to harden the developed photoresist and improve adhesion to the wafer surface.

Hard bake may not be required for processes where a soft resist is needed as in metal liftoff patterning.

The unwanted photoresist material and residues are then etched away and thereby the pattern is formed on the wafer surface. Pattern transfer to the wafer can be handled using one of three approaches: subtractive transfer (etching), additive transfer (selective deposition), and impurity doping (ion implantation). The subtractive method uses wet chemical or dry plasma etching which are the most commonly used pattern transfer approaches. Dry etch, however, provides relatively more profile control over wet etch.

To print and pattern small features on the mask, a light source with a short wavelength need to be used according to (1.1):

$$R = \frac{\lambda}{\mathrm{NA}} \tag{1.1}$$

where R is the resolution, NA is the numerical aperture that determines the range of angles over which the light source can emit light, and λ is the wavelength of the applied light.

1.3.2 Substrate technologies

Substrates are essential building structures that are used to provide mechanical support and electrical connections to the electronic components in almost any RF circuit. In the decision-making process when selecting the suitable substrate technology for a specific application, several material properties such as electrical, mechanical, chemical, and thermal parameters in addition to the manufacturing cost and compatibility with a particular fabrication technology must be considered. Whether the substrate is a co-fired ceramic (e.g., alumina), a PCB (e.g., FR4), or a semiconductor substrate, a prior knowledge of the material properties is essential for understanding how the different properties of the substrate and conductors could impact the performance of a particular design.

The co-fired ceramic substrates are available in wide varieties in terms of the dielectric material purity and many other properties. Ceramic substrates are suitable for high volume production and can be used at the high microwave frequencies (up to ~50 GHz) depending on the choice of the conductor and dielectric materials in terms of losses. Co-fired ceramics can be employed in high-density RF interconnects that support active and passive components and provide extremely reliable hermetic packaging solutions with a wide range of benefits including protection against moisture, heat, thermal expansion, mechanical shock, and corrosion. Ceramic multilayer substrates are also beneficial in realizing embedded inductors and capacitors within the laminate and therefore it is possible to reduce the number of surface mount components in a given design. The two main types of co-fired ceramics are the low temperature co-fired ceramics (LTCCs) and the high temperature co-fired ceramics (HTCCs). The temperatures of sintering at which the two types are distinguished are approximately 900°C for LTCC and 1,500°C for HTCC. The lower temperature needed for LTCC in comparison to HTCC is due to the fact that LTCC has a higher amount of glass content that enables binding the ceramic grains at a lower temperature. LTCC is also referred to as glass ceramics because glass is mixed with alumina of a specific composition in addition to small quantities of binding agents to create a homogeneous compound. Both LTCC and HTCC are manufactured to construct a multilayer ceramic substrate that is co-fired with a metal. In terms of the metallization, LTCC uses a composition of highly conductive materials such as copper, gold, or silver at temperatures less than the melting points of the metal. However, HTCC uses metals with relatively high resistivity such as tungsten, molybdenum, or moly-manganese that is co-fired at high temperatures to support applications where mechanical rigidity and long-term reliability are of high importance.

The production of multilayer-based components in LTCC or HTCC starts by casting the ceramic composites to produce uniform thicknesses of metal sheets. Then, the metal sheets are patterned and vias are created using a punching machine according to the layout of the given design. The three-dimensional (3D) structure is then constructed from the two-dimensional (2D) sheets where via holes are filled using usually a thick-film screen printing process. The fabrication of ceramic multilayer substrates is different from the semiconductor substrates in terms of the processing sequence. In semiconductor fabrication, the layers are processed sequentially on top of each other; while in the ceramic fabrication process, the layers are individually patterned and then stacked and co-fired to establish a homogeneous multilayer structure.

Key differences between the LTCC and HTCC technologies include:

- LTCC has relatively lower dielectric constant and lower losses when compared to HTCC;
- HTCC provides higher temperature and mechanical stability;
- LTCC has higher electrical conductivity of metals;
- LTCC has higher number of layers;
- LTCC has shorter manufacturing cycle time; and
- LTCC uses relatively low-cost tools.

PCB substrates using the low-cost flame retardant 4 (FR4) martial which is made of a glass-reinforced epoxy are suitable for low-frequency applications. The tangent loss of the FR4 dielectric material is a limiting factor that restrains its applications at high frequencies. Other PCB substrates such as Rogers's laminates provide low loss materials that can be used up to 20 GHz with reasonable insertion loss. Flexible and organic PCBs are also available and their usage mainly associated with lightweight mobile devices such as wearable medical equipment and in robotics. Ceramic substrates are advantageous over PCB substrates in terms of having lower thermal expansion coefficient, can operate at higher temperatures, capable of realizing hermetic packages, provide better thermal properties, better high-frequency performance, and enable high integration density by embedding passive components in a multilayer substrate resulting in smaller package size.

The following discussion highlights to some extent the different semiconductor substrate technologies which are mainly associated with the fabrication of RFICs and MICs. Semiconductor substrate materials are classified into two main groups: elemental and compound semiconductors:

- Elemental semiconductors which are found in the IV group of the periodic table of elements such as silicon (Si).
- Compound semiconductors can be further divided into three subgroups:
  - Binary compounds formed from elements of group III–V such as GaAs and AlP and also elements of group II–VI such as ZnS.
  - Ternary compounds formed from elements of group III–V such as InGaAs and AlGaAs.
  - Quaternary compounds formed from four elements of group III and V such as InGaAsP.

Due to the maturity of the conventional Si lithography process, Si wafer substrates are the primary material used in semiconductor manufacturing for the implementation of most RFICs. Another substrate that has been gaining traction in specific applications is the gallium arsenide (GaAs) which is more prominently used in MMICs. GaAs is a semi-insulating substrate and their components possess low capacitive and conductor losses. GaAs was developed and used for IC fabrication to overcome the operating frequency limitations encountered by Si-based devices when used at high frequencies. The high electron mobility of GaAs material, when compared to Si, put GaAs in a favorable position for high-speed circuits. However, the high fabrication costs have limited its economic viability. Silicon-germanium (SiGe) [6] is another semiconductor technology that combines the processing maturity, integration levels, and cost benefits of Si fabrication with the speed and performance of more expensive technologies such as GaAs. SiGe provides significant improvements in current, noise, and power capabilities of SiGe-based transistors when compared to transistors fabricated by Si technology. SiGe is well-suited for application areas that require a high frequency of operation (gigahertz range) and high-speed communications circuits such as data converters. Market growth of SiGe is expected to increase in the next few years especially with the deployment of 5G networks and the demand on large-scale internet connectivity. At present, however, SiGe is facing high competition from alternative technologies.

A pure Si substrate is a low resistive material that causes dielectric losses to the metal structure of bulk Si-based devices. Therefore, passive components realized on Si substrates possess lower *Q*-values than their counterparts implemented on GaAs substrates. The need for lower power microelectronics was the impetus to the development of several wafer substrate techniques to insulate between the bulk Si substrate and the device metal layers such as silicon on sapphire (SOS) and silicon on insulator (SOI).

1.3.2.1 Silicon on sapphire

SOS is a wafer substrate technology developed in the 1960s as a pathway to the production of high-speed and low power consuming devices [7]. One of the SOS main advantages is the excellent linearity performance which is a key requirement in high-power RFFE modules. However, the manufacturability of SOS is a limiting factor since forming a Si layer with minimal crystallization defects has been the main difficulty that hinders the potentials for high volume production and therefore the potential for commercializing SOS device products. Recent developments by Peregrine Semiconductor, US-based company, have greatly improved the crystallization characteristics of Si film on sapphire wafers and paved the way for possible future adoption and consideration by the RFIC device manufacturers. In essence, SOS substrate is produced using an epitaxial growth to deposit a Si film on a sapphire substrate. In turn, this will create crystallization defects due to the lattice unconformity with the sapphire. This defect is addressed by implanting Si ions inside the Si film to destroy the crystallization effects of the Si film at the interface with the sapphire substrate and converts it into an amorphous condition. With heat treatment and oxidization environment, the Si film is recrystallized and formed. The remaining SiO_2 film is then removed to obtain the desired SOS substrate [7].

1.3.2.2 Silicon on insulator

The basic structure of SOI substrate consists of a SiO_2 layer formed between a top thin Si layer in which devices are built and a bulk Si substrate. SOI wafer substrate technology is considered an evolutionary step in chip manufacturing that exhibits several advantages over others such as SOS. The main advantage is the reduced parasitic capacitance between the device metal layers and the substrate; therefore, improved electrical performance can be obtained. For example, SOI substrates are commonly utilized to reduce source and drain to substrate capacitance in field-effect-based transistors (FETs) used in manufacturing RF switches that in turn has a significant effect on the switch isolation, speed, and insertion loss. Moreover, SOI substrate technology offers less performance variation over temperature, compatible with most conventional fabrication processes, and doesn't require special training and equipment to manufacture.

The most common fabrication process for SOI wafers is carried out using separation by implantation of oxygen (SIMOX) to form an electrically insulating SiO_2 layer above a bulk Si support wafer. Other applications use separation by implantation of nitrogen (SIMNI) where oxygen is replaced by nitrogen or combination of oxygen and nitrogen (SIMON) without any adverse impact. All of SIMOX, SIMNI, and SIMON processes have good capability to precisely control the thickness of the oxide layer. The oxygen or nitrogen ions are implanted into the silicon by an ion beam. The energy of the ion beam determines the depth of the insulating layer as well as the thickness of the subsequent silicon overlay.

Although SOI technology is becoming the answer to several technical challenges confronting the IC industry, production of ultra-thin SOI layers to establish a low profile and thickness uniformity between the bulk Si of the device and the buried oxide layer is still an outstanding challenge. Furthermore, SOI can suffer from self-heating effects due to the low thermal conductivity of the oxide layer (0.4 W/m-K for SiO_2) if compared to a purely bulk Si (150 W/m-K for Si).

1.3.3 Microelectronics metallization and characterization

Metallization is the process of depositing a metallic material or an alloy on a wafer surface or a laminate. Thick metal is commonly used for chip-to-chip routing, intrachip connections, and the implementation of distributed matching components. The conductivity of the metal is related to its grain size where large grain size exhibits high conductivity and equivalently low resistivity.

With the numerous known interconnect reliability concerns, several metallization characteristics must be analyzed so that better trade-offs can be made when selecting the proper alloy for a given application. For instance, introducing a new metallization usually require knowledge about its key parameters such as the metal electrical conductivity, thermal conductivity, easiness and compatibility with standard microelectronics fabrication processes, resistance to electromigration, resistance to corrosion, and other mechanical properties such as the tendency to interact with Si and diffuse into it. Diffusion of metal can cause short-circuit paths to the chip or interconnects.

Sections 1.3.3.1–1.3.3.7 discuss several metallization parameters that are important in characterizing and selecting the proper material or composition for a given application.

1.3.3.1 Skin depth

The skin depth effect is arising from the fact that a time-invariant current in a conductor flows uniformly through the entire cross-sectional area of a homogeneous conductor whereas as the frequency increases, the current density has a tendency to drift toward the surface of the conductor. At high frequencies, the current is confined in a very thin layer near the conductor surface. This nonuniform distribution of current density in a conductor as a function of frequency is known as the skin effect. The phenomenon is more pronounced in ferromagnetic conductors (e.g., iron and steel) that have relative permeability ranging from few hundreds to few thousands if compared to nonferromagnetic metals (e.g., copper and aluminum) that have a relative permeability of one.

The depth of penetration, δ, at which the current density is 1/e (i.e., approximately 37%) of the current density at the surface of the conductor is equivalent to one skin depth and can be calculated using the following relation:

$$\delta = \sqrt{\frac{\rho}{\pi f \mu}} = \sqrt{\frac{1}{\pi f \mu \sigma}} \tag{1.2}$$

where μ is the permeability of the material, f is the frequency of the flowing current in hertz, p is the resistivity of the material in ohms-meter, and σ is the conductivity of the material in siemens/meter. The distribution of current density is highest near the surface of the conductor and decreases exponentially as the distance from the surface increases as illustrated in Figure 1.8 and according to (1.3a) and (1.3b):

$$J = \frac{I}{A} \tag{1.3a}$$

$$J = J_0 e^{-\left(\frac{d}{\delta}\right)} \tag{1.3b}$$

where J_0 is the current density at DC in amperes/m^2, A is the cross-sectional area of the conductor in meter squared, I is the current through the conductor in amperes,

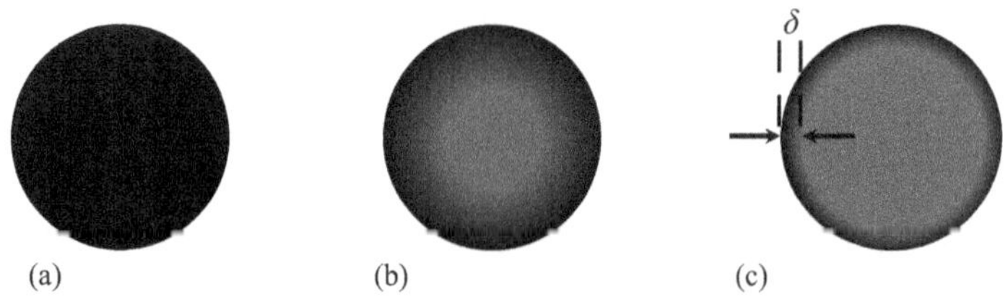

Figure 1.8 An illustration of the skin depth decrease in response to the frequency increase: (a) due to time-invariant, (b) low frequency, and (c) high-frequency source

Table 1.1 Skin depth of common metallization calculated at 1.0 GHz [8]

Material	Electrical conductivity (S/m)	Skin depth (μm)	5δ
Copper	5.8×10^7	2.11	10.55
Iron	1.0×10^7	0.016	0.08
Aluminum	3.8×10^7	2.6	13
Gold	4.5×10^7	2.5	12.5
Tungsten	1.8×10^7	3.7	18.5
Nickel	1.4×10^7	4.2	21

and d is the depth beneath the surface. From (1.3b), 63% of the current is confined within one skin depth of the surface and 99% within five times the skin depth.

Table 1.1 summarizes the thickness of commonly used alloys in microelectronic manufacturing at one and five times their skin depth.

Distributing electric power cables are typically made of copper instead of iron although iron has a lower cost. This is due to the fact that copper exhibits relatively higher conductivity and lower permeability than iron. Therefore, iron possesses a thinner skin depth and thus has higher losses than copper. Additionally, metal conductivity decreases with increasing temperatures which is another reason that makes iron not a favorable option. The skin depth of interconnecting traces depends on the interconnect technologies in use. In a sense, on-chip, thin-film, PCB, and ceramic are all interconnecting technologies that comprise different ranges of trace width and thickness yielding different levels of skin effects. This can be seen as a degree of freedom in the design of hybrid modules where decisions can be made on selecting what technology can be used to support long routing traces for low loss and high-reliability designs.

Another related phenomenon to the skin depth effect is the proximity effect. The current distribution within the conductors can be affected due to the presence of adjacent current-carrying conductors, and this effect is known as the proximity effect. The close proximity of one conductor to another at relatively high frequencies induces current due to electromagnetic induction which explains such a phenomenon. Figure 1.9 demonstrates this effect due to common- and odd-mode current sources.

1.3.3.2 Thin film stress

Thin film stress effects are one of the known reliability concerns in microelectronic devices. The implication of thin film stress can be expressed in two opposite forces: compressive stress and tensile stress. While compressive stress causes hillocks, tensile stress causes cracks as depicted in Figure 1.10(a) and (b), respectively. Both stresses can be measured using one of the surface texture measurement instruments, either using contact profilometer or noncontact interferometer. The primary cause of such stresses can be attributed to changes in temperature in a multilayered structure constructed by stacking different material types where the coefficients of thermal expansion between the material layers, for example, a thin film and a substrate, are different.

1.3.3.3 Thin film thickness measurement methods

The measurement of a thin film thickness can be carried out through destructive or nondestructive processes. For accurate measurement, a destructive approach is usually needed and involves the use of a transmission electron microscope (TEM), or a scanning electron microscope (SEM) applied through a cross-section of the thin material specimen. TEM is used for imaging thin layers in the order of 100 Angstroms, while SEM can be used with relatively thick layers in the order of 1000 Angstroms.

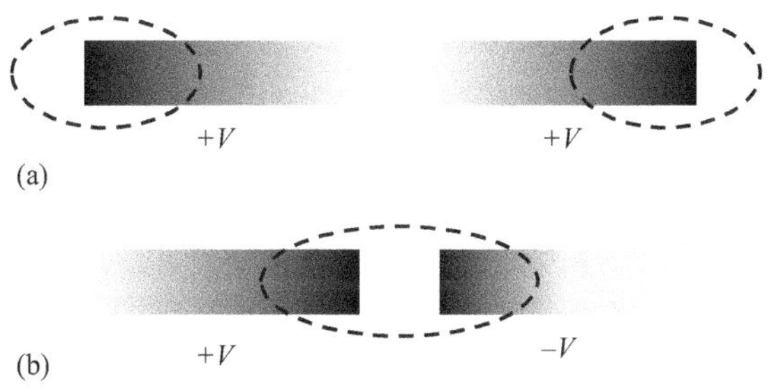

Figure 1.9 Current density distribution under skin and proximity effects. Dark gray color represents high current density (a) common-mode current and (b) differential mode current

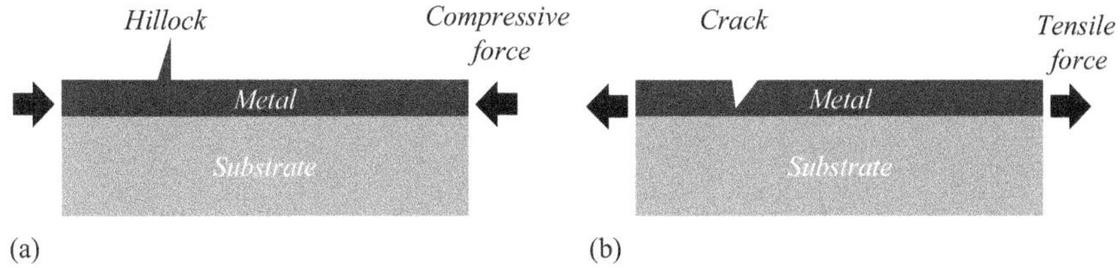

Figure 1.10 Thin film structures showing the effect of high stress: (a) compressive and (b) tensile stress

Nondestructive methods are relatively less accurate and can be carried through acoustic wave echoes by calculating the reflections at the interface of the different materials due to known acoustic wave speed as expressed in (1.4):

$$d = \frac{V \times \Delta t}{2} \tag{1.4}$$

where V is the acoustic wave-speed and Δt is the time between reflectivity peaks. Another indirect measurement can be performed through four-point probe measurements as will be discussed in Section 1.3.3.4.

1.3.3.4 Characterization of sheet and bulk resistance

Sheet and bulk resistance is one of the important parameters of a conducting material that directly impacts the circuit performance and the quality factor of its components. Precise measurement of sheet and bulk resistance is then a necessary characterization step. For instance, the resistivity of metal interconnect is inversely proportional to the interconnect cross-section and to accurately characterize the loss in a conducting film it is important to ensure constant resistivity across the deposited film. Resistors in integrated circuits are fabricated from a thin sheet of material (e.g., Nichrome) used to form either thin or thick films. Therefore, it is more convenient to measure the sheet and bulk resistivity of such materials.

Furthermore, prior knowledge of the film conductivity can be utilized along with the sheet resistance measurement to monitor the deposition process uniformity as an indirect way of measuring the film thickness.

Calculation of sheet and bulk resistance can be carried out using a two- or four-point probe measurement of the material resistivity. The two-point measurement approach, however, is prone to errors due to the unwanted contribution of the electrical contact and lead resistance of the probe to the measurement results. Alternatively, a four-point probe measurement setup [9] provides an accurate measurement tool, especially, for low resistive measurements. The principle of operation starts by passing a known current to the two outermost points and measuring the voltage drop across the sample at the inner two points via a high impedance voltmeter as depicted in Figure 1.11. The conductive tips of the probes are usually made of tungsten and are equally spaced so that the sheet resistance can be accurately calculated. The assumption here is that a finite sample of the conducting sheet has a thickness much less than the spacing (S) between the probe tips and the metal film is semi-infinite in the lateral dimension. To obtain accurate measurements, the edges of the sheet sample must be distanced at least four times the spacing distance from the measurement point. This is to ensure that lines of current sources along the direction of the probe tips are not disturbed by the sample boundaries.

1.3.3.5 Electromigration

The rising demand for reduced size integrated circuits is accompanied by an increase in their current densities. This implies that the implications of electromigration will continue to aggravate, especially in the case of interconnects with

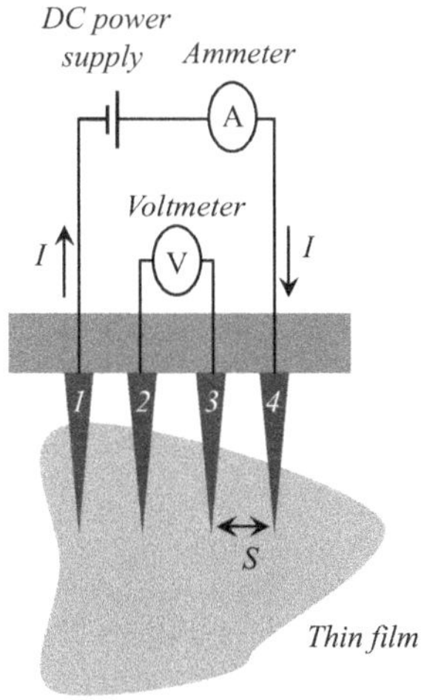

Figure 1.11 Four-point probe measurement setup

narrow line widths and higher aspect ratio vias. Electromigration, when occurs, can have a catastrophic impact on the production yield in high-volume manufacturing.

Electromigration is the transport of metal atoms due to the momentum transfer between the metal electrons pushing the metal atoms [10]. In a sense, the collision of electrons with atoms under the influence of an electric field leads to defects such as vacancies at grain boundaries in the metal conductor. High current densities have undesired effects on the conductor due to the finite resistance of the conductor material causing a rise of the conductor's temperature or what is known as Joule heating and eventually leads to destructive damage in the form of voids or hillocks. While voids can lead to the formation of an electrically open circuit, hillocks can have sufficient depth to short out a nearby junction.

In microelectronic designs, thin-film metallic interconnects are used to transport currents between the various components on a chip or within a module. Therefore, conductors used in high power applications are selected so that they have maximum current ratings to eliminate reliability concerns arising from the implications of Joule heating. Electromigration depends on the conductor material shape, temperature, applied current density, and the environment housing the conductor. The maximum current density that can be supported by an interconnect conductor can thus be a limiting factor due to reliability constraints. Electromigration limits the maximum current densities that can flow through a thin film interconnect due to reaching the maximum melting temperature.

Copper (Cu) thin films have higher electromigration activation energy levels due to its high electrical and thermal conductivity and higher melting temperature ($1e^6$ A/cm^2) when compared to aluminum (Al) which has melting temperature of $2e^5$ A/cm^2, thus Cu can relatively offer resistance to electromigration. Both Cu and Al are polycrystalline, meaning they are structured as grains of the lattice. At the application of current, the electrons collide with the atoms in the lattice causing them to migrate in the direction of the electron flow.

Depending on the material properties, one or more of the following migration mechanisms can be a dominant factor affecting electromigration as they are not independent of each other [11]:

- Thermo-migration due to thermal gradient,
- Stress-migration due to mechanical stress, and
- Electromigration due to applied electric field.

Black's empirical equation in (1.5) estimates the mean time to failure (MTTF) [12] as a function of current density and temperature. Measurements of lifetimes happen at elevated temperatures and current densities to accelerate the aging process and to estimate the life expectancy of a part in a reasonable period of time rather than an extended period of time under normal operating conditions. According to Black's equation which is derived from the Arrhenius equation [12], increasing the temperature leads to an increase in the reaction rates that in return decreases the time needed to perform the failure analysis. This reliability stress test is referred to as high temperature operating life (HTOL):

$$\mathrm{MTTF} = \frac{A}{J^n} \times \exp\left(\frac{E_a}{k \cdot T}\right) \tag{1.5}$$

where A is the cross-section area of the conductor, J is the current density flowing in the conductor, n is a scaling factor, T is the temperature in Kelvin, k is Boltzmann constant in electron-volts/Kelvin, and E_a is the effective activation energy of the electromigration process in electron-volts that is the minimum energy required to start a process reaction. While the relation in (1.5) is used to calculate MTTF of a device under constant current stress, in general, if N product samples fail after successively longer times $t_1, t_2, \ldots, t_n$, the average MTTF can be expressed as:

$$\mathrm{MTTF} = \frac{t_1 + t_2 + \cdots + t_N}{N} \tag{1.6}$$

Another commonly used time measure for statistical failure analysis is the median time to failure which is the time for 50% of the samples to fail.

1.3.3.6 Electroless and electrolytic plating

Material coatings are used in many instances, for example, in forming metal enclosures encapsulating RF modules to prevent EMI emissions into the neighboring modules in a complex system and to reduce the module susceptibility when subjected to external EM fields. Additionally, material coating provides mechanical strength and resistant to corrosion and friction. Typically, composite materials that can provide shielding to both magnetic and electric fields are used for optimal shielding effectiveness.

Both electroless and electrolytic are plating techniques which can be used for material coating. An electroless coating is a deposition process that uses chemical reaction without requiring an external source of electric current, whereas an electrolytic deposition process requires a current source to coat the surface of a material.

Electroless plating is advantageous over electrolytic plating because of its uniform plate distribution which is considered an important feature, especially in microscale packaged devices. Electroless plating also helps in avoiding solder joint embrittlement. In contrast, electrolytic plating causes nonuniformity in the deposit. This is due to the fact that high currents tend to reside on the edges and corners of the part being coated and thereby current distribution may not be uniform.

1.3.3.7 Metal selection and design considerations

Metal stack composition and selection is one of the critical factors in RFICs that impacts the device technology readiness in terms of reliability and performance. Packaging engineers usually perform thorough analysis before making decisions on the material selection and material composition used in a module or a device.

Aluminum-copper (AlCu) alloy is one of the commonly used metal compositions. Al metal alone is very susceptible to electromigration damage, while pure Cu possesses high electromigration resistance, and hence better reliability. Additionally, pure Al is more resistive than pure Cu. However, Cu is known for its high tendency to diffuse into silicon and has poor adhesion with silicon dioxide. When Al is alloyed with Cu, electromigration resistance of Al significantly improves as Cu serves as an adhesive filler between the Al grains that prevent them from migrating. Alternatively, tantalum (Ta) metal can also be used as a barrier layer to prevent copper diffusion into surrounding semiconductor structures and titanium (Ti) to block Al from diffusing into the Si layer.

The coefficient of thermal expansion (CTE) is a material parameter that measures how a matter changes in its shape in response to a change in temperature. The lower the CTE, the preferable the material is. For thermal management purposes, it is desirable to select materials that have high thermal conductivity and low CTE since device lifetime can be decreased drastically because of the large thermal stresses which may alter the electrical performance of a device. Table 1.2 lists the CTE and thermal conductivity of some of the commonly used alloys. Materials used for heat dissipation and spreaders have to be highly conductive materials.

Selection of a conductive film can also depend on many other mechanical factors such as easiness to form, to etch, and to pattern using standard fabrication

Table 1.2 Linear thermal expansion coefficients of commonly used metals [13]

Material	**CTE (ppm/°C)**	**Thermal conductivity W/(m-K)**
Copper	16.5	390
Iron	11.7	80
Aluminum	23.6	200
Gold	14.2	310
Tungsten	4.6	174
Nickel	13.3	91

tools, surface roughness, adhesion, and stress to ensure stability throughout the fabrication process.

1.3.4 Dielectric permittivity characterization

Dielectric materials (either thin- or thick-films) are key elements of any electronic circuit that are used to protect against the occurrence of voltage breakdown between the active signal traces and provide a base to mount on the circuit components. High permittivity dielectrics help in reducing the size (i.e., increase capacitances) of distributed circuits due to the increase in the electric flux density. Dielectrics are classified based on their electrical parameters into dielectric loss tangent, dielectric constant, and dielectric strength. Materials that exhibit low electrical loss, high thermal conductance, and low thermal expansion are highly desirable. For instance, low electrical loss materials provide low power dissipation, low Joule heating and result in reduced signal attenuation. In most of the RF circuit designs, the selection of a dielectric material for high-frequency applications should start by determining the dielectric with the lowest loss and then finding the required dielectric constant and breakdown voltage.

The speed of signal propagation in high-frequency, high-speed circuits are influenced by the signal-propagation delay that depends on both the material dielectric constant and the conductive structure. Assuming a lossless transmission line with a transverse electromagnetic (TEM) propagation mode, its time delay can be expressed as:

$$t_d = \frac{l\sqrt{\varepsilon_r' \mu_r'}}{c} \tag{1.7}$$

where l is the line length of the conductor, c is speed of light in vacuum, μ_r' is the real part of the complex permeability given by $\mu = \mu_0 \, (\mu_r' - j\mu_r'')$, ε_r' is the real part of the complex permittivity given by $\varepsilon = \varepsilon_o \, (\varepsilon_r - j\varepsilon_r'')$, μ_0 and ε_o are the permeability and permittivity of vacuum, respectively. As can be concluded from (1.7), the propagation delay can be reduced by using a substrate with a low value of the real part of the complex permittivity. In other words, wave propagation speed and impedance can be modified when using different materials comprising different relative permittivity. The imaginary parts of the complex permeability and permittivity (i.e., μ_r'' and ε_r'') are there to express the associated loss in the used material. The dielectric loss can be expressed through the tangent loss ($\tan\delta = \varepsilon_r''/\varepsilon_r'$) which is the ratio of the imaginary part to the real part of the complex permittivity. Hence, the magnitudes of ε' and ε'' depend on the frequency, ω, of the excitation wave.

In the development of RFIC chips and modules, it is possible that new composite materials are researched in attempts to discover enhanced material properties. Therefore, measurement methods that can accurately determine the permittivity and permeability of a substrate material are necessary. The fact that each band of frequencies exhibits different loss levels than the other and most dielectric materials are anisotropic, it is challenging to characterize the material properties using a single method. In practice, there are different methods that can deal with such challenges such as isotropic or anisotropic, bulk or film, temperature dependency, and materials with different surface roughness. Among those techniques, resonant-based methods (such as re-entrant cavity, split-cylinder resonators, cavity resonators, and Fabry–Perot resonators) [14] are the most suitable methods for anisotropic dielectrics, where permittivity or permeability is determined from measurements of the resonance frequency and quality factor. For a thin anisotropic material, two methods of characterization are required, one for the permittivity normal to the material face and another for the in-plane permittivity. Other techniques to determine the complex permittivity of a dielectric substrate can also be carried out using transmission line methods. In [15], coplanar waveguide (CPW) transmission line was used to match the impedance between two different substrates (a reference substrate with known low loss and constant permittivity over frequency and the substrate whose permittivity is to be determined) using identical conductor geometries fabricated on them. By measuring the frequency-dependent propagation constant (γ), of the reference and the test CPW lines using a multiline TRL calibration method, one can find the total capacitance and conductance per unit length and consequently the relative permittivity and loss tangent of the substrate under test.

1.3.5 Cleanroom classifications

A cleanroom is a controlled environment that has a low level of air contamination such as dust, chemical vapors, and/or aerosol particles and is used in almost every precision microelectronics and nanotechnology manufacturing facilities where small particles can adversely affect the quality and purity of a manufacturing process. Air filtration methods are used in cleanrooms to prevent contamination and to remove particles from the air.

There are two standards of measurement that are vastly used for cleanroom classification. Each of which comprises different classes of cleanrooms to indicate the level of air cleanliness. The US FED STD 209E standard which counts the number of half-sized micron (0.5 μm) particles or larger permitted in a cubic foot of air. This standard, however, was replaced in 2001 by an American National Standard which is the ISO 14644-1 to account for particles smaller than 0.5 microns by extending the particle size range to start from 0.1 μm up to 5 μm per cubic meter of air for more precise measurement of air cleanliness. Table 1.3 summarizes the standardized levels of air cleanliness in terms of concentration of airborne particles in cleanrooms specified by the number of particles per cubic meter (m^3) of air at a specified particle size measured in microns [16]. Choosing a cleanroom classification is dependable upon which particle size begins to influence the fabrication quality of a product.

The lowest level of cleanroom standards is an ISO 9 standard that corresponds to the ambient air in a typical city environment that contains approximately 35,000,000 particles/m^3, for particles that are 0.5 μm and larger in diameter.

Table 1.3 ISO 14644-1 cleanroom standards and its equivalent US FED STD 209E

ISO class	**Maximum particles/m^3**						**FED STD 209E (Old Standard)**	
	≥ 0.1 μm	**≥ 0.2 μm**	**≥ 0.3 μm**	**≥ 0.5 μm**	**≥ 1 μm**	**≥ 5 μm**	**English unit (ft^3)**	**Metric unit (m^3)**
ISO 1	10	2	–	–	–	–	–	–
ISO 2	100	24	10	4	–	–	–	–
ISO 3	1,000	237	102	35	8	–	Class 1	M 1.5
ISO 4	10,000	2,370	1,020	352	83	–	Class 10	M 2.5
ISO 5	100,000	23,700	10,200	3,520	832	29	Class 100	M 3.5
ISO 6	1×10^6	237,000	102,000	35,200	8,320	293	Class 1000	M 4.5
ISO 7	1×10^7	2.37×10^6	1,020,000	352,000	83,200	2,930	Class 10,000	M 5.5
ISO 8	1×10^8	2.37×10^7	1.02×10^7	3,520,000	832,000	29,300	Class 100,000	M 6.5
ISO 9	1×10^9	2.37×10^8	1.02×10^8	35,200,000	8,320,000	293,000	–	–

1.4 RF and microwave module types

RF receiver and transmitter architectures share common active and passive components such as amplifiers, switches, filters, and matching networks. The way module components are manufactured and assembled results in different types of modules. High-volume RF and microwave modules are based on one of the following module assembly and integration technologies: monolithic microwave integrated circuits (MMICs), hybrid microwave integrated circuits (HMICs), or multichip modules (MCMs).

The module assembly process encompasses several steps including substrate fabrication, die and component attach, molding, and shielding. Each of which has its own set of challenges, especially when operating at microwave frequencies that require special attention at the design stage. For instance, in the module assembly process, to place parts on a module with a high level of accuracy, it is recommended to use automated fine-pitch placement machines and to conduct placement accuracy studies to extract information regarding the proper fudge factor needed for a particular package type. After attaching the components onto the module surface, it is also recommended to verify if there is any unintended open or short-circuits through nondestructive optical inspection or X-ray inspection.

Besides the fabrication and assembly process challenges at high-volume manufacturing, there come other critical business commitments including capacity planning, material supply, personnel staffing, and factory utilization.

1.4.1 Monolithic modules

MMICs allow integration of all components on a monolithic substrate that in turn enhances the overall speed of the module circuits while decreasing the power consumption. The monolithic method is a reliable process suitable for high-frequency applications where distributed passives can be realized in a reasonable form factor enabling seamless device integration on a single-piece semiconductor material (e.g., GaAs). Here, all the components are fabricated simultaneously by various process methods which include diffusion or ion implantation. GaAs is one process technology suitable for monolithic fabrication that allows millions of devices to be interconnected seamlessly. Monolithic technologies are advantageous in being reliable and well-suited for high-volume manufacturing by avoiding bond wires for die interconnects. Moreover, monolithic microwave technology can be considered as the foundation toward the realization of SoCs.

1.4.2 Hybrid modules

A hybrid module consists of a dielectric substrate (e.g., alumina) carrying one or more chips fabricated using different process technologies to establish the targeted function of the module. These chips may themselves be monolithic integrated circuits, active or passive devices that are produced separately using diverse technologies. In addition, SMT devices may be soldered to the circuit tracks to form a complete circuit. The chips and components can be interconnected using bonding wires deposited on a dielectric substrate or interconnections can be performed via TEM mode transmission lines (e.g., striplines). A hybrid module can use mixed technologies, such as Si, InP, and GaAs chips based on either thick or thin films.

1.4.2.1 Thick-film ICs

Thick-film ICs can be fabricated using a screen printing process which is a printing technique that uses a mesh (designed to resemble the pattern that needs to be printed) to transfer ink (a paste that could be conductive or resistive) onto a substrate. The multilayer interconnects in a conventional thick-film technology are formed by successive screen printing of conductors and dielectric layers on a base substrate. Once printed, the materials will be in the form of inks that require each layer to be dried and then fired to produce it in its ready to use form. Thick-film MICs are low cost and mostly utilized for low-frequency microwave ICs [17]. Substrate materials such as alumina (Al_2O_3), beryllia (BeO), and aluminum nitride (AlN) are the most commonly used base material.

1.4.2.2 Thin-film ICs

There is high interest in continually developing thin-film fabrication technologies that can be used to meet the demand for reliable, compact, and high-frequency RF/MICs products. Thin-film manufacturing process is precisely controlled and more importantly is a repeatable process. Thin-films require the use of a cleanroom environment and thereby it is relatively a costly process when compared to thick-films. The film material is selected so that it has good adhesion to the substrate. This is considered one of the important selection criteria of the conductor material used as the first layer of metal film on the substrate [17]. Thin-film technology has been known, for some time, as *miniature hybrid MIC technology* which is also a multilevel fabrication process used to implement RF passives fabricated on a substrate and externally attached to solid-state devices.

1.4.3 Multichip modules

The complexity of RFIC multichip modules arises from the need to integrate many subcomponents and multiple dies exhibiting a variety of technologies and the need to be housed in a single package since they cannot be fabricated under a single die [18]. The relentless price and performance requirements are the driving force for the microelectronics manufacturers to switch to new technologies and tools enabling them to survive a highly competitive RFFE market. In this context, module components and devices are developed using different technologies and processes to achieve the best possible device level performance. A single MCM module can encompass several bare IC dies (built using different technologies such as CMOS controllers, GaAs power amplifiers, SOI switches, SMT capacitors and inductors, and flip-chip or CSP filters) mounted on a custom-designed common multilayer substrate that provides mechanical support for the chips as well as the interconnection between the integrated devices as depicted in Figure 1.12.

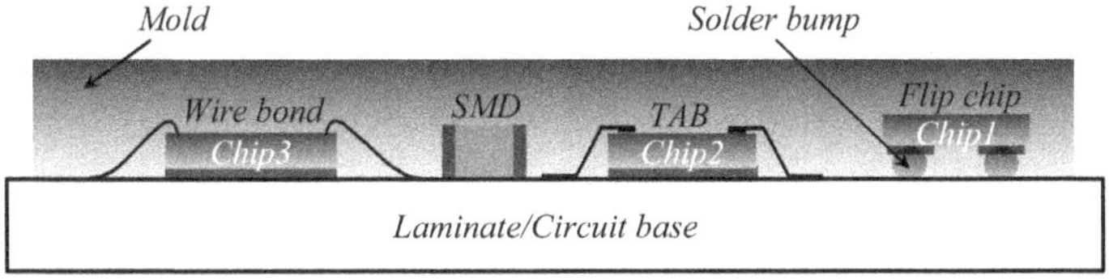

Figure 1.12 Illustrative example of an MCM comprising different die-attach technologies

MCMs provide several advantages over other module and PCB implementation types in terms of improved electrical performance through shorter interconnect lengths between the integrated dies, higher integration density, lower cost, lower power consumption, and more flexibility on designing the EM interaction between the integrated components and the module PCB board. In addition, an MCM allows passive components to be embedded on its multilayer substrates.

Main types of MCM packaging technology with respect to the used high density interconnect (HDI) substrate include:

- MCM-L: Laminated MCM with a multilayer substrate similar to the PCB structure;
- MCM-C: Co-fired ceramic MCM, usually LTCC; and
- MCM-D: Deposited MCM that uses deposition process to the laminate metals and dielectrics in a similar fashion to that utilized in the semiconductor thin-film processing.

References

[1] Tsutsumi J., and Matsumoto K. "Super-isolation duplexer aiming for removing Rx inter-stage filter in W-CDMA handsets." *IEEE 38th European Microwave Conference*; Amsterdam, the Netherlands, Oct 2008. pp. 1066–69

[2] Tummala R., and Madisetti V. "System on chip or system on package?" *IEEE Design & Test of Computers*. 1999;**16**(2):48–56

[3] Chiang T.-Y., Souri S.J., Chui C.O., and Saraswat K.C. "Thermal analysis of heterogeneous 3DICs with various integration scenarios." *IEEE International Electron Devices Meeting*; Washington, DC, Dec 2001. pp. 681–84

[4] Tsai W.S., Huang C.Y., Chung C.K., Yu K.H., and Lin C.F. "Generational changes of flip chip interconnection technology." *12th International Microsystems, Packaging, Assembly and Circuits Technology Conference (IMPACT)*; Taipei, 2017. pp. 306–10

[5] Jaeger R. *Introduction to Microelectronic Fabrication: Volume 5 of Modular Series on Solid State Devices*. 2nd edn. Upper Saddle River, NJ: Pearson; 2001

[6] Cressler J.D. "SiGe HBT technology: a new contender for Si-based RF and microwave circuit applications." *IEEE Transactions on Microwave Theory and Techniques*. 1998;**46**(5):572–89

[7] Nakamura T., Matsuhashi H., and Nagatomo Y. "Silicon on sapphire (SOS) device technology." *Oki Technical Review*. 2004;**71**(4):66–69

[8] Rogers J., and Plett C. *Radio Frequency Integrated Circuit Design*. Norwood, MA: Artech House, Inc.; 2003, p. 134

[9] Smits F.M. "Measurement of sheet resistivities with the four-point probe." *The Bell System Technical Journal*. 1958;**37**(3):711–18

[10] Lienig J., and Jerke G. "Electromigration-aware physical design of integrated circuits." *18th International Conference on VLSI Design held jointly with 4th International Conference on Embedded Systems Design*; Kolkata, India, Jan 2005. pp. 77–82

[11] Heryanto A., Pey K.L., Lim Y.K. *et al.*, "Study of stress migration and electromigration interaction in copper/low-κ interconnects." *2010 IEEE International Reliability Physics Symposium*; Anaheim, CA, 2010. pp. 586–90

[12] Black J.R. "Electromigration failure modes in aluminum metallization for semiconductor devices." *Proceedings of the IEEE*. 1969;**57**(9):1587–94

[13] Agilent (Keysight) Technologies. *Laser and Optics User's Manual*. Santa Clara, CA: Agilent Technologies; 2002, Chapter 17

[14] Baker-Jarvis J., Janezic M., Riddle B., Holloway C.L., Paulter N.G., and Blendell J.E. "Dielectric and conductor-loss characterization and measurements on electronic packing materials." *National Institute of Standards and Technology (NIST) Technical Note 1520*, 2001

[15] Janezic, M.D., and Williams, D.F. "Permittivity characterization from transmission-line measurements." *1997 IEEE MTT-S International Microwave Symposium Digest*. 1997;**3**:1343–46

[16] "ISO 14644-1:2015 – Cleanrooms and associated controlled environments – Part 1: Classification of air cleanliness by particle concentration." ISO. Available at: https://www.iso.org/standard/53394.html [Accessed Sep 12, 2018]

[17] Bahl I.J. *Lumped Elements for RF and Microwave Circuits*. Norwood, MA: Artech House; 2003

[18] Doane D.A., and Franzon P.D. *Multichip Module Technologies and Alternatives: The Basics*. New York, NY: Van Nostrand Reinhold; 1993

Chapter 2

Lumped and distributed passive elements

2.1 Introduction

Lumped and distributed elements as passive components are the constituting parts of almost any radio frequency (RF) and microwave circuit. When the largest physical dimension (l) of an RF component is much shorter than the operating wavelength (λ), typically $l < \lambda/20$, the component is referred to as lumped element since wave phase variations across the element are negligible. The largest dimension of a distributed element, however, is comparable to the operating wavelength and is usually characterized by the element's length, characteristic impedance, and propagation constant. One of the main advantages of using distributed elements over their lumped counterparts is the relatively high-quality factor (Q-factor). Lumped elements, however, comprise relatively wider bandwidth and small form factor which implies reduced coupling between elements located in proximity.

The use of lumped elements in the design of microwave integrated circuits (MICs) has been reported since the 1960s, primarily to achieve reduced size designs at low frequencies [1–3]. Soon after the start of monolithic microwave integrated circuits (MMICs), the use of lumped elements exaggerated for high-frequency applications in the late 1970s [4]. The impetus for using lumped elements was their low cost, large bandwidth, large impedance transformation ratio, negligible phase delay, and compact size in relative to the operating wavelength. Moreover, designs based on lumped elements result in reduced chip size and thereby enable more chips per wafer and reduced manufacturing costs.

Integrated RF components in an RF and microwave module design normally require the knowledge of their performance at the harmonics of the fundamental design frequency which is a critical design consideration when dealing with non-linear designs. It is also important to understand and choose the analysis method that is applicable for each component type. Analysis of lumped components is generally carried out via ordinary differential equations (ODEs) since lumped elements have dependent variables that are function of time. Whereas distributed elements are circuit components with dependent variables that are function of time and space and their analysis entails solving partial differential equations (PDEs). Figure 2.1 lists the different analysis methods that can be used depending on the electrical size of the component.

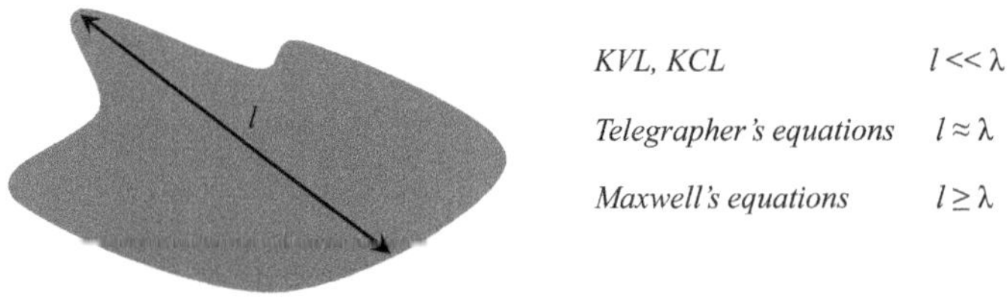

Figure 2.1 Available analysis methods depending on the electrical size of the RF component

The fabrication of lumped elements can be realized using various technologies including co-fired ceramics, thin- and thick-films, while common distributed elements exist as forms of transmission lines (TLs) such as open- and short-circuited stubs, and quarter wavelength transformers that can be implemented using photolithography-based methods. The availability of high dielectric constant substrate materials has extended the usage of distributed circuits to low microwave frequencies.

In microsystems and micro RF structures that operate at RF and microwave frequencies, it is incorrect to treat the circuit interconnect as a transmission line since the signal level is relatively constant and does not change along the length of interconnect. In such case, an interconnect can be described simply as a lumped element, mostly as an inductor with a value usually in the nanohenry range.

2.2 Lumped elements

Generally, linear passive circuits consist of three main elements: resistor, inductor, and capacitor (RLC) as illustrated in Figure 2.2. These elements can be categorized by their electrical size into lumped and distributed elements. A component electrical length is a unit-less measure of the component physical length in terms of a certain wavelength.

Sections 2.2.1–2.2.3 discuss each of these three elements in terms of their I–V mathematical relations, equivalent circuit models, and their implementation methods.

2.2.1 Resistors

Resistors are used in several RF and microwave circuit designs, for example, to provide impedance transformation or isolation as in power dividers, circuit termination loads, attenuators, and for biasing and stabilizing active devices. Electrical resistance is a material property that measures the material opposition to the flow of electric current or alternatively, it determines the rate at which electrical energy is converted into heat energy at a given current. A resistor exhibits a linear I–V characteristic governed by Ohms law as:

$$R = \frac{V}{I} \tag{2.1}$$

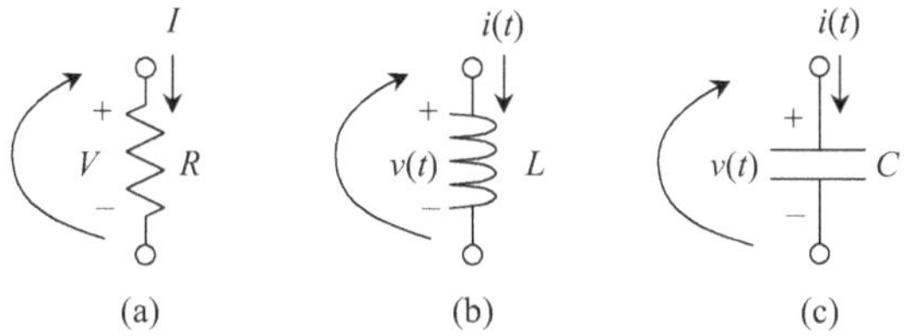

Figure 2.2 *Circuit schematic of passive lumped elements: (a) resistor, (b) inductor, and (c) capacitor*

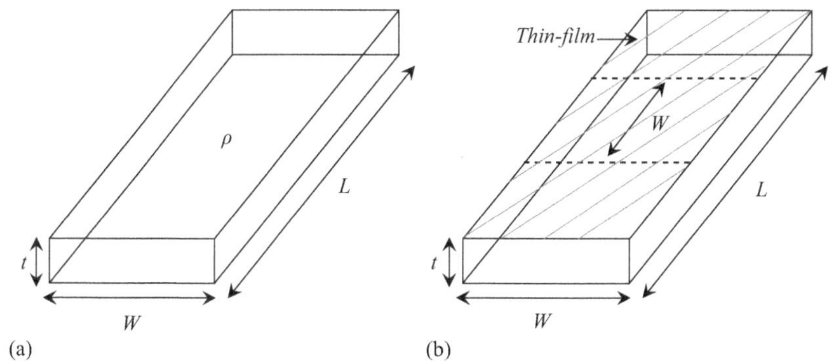

Figure 2.3 *(a) A slab of a resistive material and (b) a sheet of a resistive material showing the number of squares of width along its length*

where V is the voltage applied between the resistor terminals causing charged particles to flow and I is the current representing the number of charged particles passing through the resistor per unit time due to the applied voltage. Figure 2.3(a) shows a slab of a resistive material and Figure 2.3(b) illustrates a thin-film or a sheet of resistive material.

Sheet resistance, volume resistivity, and material dimensions are all related. In a sense, multiplying the sheet resistance by the thickness of the material gives the volume resistivity. Similarly, the sheet resistivity can be obtained by dividing the volume resistivity by the material thickness. These relations are defined as:

$$\rho = R_{\text{sheet}} t = R\frac{A}{L} = R\frac{tW}{L} \tag{2.2}$$

$$R_{\text{sheet}} = \frac{\rho}{t} = \frac{1}{\sigma t} \tag{2.3}$$

where ρ is the volume or bulk resistivity (ohm-m), A is the cross-sectional area (m^2), W is the width of the slab material (m), R is the resistance (ohm), R_{sheet} is the sheet resistance (ohms/square), σ is the material conductivity (S/m), and t is the material thickness (m).

From (2.2), it can be seen that the resistance of a material increases with increasing the length (L) but decreases with increasing the cross-sectional area. Sheet resistance is defined as the lateral resistance at one square of width of the thin lossy material. Therefore, the number of squares of width along the material total length determines its total resistance in ohms. Consider Figure 2.3(b) as an example which depicts a planar resistor comprising three squares of width along its length, the total resistance (R) is then equal to three times the sheet resistance ($3R_{\text{sheet}}$).

In general, the electrical conductivity of the solid materials can be classified into three categories:

- insulators: $\rho > 10^2$ Ω.cm;
- conductors: $\rho < 10^{-3}$ Ω.cm; and
- semiconductors: 10^{-3} Ω.cm $< \rho < 10^2$ Ω.cm.

2.2.1.1 Resistor modeling

An ideal resistor is frequency-independent; however, a practical resistor operating at high frequencies will exhibit additional parasitic inductances and capacitances contributing to the actual resistance value R. The two-port model illustrated in Figure 2.4 represents an equivalent circuit model of a chip resistor assuming symmetric contacts. The inductances L_s and L_c are attributed to the finite length of the resistor body and metallic contacts, respectively. C_g and C_p describe the parasitic capacitances between the metal contacts and the ground plane and between the contacts themselves, respectively. R is the total equivalent resistance.

2.2.1.2 Resistor implementation

The two main types of chip resistors are the thin- and thick-film resistors. In all implementation types, resistors comprise a resistive layer supported by a substrate. A thin-film resistor is realized using a lossy metal such as Ta, Ti, or NiCr and has a thickness in the order of 0.1 um or less. Thin-films are usually produced by vacuum deposition techniques onto an insulating substrate and are suitable for trimming to adjust the value of resistance through a subtractive process using, for example, laser trimming methods. However, a thick-film resistor is achieved through an additive fabrication process that uses a bulk semiconductor film such as doped Si or GaAs with thicknesses in the order of 100 um. Hence, thick-film resistors are compatible with MIC fabrication processes. The realization of thick-film resistors starts by depositing a resistive material past onto a substrate then curing it at high temperature using screen and stencil printing to form a glass-like material.

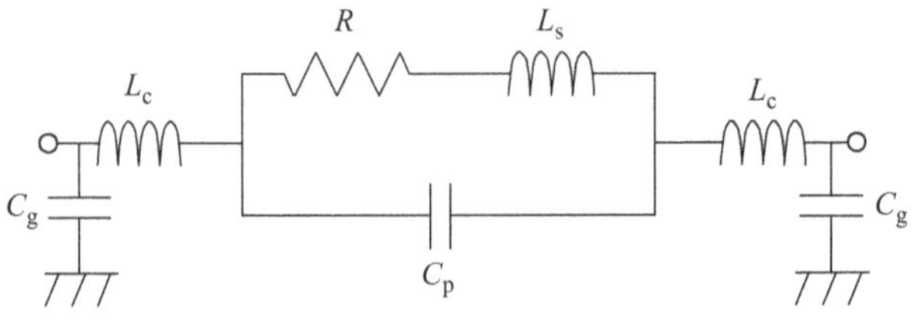

Figure 2.4 Lumped equivalent two-port circuit model of a chip resistor

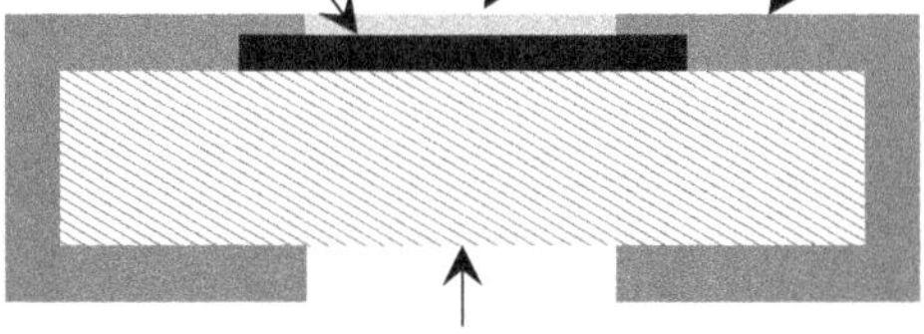

Figure 2.5 Structure of a thin- or thick-film chip resistor

Temperature coefficient (TC) of a resistor is a measure of the resistance variation with respect to temperature, and it depends on the material selection used in making the resistor. A resistor with a positive TC value implies that the resistance value increases with increased temperature and the opposite is true. Between the two types, thin-film-based resistors are commonly used as a surface mounted device (SMD) for precision applications because they exhibit lower TC and lower thermal noise when compared to bulk/thick-film resistors. However, the low-cost fabrication of thick-film resistors makes them strong candidates for high volume manufacturing, especially, for circuit designs that don't rely on critical resistance values.

Figure 2.5 depicts the structure of a thick- or thin-film chip resistor with conductive end contacts and a ceramic base such as alumina used to support the film material. Small resistance values of integrated resistors realized using monolithic fabrication methods are printed as straight or serpentine structures of the resistive material. The implementation of relatively larger resistance values in a small substrate area usually takes the shape of a meander line.

2.2.1.3 Attenuators

One of the direct applications of resistors is the attenuator. Attenuators are simple voltage divider networks designed to reduce the amplitude of a signal passing through it, ideally, without significantly influencing the signal integrity. RF attenuators are generally used in high power RF circuitry and exist in six different types: fixed, continuously variable, step, programmable, dc bias passing, and dc blocking. Fixed attenuators provide fixed attenuation level to the input signal. A fixed attenuation network has resistor values governed by a set of formulas (depending on the configuration used) that can result in a specific characteristic impedance over a given frequency range. Continuously variable attenuators provide a degree of freedom by allowing the manual adjustment of the attenuation value over a specific range. Varying the attenuation level can be carried out using a solid-state device such as a PIN diode or a field effect transistor (FET). In step attenuators, several fixed attenuators are used with a switching mechanism to switch between them. They comprise a set of resistive networks to generate a particular attenuation level at each state. The programmable attenuator can be thought of as a digital step attenuator, generally, controlled by a computer logic circuit. DC bias passing and DC blocking attenuators both have relatively an opposite attenuation function. DC bias attenuator reroutes or bypasses DC signal from input to output while attenuating the RF signal in the main path by including capacitive elements at the attenuator input and output terminals. DC blocking attenuators completely block DC signal to pass from the input to the output.

Design specifications of an attenuator include the signal attenuation level measured in decibels, power handling, bandwidth, tuning resolution and accuracy, repeatability, and the impedance presented by the attenuator to the other circuitries.

The implementation of attenuators can be in different forms of balanced symmetric and unbalanced asymmetric configurations. Unbalanced configurations include T, Π, and L resistive networks. The balanced configurations include T (known as the H attenuator) and Π (known as the O attenuator). Attenuators can be realized as surface mount, waveguide, or coaxial. Also, they can be classified into unidirectional or bidirectional.

2.2.2 Capacitors

A capacitor is as an electric-field storage element measured in farads (F). A capacitor is an extensively used element in RF circuitry for resonant frequency tuning, bypassing, inter-stage coupling, DC blocking or decoupling element, and impedance matching. The basic structure of a capacitor consists of two parallel electrodes separated by a dielectric medium where electric energy is stored due to applying a potential difference across the capacitor plates. The capacitance value depends on the plate dimensions, separation, and permittivity of the material sandwiched between the two plates. Capacitance (C) in terms of the total charges (Q) at each electrode and the voltage (V) applied across them is defined as:

$$C = \frac{Q}{V} \tag{2.4}$$

The alternating current across the capacitor terminals is equal to the rate of change of the charge on the capacitor plates or the voltage change across the capacitor terminals as expressed mathematically by:

$$i(t) = \frac{dq}{dt} = C\frac{dv(t)}{dt} \tag{2.5}$$

Considering the voltage in a phasor form $v(t) = e^{j\omega t}$, (2.5) can be rewritten as:

$$i = C\frac{d(e^{j\omega t})}{dt} = j\omega Ce^{j\omega t} = j\omega Cv \tag{2.6}$$

And the voltage across the capacitor is expressed as:

$$v(t) = \frac{1}{C}\int i(t)\, dt \tag{2.7}$$

Assuming v is the root mean square (RMS) value of the voltage across the capacitor terminals, the stored electric energy (W_e) is given by:

$$W_e = \frac{1}{2}Cv^2 \tag{2.8}$$

According to Gauss's law which states that the total electric flux emerging out of a closed surface is equal to the charge enclosed divided by the material permittivity ($\varepsilon = \varepsilon_0 \cdot \varepsilon_r$), one can derive an equation for the relationship between the capacitance, the electrodes dimension, and the permittivity of the dielectric material located between the electrodes. Considering the following surface integral form of Gauss's law:

$$\oint_S E \cdot dA = \frac{Q}{\varepsilon_0 \varepsilon_r} \tag{2.9}$$

where ε_r is the material relative permittivity to the free space permittivity. ε_0 is the free space permittivity and equals 8.854×10^{-12} f/m.

Solving (2.9) for the electric field (E) lines traveling in one direction between the two plates as illustrated in Figure 2.6, we can obtain:

$$Q = \varepsilon_0 \varepsilon_r E A \tag{2.10}$$

where A is the area of the plates. The voltage across the plates that is needed to bring a positive unit charge from the negatively charged plate to the positively charged plate through the separation distance d can be defined as the gradient of electric potential as:

$$E = -\nabla V = -\frac{\partial V}{\partial Z} \Rightarrow V = -Ed \tag{2.11}$$

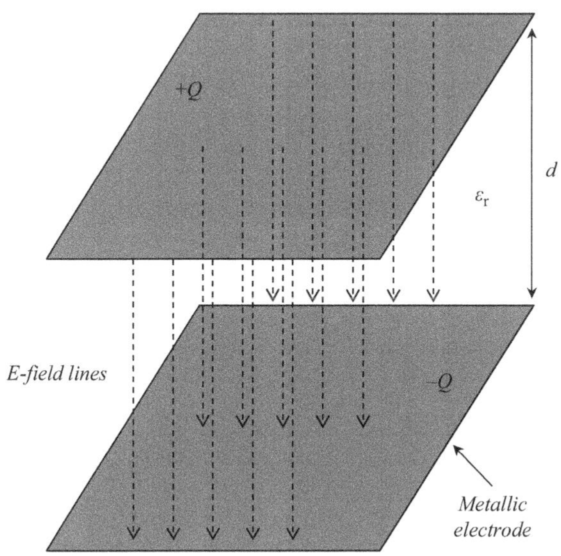

Figure 2.6 Basic structure of a parallel plate capacitor separated by a dielectric material showing the electric field lines traveling from a positively charged plate

Thus, from (2.4), (2.10), and (2.11), the capacitance can be expressed in terms of the area of the electrodes and separation between them as:

$$C = \varepsilon_0 \varepsilon_r \frac{A}{d} \tag{2.12}$$

The reactive impedance of inductors and capacitors possess opposite signs. The reactance of a capacitor is monotonically proportional to the signal frequency and its impedance imaginary part has a negative sign. The purely reactive impedance of a capacitor is given by:

$$Z_C = \frac{v}{i} = \frac{1}{j\omega C} \tag{2.13}$$

where $\omega = 2\pi f$ is the angular frequency, i and v are the current and the voltage phasors, and v/i is a complex number. The imaginary sign (j) is to indicate the $\pi/2$ phase shift between the voltage and the current for a sinusoidal input. Here, the phase of the electric current leads the phase of the voltage.

2.2.2.1 Capacitor modeling

The equivalent lumped circuit model of a chip capacitor illustrated in Figure 2.7 shows that both series and parallel resonance frequencies can exist. The model accounts for many parasitic components which become predominant at high frequencies. The parasitic inductance (L_s) is due to the finite chip terminal size, and the series resistance (R_s) is due to the contact and electrode resistances. The model also accounts for dielectric loss through the parallel resistor (R_p) and parasitic parallel capacitance (C_p) between the electrodes. From the capacitor circuit model, one can conclude that a capacitor behaves as an inductor above the series resonant frequency, also known as the self-resonant frequency (SRF).

Chip capacitors find many applications in multichip modules as a surface mount device that can be used for circuit resonance tuning or matching. Similarly, monolithic capacitors are an integral part of the MMICs process and are utilized in the design of several microwave circuits such as filters, baluns, and tunable circuits. Monolithic capacitors can be implemented as metal-insulator-metal (MIM) or as interdigital capacitors. The equivalent circuit model for a MIM capacitor is depicted in Figure 2.8. The MIM capacitor is formed usually of a thin low loss dielectric

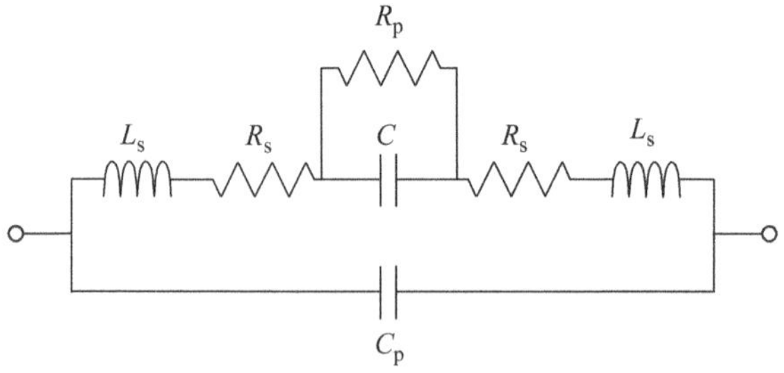

Figure 2.7 Equivalent circuit model of a chip capacitor

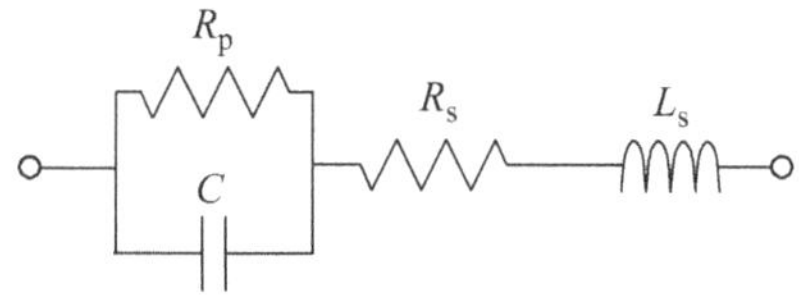

Figure 2.8 Equivalent circuit model of an MIM capacitor

material sandwiched between two plates. In semiconductor microelectronic devices, the commonly used dielectric material is either silicon dioxide (SiO_2) or silicon nitride (SiN).

2.2.2.2 Capacitor implementation

There exists a myriad of MIM chip and planar capacitors. Chip capacitors are available as multilayer, multiplate, and a pair of parallel plate capacitors; whereas planar capacitors can be realized as interdigitated capacitors composed of closely spaced fingers.

In the interdigital capacitor depicted in Figure 2.9(a), the capacitance per unit length (C) along its longitudinal dimension (W) can be approximately calculated from [1]:

$$C = \frac{\varepsilon_r + 1}{W} l[(n-3)A_1 + A_2] \tag{2.14}$$

where n is the number of fingers, l is the length of the finger in μm. W is the longitudinal dimension in μm. A_1 and A_2 are the interior and the two exterior capacitances per unit length of the fingers, respectively. This approximate relation is based on grouping several microstrip subsections (interior fingers, two outermost fingers, gaps, end-discontinuities, T-junctions, etc.) and doesn't include the parasitic effects between the fingers.

The total capacitance in pico-farad can be reduced to:

$$C_{\text{total}} = (\varepsilon_r + 1) l[(n-3)A_1 + A_2] \tag{2.15}$$

Figure 2.9(b) shows a configuration of a multilayer plate capacitor that assumes uniform separation and uniform overlap between the plates. In this structure, the total capacitance can be calculated in a similar fashion to (2.12) as:

$$C = \varepsilon_0 \varepsilon_r \frac{A(n-1)}{d} \tag{2.16}$$

where n is the number of plates in the multilayer capacitor.

The total capacitance of a parallel-plate capacitor shown in Figure 2.9(c) filed with different dielectrics stacked in series and sharing a parallel-plate area (A) is calculated through the following formula assuming uniform dielectric thicknesses:

$$C = k\varepsilon_0 \varepsilon_{r,\text{equiv}} \frac{A}{d_{\text{total}}} \tag{2.17}$$

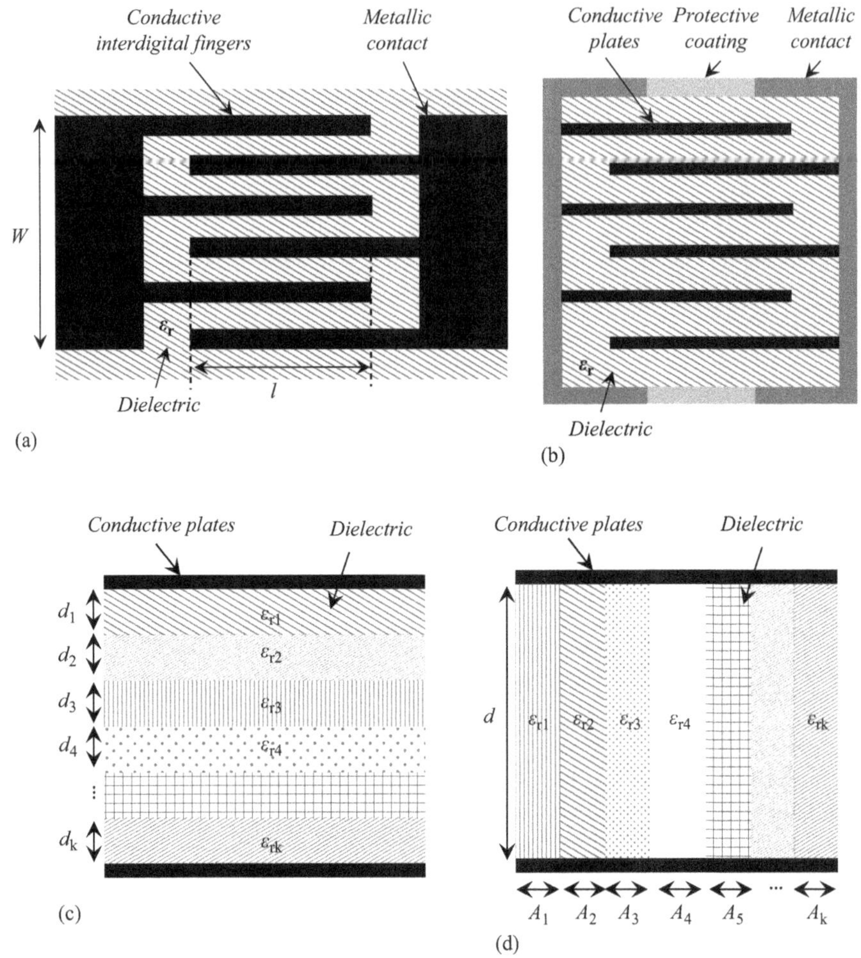

Figure 2.9 Different structures of lumped capacitors: (a) interdigital, (b) multiplate, (c) multidielectric—stacked in series, and (d) multidielectric—stacked in parallel

The equivalent dielectric constant $\varepsilon_{r,\text{equiv}}$ and the total stack thickness are given as:

$$\frac{1}{\varepsilon_{r,\text{equiv}}} = \frac{1}{\varepsilon_{r,1}} + \frac{1}{\varepsilon_{r,2}} + \cdots + \frac{1}{\varepsilon_{r,k}} \tag{2.18}$$

$$d_{\text{total}} = d_1 + d_2 + \cdots + d_k \tag{2.19}$$

where d_{k} and $\varepsilon_{r,k}$ are the thickness and dielectric constant, respectively, of the kth layer.

The capacitance of a parallel-plate capacitor filed with different dielectrics stacked in parallel as illustrated in Figure 2.9(d) is calculated through the following formula assuming uniform area of the successive dielectric segments as:

$$C = \frac{1}{k}\varepsilon_0\varepsilon_{r,\text{equiv}}\frac{A}{d} \tag{2.20}$$

and

$$\varepsilon_{r,\text{equiv}} = \varepsilon_{r,1} + \varepsilon_{r,2} + \varepsilon_{r,3} + \cdots + \varepsilon_{r,k} \tag{2.21}$$

Capacitive gap coupled conductors also provide another form of capacitors when the short dimension edges of the conductive lines are coupled with each other or with a ground line to construct a series or a shunt capacitor, respectively. At the gap between the conductors, the electric field lines dominate and the magnetic field is very weak. The capacitance to the gap is dependable on the gap size, the electrical length of the conductive line, and its impedance defining its linewidth.

High capacitance values can be attained using several implementation techniques. For example, by using multilayer MIM and interdigital configurations, incorporating high dielectric constant materials and ultra-thin-film dielectrics, or by shaping the geometry of the plates using fractal shapes where one can increase the total capacitance of a capacitor by virtue of the lateral fringing field capacitance. In addition, the use of ferroelectric materials such as a composition of barium strontium titanate (BST) provides a high dielectric constant material that can also be tunable.

Improving the quality factor of a capacitor can be obtained by using low loss dielectric material, thick and high conductivity metal for the electrodes, and by using suspended substrate to reduce parasitic loss between the bottom electrode and the semiconductor substrate as the case of a monolithic integrated capacitor on a Si-substrate. Capacitors with *Q*-factor above 100 can be realized using SiO_2, Si_3N_4, and Ta_2O_5 dielectric materials [1].

2.2.2.3 Electrostatic discharge in capacitors

Electrostatic discharge (ESD) is one of the important reliability considerations associated with the manufacturing, assembling, handling, and testing of microelectronic integrated circuits and components. ESD can have a catastrophic impact on the performance of an integrated circuit and its production yield, especially, for small size devices with very thin conducting paths. This impact can be either in the form of instantaneous or latent failure where a device performance degrades over time. Applying voltage across the capacitor plates leads to the accumulation of static charges on the plate surfaces and upon reaching the dielectric breakdown voltage, the static charges start to transfer between the two plates causing an ESD. The capability of capacitors to withstand ESD pulses depends on the plate area, the geometry of the electrodes and the number of sharp edges on the plates, environmental conditions such as humidity, temperature and contamination, and the dielectric field strength of the medium between the plates. All these elements contribute to the determination of the maximum voltage rating of a capacitor component. The dielectric field strength is an intrinsic property of the material that represents the maximum electric field a material can withstand without breaking down and form an electrically conductive path. This path of voltage breakdown causes a permanent modification to the dielectric material in the failing path.

The assessment of ESD can be carried out through one of the following standardized models such as the human body model (HBM), charged device model (CDM), machine model (MM), and system level ESD. Their test procedures are defined by the reliability requirements per *Joint Electron Device Engineering Council* (JEDEC) standards.

The HBM is used to simulate the electrostatic discharge effect from the human body into the device under test (DUT). The model uses a series RC network consisting of a 100 pF capacitor and a 1.5 KΩ resistor for discharging and a 1 MΩ is used for charging as shown in Figure 2.10(a).

The CDM emulates the effect of discharge from an electrical circuit into the DUT; hence in practice, this might occur when there is direct metal contact between the device and the manufacturing environment such as touching test handlers or other handlers in the wafer fab environment. CDM is a more rigorous model than the HBM as the current levels are higher than those assumed for HBM. This is represented in the CDM by excluding the current limiting resistor in the discharge path as depicted in Figure 2.10(b).

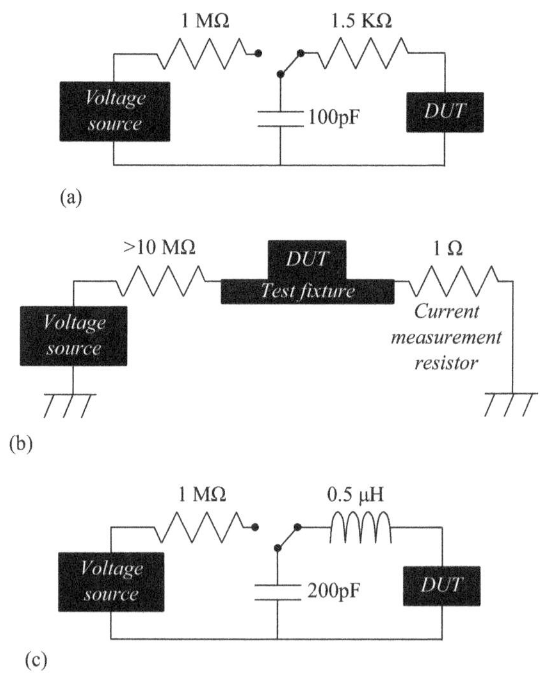

Figure 2.10 Different standard ESD models: (a) human body model, (b) charged device model, and (c) machine model

While the HBM and the CDM are component-level models, the third test is more concerned of the system level ESD which simulates the discharge effect of an object into an end-equipment or an electronic module.

The MM is used to simulate the discharging effect of a machine (e.g., handling or assembly machines) into the DUT. The model is similar to the HBM circuit setup; however, the charging capacitor value is 200 pF and the series resistance is replaced by an inductor of 0.5 μH for the discharging path as illustrated in Figure 2.10(c). The use of an inductor results in an oscillating current reaching the DUT. While HBM and MM provide similar failure mechanism due to their relatively similar pulse rise time in the order of 10 ns, CDM follows a different mechanism as it exhibits faster rise time in the order of 100 ps. The MM test is less commonly used than the HBM test, as usually passing the HBM test guarantees passing MM but the opposite is invalid.

Generally, the dielectric strength of a material decreases with increased frequency, material thickness, dielectric constant, and operating temperature. However, the dielectric strength increases with the increase in the capacitor electrode area. In order to protect against ESD, replacing the dielectric material between the capacitor plates with a higher field strength material may seem to be an option. However, it is a nonpreferable solution because it requires a change in the device fabrication process. Small size capacitors are sensitive to ESD damage. Practically cascading capacitors in series to increase the electrode area instead of using a single capacitor help to protect against ESD. Additionally, changing the electrode shape in MIM capacitors improves the ESD performance by avoiding sharp edges, for example, by using circular shape electrodes instead of rectangular or square shapes.

2.2.2.4 On-module electrostatic discharge protection

The fact that ESD could have a voltage of several thousand volts and could cause severe damage to the internal components of an IC within an RF module, effective ESD protection devices are then necessary to reduce the peak voltages before they strike the sensitive ICs. The methods to realize ESD protection on-module differ from each other with respect to their principle of operation, the characteristics of the different ESD events, and the circuit mode of operation (i.e., low speed or high-speed circuit). Usually, in RFIC transceiver modules, the antenna pin is the most ESD sensitive I/O port as it connects the module circuits to the outside world. On-chip ESD protection is somewhat not feasible due to the small form factor of the microscale chips, and the type of the chip stack material that may not be applicable for realizing an ESD protection device.

An ESD protection component when placed in a shunt configuration to ground at the module's I/O pin presents a high impedance path to the current flowing in the circuit while it allows a low-resistance path to ground during an ESD event. These components need to be placed as close as possible to the module sensitive I/O pins for effective ESD protection. Proper placement of an ESD protection device on the module can be established by avoiding lengthy traces to eliminate unwanted inductance that can obstruct the ESD current path to ground since any RF inductance increases the RF impedance. It is also desirable to have a low capacitance ESD protection at gigahertz applications to avoid impedance loading effects. Some products require two-stage protection: one at the module pins that are meant to suppress direct ESD energy and a second near the IC peripheral to mitigate any residual ESD stresses left after and couldn't be sufficiently suppressed by the first stage.

ESD protection devices could be active or passive components that include resistors, decoupling capacitors, inductors, ferrite beads, transient voltage suppressors (TVSs), spark gaps, or LC filters. The LC filters are effective in blocking transients such as ESD pulse amplitudes and pulsed electromagnetic interference (PEMI) events. The air breakdown spark gaps can be realized using metallization patterns formed on the module or on the chip package. However, using spark gaps as ESD protection are limited in terms of their repeatability, reliability, and response time to ESD. One of the simplest means of protecting against ESD current pulse is having a large shunt inductor connected to the sensitive pins of the module. The TVS devices are used to address system-level ESD events and considered as a low-cost solution to protect sensitive ICs connected through external plug pins. As the name indicates, they work as voltage clamping devices that suppress any high voltages above the breakdown voltage. TVS devices include diodes, polymer voltage suppression (PVS), varistors, spark gaps, or a combination of them. The PVS is a self-resetting polymer with structures of low profile film materials capable of increasing its resistance as a function of current. PVS is also capable of resetting after the occurrence of ESD current. The varistors are voltage-dependent resistors realized usually based on ceramic technology. A varistor has an initial high impedance value and changes to a low impedance value once an ESD occurs to allow ESD current to flow to ground and provide overvoltage protection to the IC that is connected in parallel to it. The TVS diode possesses relatively faster response time and comes in two types: bidirectional and unidirectional. A bidirectional diode is constructed using two opposing Zener diodes in a series configuration.

Note that any ESD protection strategy causes additional insertion loss to the RF chain and therefore leads to RF performance challenges. The selection of the proper ESD protection method and co-designing the RF chain with it are then important for realizing a competitive performance.

2.2.3 Inductors

An inductor is a magnetic field storage element measured in units called Henrys (H). Inductors are one of the significant elements that take place in many MIC designs as matching components, phasing elements, and RF chokes. Their realization can be in the form of embedded inductors in a multilayer substrate or as a surface mount device. Considering a loop of current (I) and a magnetic flux (ϕ) that passes through the loop area S, the inductance (L) can be calculated as:

$$L = \frac{\phi}{I}, \quad \phi = \oint_S B \times dS \tag{2.22}$$

The current and voltage relation across the inductor terminals are expressed mathematically in (2.23) and (2.24), respectively:

$$i(t) = \frac{1}{L}\int v(t)\, dt \tag{2.23}$$

$$v(t) = L\frac{di(t)}{dt} \tag{2.24}$$

From (2.24), the voltage across an inductor is proportional to the time rate of change of the current passing through it. Thus, a time-invariant current will result in a zero voltage across the inductor. The reactance of an inductor is proportional to the signal frequency and its impedance imaginary part has a positive sign. The purely reactive impedance of an inductor is given by:

$$Z_L = j\omega L \tag{2.25}$$

Assuming i is the RMS value of the voltage across the inductor, the magnetic energy stored in an inductor is given by:

$$W_m = \frac{1}{2}Li^2 \tag{2.26}$$

2.2.3.1 Inductor modeling

Figure 2.11 illustrates two simple equivalent circuit models of an integrated inductor that is capable of providing accurate inductor behavior models over a limited range of frequency.

Among the several existing inductor configurations, the focus in this section will be on modeling a spiral inductor on a silicon substrate as they are commonly used in the microelectronics semiconductor devices. In the literature, there exist several models for spiral inductors patterned on a semiconductor substrate comprising frequency independent and frequency dependent models. For instance, several frequency independent models have been reported in literature including symmetric and asymmetric π-networks, series RL, and series RL with a parallel C.

A conventional spiral inductor is essentially asymmetry in terms of the inductor layout and therefore an asymmetric π-model would be more realistic to represent it. However, symmetric π-models also exist for analysis simplifications. In Figure 2.12, the capacitance C_0 models the overall distributed turn-to-turn capacitance of the windings and R_s and L represent the total resistance and inductance of the inductor, respectively. C_{ox} represents the capacitance between the spiral metal and the oxide layer on top of the Si substrate and is usually formed as SiO_2 layer. $C_{sub1,2}$ and $R_{sub1,2}$ model the capacitance and resistive loss, respectively, due to the Si substrate. In this model above the self-resonant frequency, the

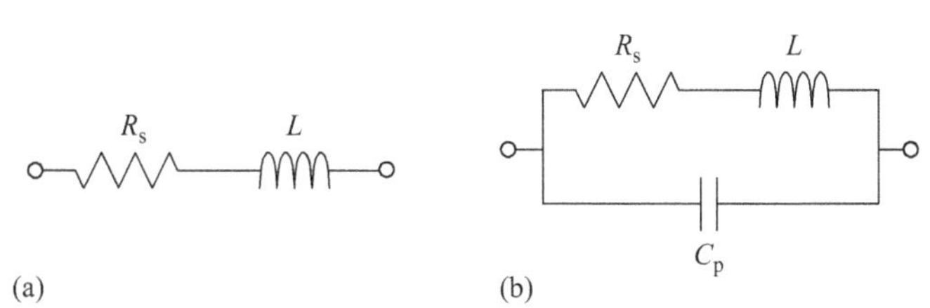

Figure 2.11 Simple equivalent circuit models of an inductor: (a) series RL and (b) series RL in parallel with C

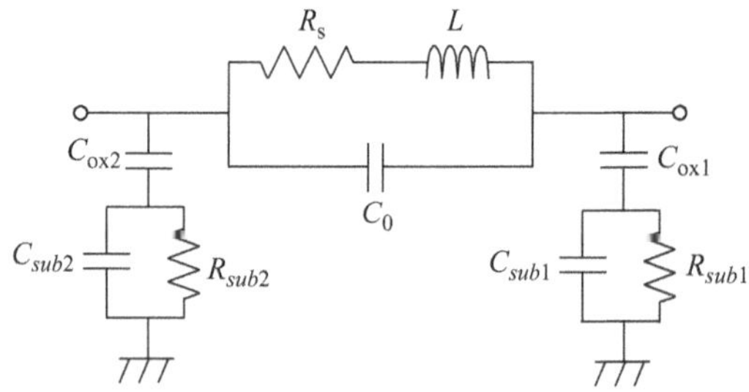

Figure 2.12 π-equivalent lumped circuit model of a spiral inductor on a Si substrate

parasitic capacitance begins to dominate the behavior of the equivalent circuit and the impedance starts to drop by increasing the frequency.

In general, frequency independent models such as the π-network shown in Figure 2.12 are limited in terms of accuracy and the frequency range they can cover. On the other hand, a frequency dependent model provides an accurate alternative; however, such models require the calculation of the model elements at each frequency point of interest. Rautio's synthesis approach [5] is one method that can be utilized to construct a frequency dependent π-model of a spiral inductor on a Si substrate. In essence, the synthesis approach uses *S*-parameter data either from measurement or simulation to construct the lumped branch model depicted in Figure 2.13(a) using closed-form equations explained in [5]. Then, the construction of the π-network depicted in Figure 2.13(b) can be used to represent an accurate model of the spiral inductor.

2.2.3.2 Inductor implementation

An inductor can be implemented in different configurations either as planar or three-dimensional (3D) structures. Each type has its own advantages and disadvantages in terms of *Q*-factor, total inductance per unit area, and SRF. A planar inductor is usually realized as a single microstrip line, a meander line, a single loop, or a spiral on a printed circuit board (PCB).

Typically, the single microstrip line, meander, and single loop inductors provide inductances in the range of 0.5 nH for on-chip realization and are seldom used. A spiral inductor depicted in Figure 2.14(a) can provide more inductance per unit area due to the relatively higher mutual coupling between the winding line that adds up to the self-inductance. However, spirals exhibit lower SRF when compared to the inductor configurations in Figure 2.14(b) and (c). Rectangular and hexagonal spirals are the most commonly used structures in RFIC designs if compared to other possible shapes (e.g., circular spirals) due to its simple layout and for designs that can tolerate slight degradation in the *Q*-factor of the inductor.

3D inductors are available in the form of helical and solenoid inductors as illustrated in Figure 2.15. They are advantageous when implemented on a multi-layered substrate as they can utilize small layout area when compared to spiral inductors of equivalent inductance. However, the coil of a 3D inductor is more

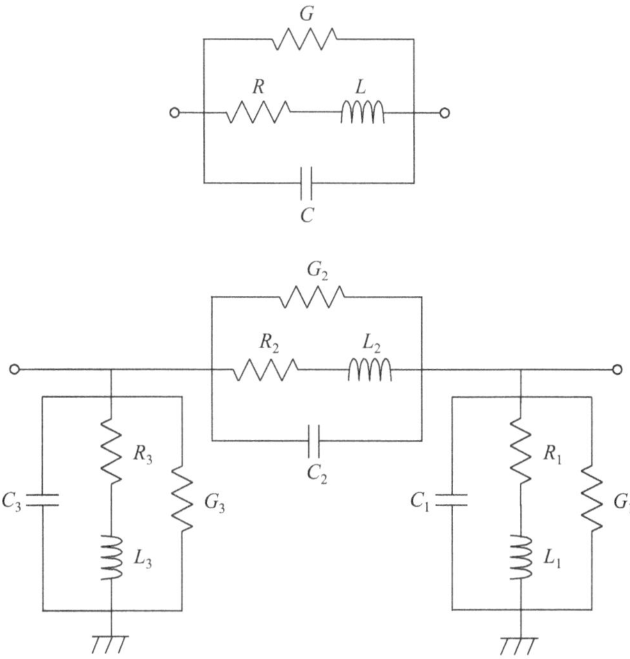

Figure 2.13 *(a) Rautio's frequency dependent lumped circuit model of a two-port network, and (b) frequency dependent π-equivalent lumped circuit models of a spiral inductor on a Si substrate*

Figure 2.14 *Different configuration of planar inductors: (a) spiral, (b) loop, and (c) meander*

exposed than a planer inductor to EM parasitic effects from other components sharing the same EM environment. In some cases of RF transceiver module designs, such inductors can lead to isolation problems if not shielded properly.

2.2.3.3 Mutual inductance

According to Faraday's law, a conductor carrying an alternating current produces a varying magnetic field that induces voltage and current on a secondary conductor in proximity. This effect of changing the current of one conductor altering the current and voltage in an adjacent conductor is known as mutual inductance (M). Having two inductors located in close proximity make their magnetic flux lines to intersect

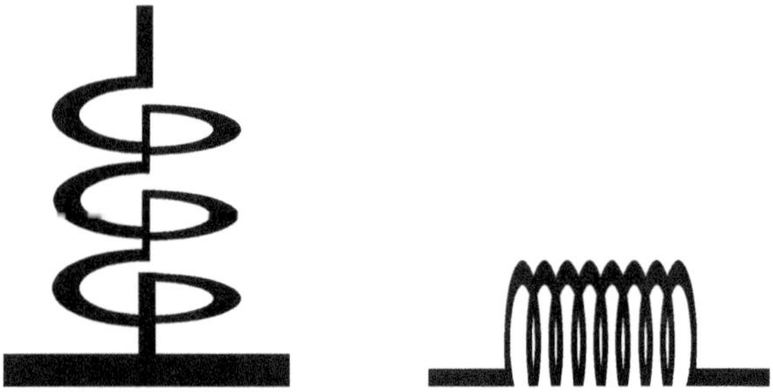

(a) (b)

Figure 2.15 Three-dimensional inductors: (a) a multilayer three quarter turns helical and (b) a solenoid

which result in a mutual coupling. Depending on the current direction in each inductor, the magnetic flux between the inductors may have aiding or canceling effect on the total inductance. If the currents on the two conductors are flowing in the same direction, the inductance of each coil increases, whereas the opposite holds true for currents flowing in opposite directions. The amount of increase or decrease in inductance is quantified as the mutual inductance and is defined as:

$$M = k\sqrt{L_1 L_2} = \frac{L_A - L_B}{2} \tag{2.27}$$

where k is the coupling factor which determines the percentage of magnetic flux coupled from one coil to another, L_1 and L_2 are the self-inductance of the two inductors. Here, $k = 1$ means that the magnetic flux produced by one conductor is fully coupled to an adjacent conductor (i.e., an ideal transformer). Practically, the magnetic flux is passed partially to the other coil and k usually has values ranging between zero to less than one. Coupled inductors and similarly coupled microwave resonant structures can be categorized based on their coupling factor into two groups. For $k < 0.5$ the structure is said to be loosely coupled and for $k > 0.5$ is referred to as tightly coupled. The mutual inductance can also be expressed in terms of the common and differential mode excitation of the two inductors. In (2.27), L_A and L_B are the total inductance of the two conductors when the current through them flows in the same and opposite direction, respectively. Under the mutual coupling condition, the self-inductance of each of the coils at the two different current flowing directions is given by:

$$L_1 + M, \quad L_2 + M \text{ (same current direction)} \tag{2.28}$$

$$L_1 - M, \quad L_2 - M \text{ (opposite current direction)} \tag{2.29}$$

Using (2.28) and (2.29), in the case of two inductors connected in series and mutually coupled, the total equivalent inductance is given by:

$$L_{eq} = L_1 + L_2 \pm 2M \tag{2.30}$$

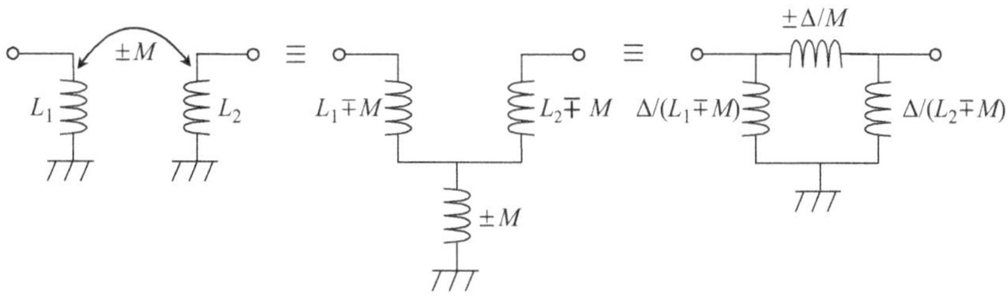

Figure 2.16 π- and T-equivalent circuits of a pair of mutually coupled inductors and the corresponding signs of $M \times \Delta = L_1L_2 - M^2$

In the case of two inductors connected in parallel and mutually coupled, the total equivalent inductance is:

$$L_{eq} = \frac{L_1L_2 - M^2}{L_1 + L_2 \pm 2M} \tag{2.31}$$

The ± sign is determined by the current direction flowing in each inductor.

Mutually coupled inductors can be transformed into an equivalent Π- or T-noncoupled inductors as in the configurations depicted in Figure 2.16. Similarly, the sign of the mutual inductance depends on the current flow direction in the mutually coupled inductors [6].

Designing a circuit that includes printed inductors can benefit from this equivalency. For example, series and shunt inductors in a circuit can be designed as mutually coupled pairs or more to reduce the total inductance that needs to be implemented in a multilayer laminate by forming a T-inductor connection.

In mutually coupled inductors and transformers, the dot notation is used to indicate the different winding direction which is important in determining the polarity or phase of the induced voltage from one inductor with respect to the other. Dot convention is also necessary for writing the appropriate circuit equations. The rule for using the dot convention in a magnetically coupled coils is that when the direction of the reference current enters the dotted terminal at one of the inductors, the polarity of the induced voltage on the other coil at the dotted terminal is positive and vice versa. In other words, the in-phase coil terminals are usually marked with a dot.

2.2.3.4 Via holes

In RFIC circuit and module designed with very tight specifications, it is important to attain a low inductance grounds to be able to achieve the anticipated electrical performance. The design of most RF circuits begins with an initial optimization process using low order models of the circuit components that are obtained from empirical or analytical models. In either case, initial optimization results are viewed with the assumption that the ground plane has no loss and zero inductance in the RF return path of the signal. Such an approach will demonstrate the best circuit performance that is not necessarily realizable. In reality, a ground plane shared between the different circuits possesses an amount of inductance even though careful attention is taken at the design stage to establish good grounding (i.e., by extending the ground plane or adding a large number of vias in a multilayer structure to reduce the ground inductance).

The use of vias to connect the terminals of a circuit to the ground plane and other components is becoming inevitable, especially in high-density RF/MICs. This is because using alternative methods such as wire bonds and ribbon bonds are not efficient in terms of realizing compact, flexible, and complex layouts. In this context, several studies have been carried out to understand and model the impact of vias on the circuit performance by looking, for example, into the amount of inductance and parasitic capacitances contribution when passing through a multilayer metallization structure. A via hole filled with metal provides a connection to the metal layers in a multilayer fabrication technology, either to create a connection to ground or a connection between the circuit elements/traces. The fabrication of vias is compatible with MMIC technology. They can be realized by laser drilling, dry, or wet etching the dielectric material to create a hole and subsequently filling it with a metal paste to make the interconnection from layer to layer or layer to ground. Vias also serve other purposes such as providing good thermal paths (low thermal resistance path) and heat-sinks for the active devices in a module. This in return assists in lowering the noise floor and improving the reliability of the active devices. On-chip via holes or pillars also provide thermal paths to the heat-sinks in addition to providing mechanical support to the device structure.

Among the several models that are available to describe the electrical behavior of a via over a wide range of frequencies, the analytical models are sufficiently accurate for vias with dimensions much smaller than the operating wavelength as the case in microelectronics RFICs and modules operating in the gigahertz range. Although full-wave electromagnetic (EM) modeling provides accurate performance results, it requires long computational time and may not be a time-efficient option for providing quick solutions at the initial design stage of a circuit or a system. Analytical models can be made easily available within the circuit simulator environment to provide instantaneous model results.

Typically the structure of a via consists of a capture pad on the metal layer to allow traces to connect to it or an SMD to land on it. A via model has a series inductance (L_{via}) that is due to the physical length of the via in parallel with a parasitic capacitance (C_{via}) to represent the capacitance between the capture pad of the via and the ground, and the cylinder body of the via to the metal layers crossed by the via (also called anti-pad capacitance or coaxial capacitance). The anti-pad capacitance surrounding the via hole is usually small and is ignored in this discussion. The influence of the series inductance is more significant than the parasitic capacitance in the case where a via runs through few thin metallization layers or when a via is connected directly to the ground plane without crossing any metal layers. A series resistance (R_{via}) is required to indicate the conductor loss which is dependable on the radius of the via and the skin depth of the used conductor.

The model in Figure 2.17(b) can be used to represent the loss, inductance, and additional parasitic capacitances of the via structure illustrated in Figure 2.17(a).

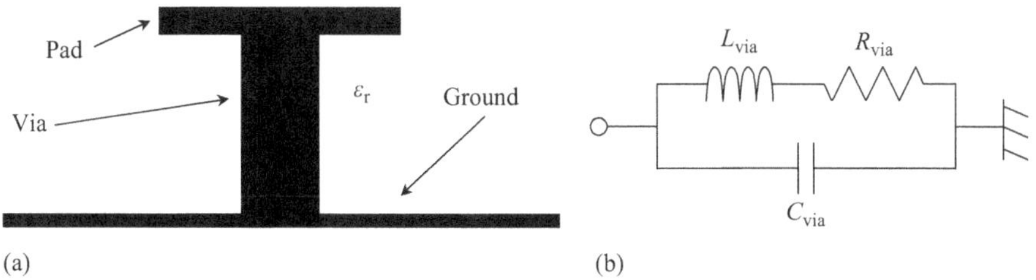

Figure 2.17 (a) Structure of a via hole to ground and (b) its equivalent circuit model

An analytical expression for the inductance, L_{via}, of a cylindrical via expressed in pico-henries (pH) is given by [7]:

$$L_{via} = \frac{\mu_o}{2\pi}\left[h \cdot \ln\left(\frac{h + \sqrt{h^2 + r^2}}{r}\right) + \frac{3}{2}\left(r - \sqrt{r^2 + h^2}\right)\right] \tag{2.32}$$

where r and h are the radius and height of the via, respectively, in microns. The resistance of the via can be approximated using the following formulas:

$$R_{via} = R_{dc}\sqrt{1 + \frac{f}{f_\delta}} \tag{2.33}$$

$$R_{dc} = \frac{h}{\sigma\pi\left(r^2 - (r - t)^2\right)} \tag{2.34}$$

$$f_\delta = \frac{1}{\sigma\pi\mu_o t^2} \tag{2.35}$$

where f is the operating frequency, t is the metal thickness, σ is the conductivity of the metal, μ_o is the free-space permeability, and f_δ is the skin effect corner frequency.

When aiming for a very low inductance via, one can use a double or a set of vias connected in parallel instead of a single via. The inductance of a double via will be lower than that for a single via hole. However, it may not result in half of the inductance value of a single via because of the mutual coupling between them that also depends on the separation and the length of the vias.

A metal filled via is advantageous over a plated via because of its lower thermal resistance. Also, a via hole filled with metal has a lower DC resistance than a plated via of the same length and diameter. Slot vias is another type of vias that can provide less inductance when compared to the double via configuration; however, they are not commonly used because they are difficult to manufacture than cylindrical vias.

2.2.3.5 Crossovers and airbridges

Routing of interconnect traces in high-density packaged electronic circuits and RF modules is a challenging task for both design and layout engineers. In many occasions, signal channels must geometrically cross each other as in Butler-matrix beam-switching antennas, transformers, spiral inductors, MIM capacitors, Lange couplers, and some transistor configurations. The complexity of layout designs combined by the limitation of the available footprint area to allow flexible routings places a high demand on designing trace crossovers, bond wires, or airbridges that can support high isolation between the crossing signal traces over a wide frequency range. An airbridge uses air as a dielectric between the two crossing conductors and realizing it requires essentially two metallization layers. The common drawbacks of the wire bonds, airbridges, and wired vias structures are their limited bandwidth and are not convenient for certain low-cost layout designs.

Alternatively, crossovers can do the same function using a single metal layer and thereby offers a low-cost solution. A single layer four-port network can allow two signal paths to physically crossover while maintaining high isolation. In this aspect, different configurations of single-layer crossovers, such as cascaded couplers, ring and multisection branch-line couplers, have been proposed [8,9]. Planar crossovers require relatively larger space to realize; however, in many applications the increase in size can be justified over having an expensive multilayer structure, especially, in mass-produced products. Figure 2.18(a) illustrates the configuration of an ideal symmetric crossover comprised of a four-port network that has a scattering matrix described in (2.36) which allows a pair of intersecting signal lines to cross each other while maintaining ideal isolation between them. Figure 2.18(b) shows a possible microstrip or stripline realization of a crossover comprised of two branch-arm hybrids cascaded so that the diagonal ports passes the signal while maintaining good isolation between them [9]:

$$S_{\text{crossover}} = e^{j\theta}\begin{bmatrix} 0 & 0 & 1 & 0 \\ 0 & 0 & 0 & 1 \\ 1 & 0 & 0 & 0 \\ 0 & 1 & 0 & 0 \end{bmatrix} \tag{2.36}$$

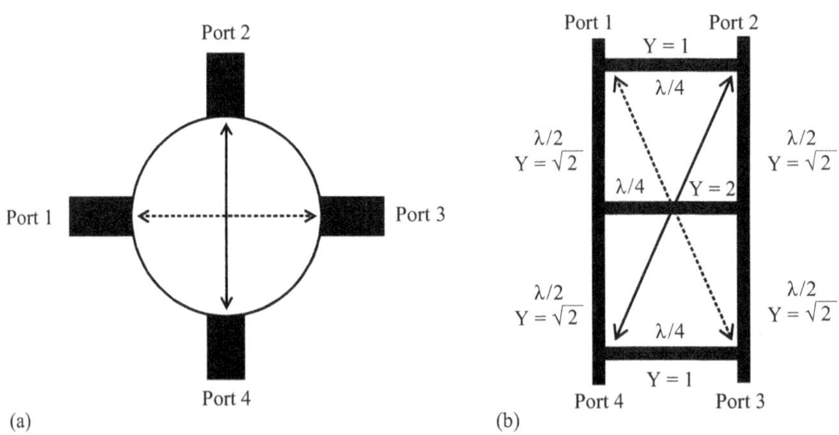

Figure 2.18 (a) A schematic of a symmetric four-port single-layer crossover, and (b) the realization of a crossover structure using two cascaded branch-arm hybrid couplers

2.3 Distributed elements

Although lumped components provide relatively broader bandwidth than their corresponding distributed elements, distributed elements are used to achieve high Q-factor reactive elements when the circuit operates at sufficiently high frequency. For example, in the synthesis of microwave filters, a lumped element topology derived from a transfer function is eventually transformed into its equivalent distributed elements depending on the implementation technology of interest. Table 2.1 summarizes commonly used distributed elements and their corresponding lumped equivalents that are perfectly accurate only at a single design frequency (f_0). This includes interconnecting TLs or discontinuities, quarter wavelength transformers, open- and short-circuited stubs as dual to lumped element capacitors and inductors, respectively, and TL gap that represents a series branch capacitor.

A transmission line is a means of transferring and guiding wave signals. In electrically large TLs, it is difficult to describe the voltages and currents along the TL using a simple lumped-element equivalent circuit. Thus, a distributed equivalent circuit which describes each short segment of the line by a lumped element network (unit cell) is used. A lossless transmission line will comprise only distributed LC elements. A distributed line realized as stripline, coplanar waveguide (CPW), or microstrip line and possesses a narrow linewidth is equivalent to lumped inductor if the electrical length is less than 90° accompanied with a small capacitance. The opposite also holds true for a wide width TL, which is equivalent to a capacitor with a small amount of inductance. This relation can be illustrated by examining the TL input impedance (Z_{in}) function expressed as:

$$Z_{\text{in}} = Z_0 \frac{Z_L - jZ_0 \tan(\theta)}{Z_0 + jZ_L \tan(\theta)} \tag{2.37}$$

where Z_0 is the characteristic impedance of the line and Z_L is the load impedance placed at an electrical distance θ from the source. Now, if we consider Z_L as an open circuit, the input impedance of the line is defined as:

$$Z_{\text{in\_open}} = -jZ_0 \cot(\theta) \tag{2.38}$$

In the case where the TL is terminated with a short circuit, the input impedance is then given as:

$$Z_{\text{in\_short}} = jZ_0 \tan(\theta) \tag{2.39}$$

Hence for $\theta < 90°$, the open circuit case in (2.38) has a negative value that indicates the capacitive behavior, while the impedance equation of the short circuit case in (2.39) has a positive value and indicates an inductive behavior. This implies that a lumped inductor or a capacitor in a circuit can be replaced with its equivalent transmission line of characteristic impedance Z_0 and electrical length θ. The shorter the transmission line forming a reactance element, the better the element behavior over a wide frequency range. This is apparent in the transformation of the first four elements in Table 2.1 where $\lambda/8$ open and short-circuited stubs are

Table 2.1 *Commonly used distributed elements and their corresponding equivalent lumped circuit*

Shunt short-circuited stub	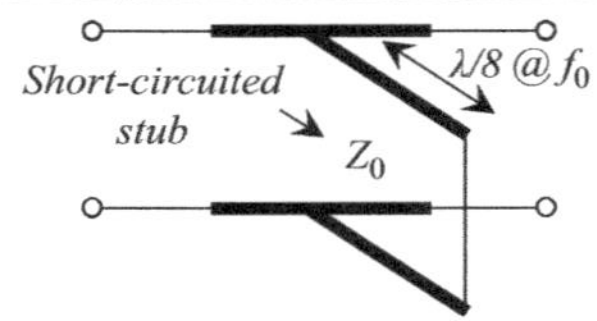	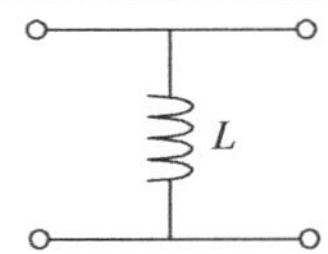
Series short-circuited stub	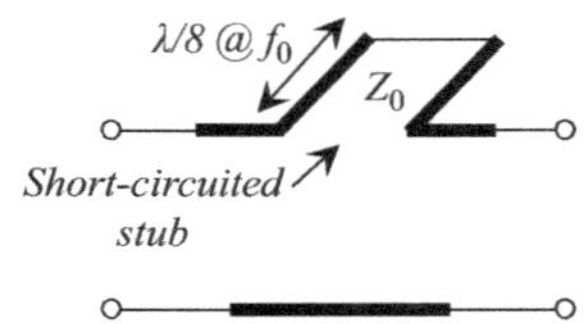	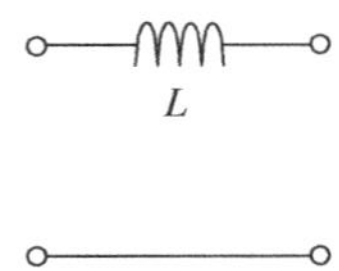
Shunt open-circuited stub	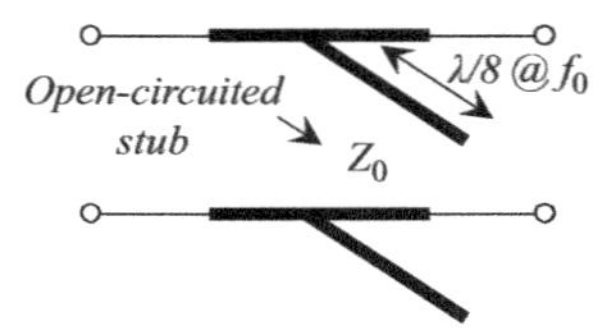	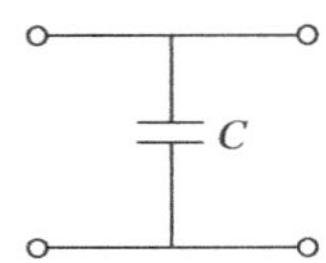
Series open-circuited stub	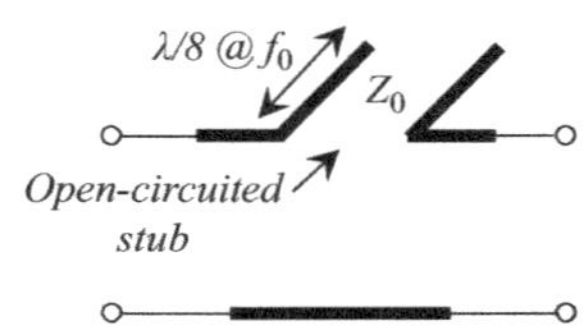	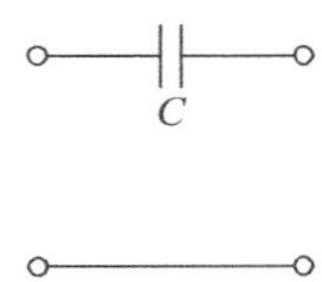

Series capacitance

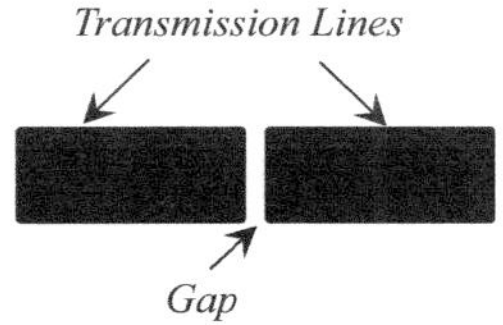

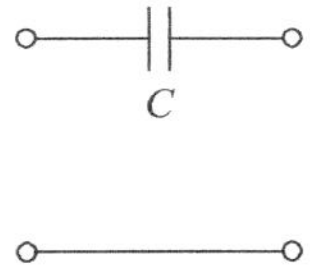

Impedance inverter or quarter wavelength transformer

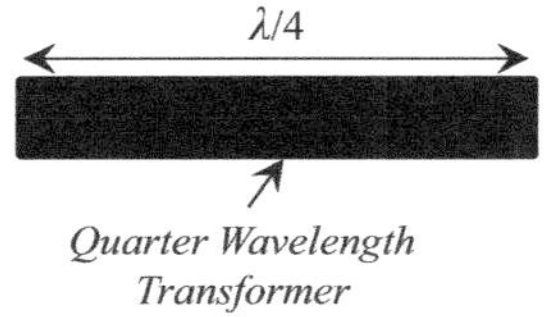

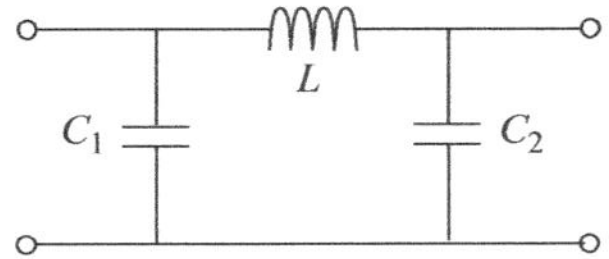

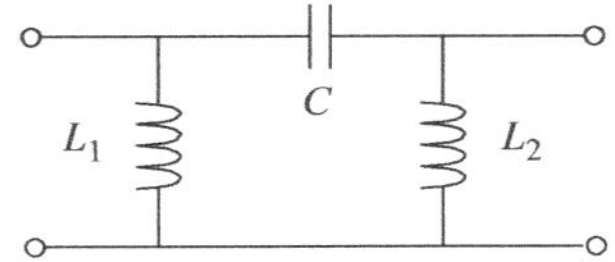

Stepped impedance (low-high-low)

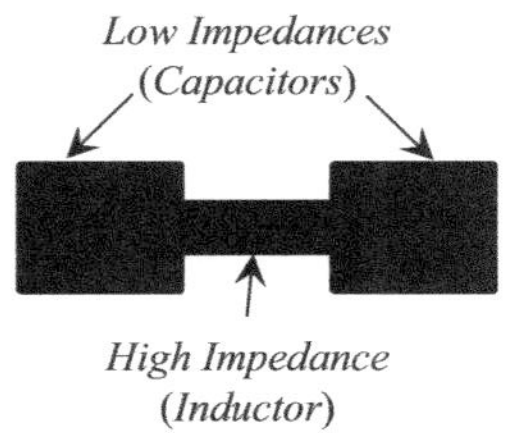

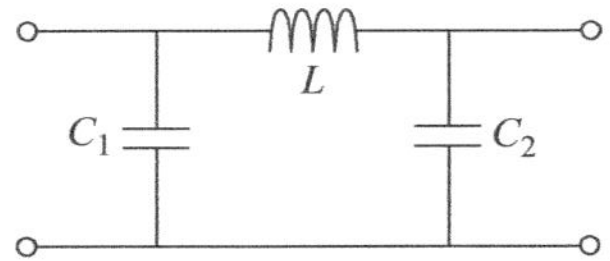

Stepped impedance (high-low-high)

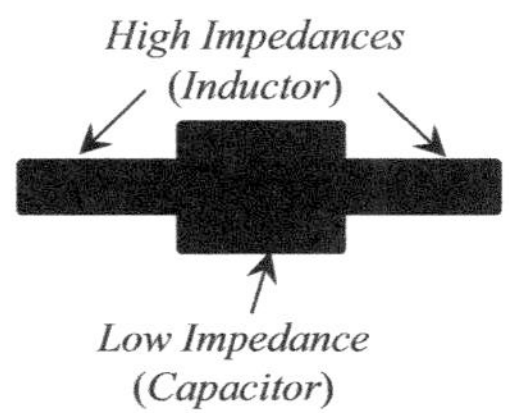

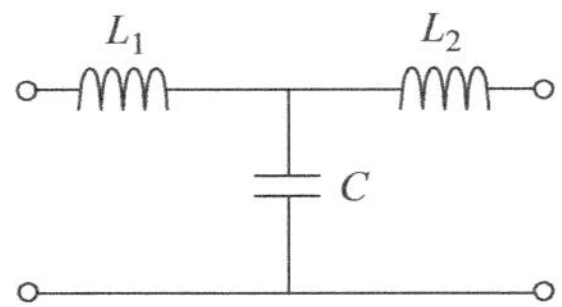

exploited to represent capacitor and inductor elements, respectively. This transformation between lumped and distributed components is known as Richards' transformation [10] that is accurate over a certain frequency range. Note that the reactance slope of the lumped element is different than its equivalent distributed element. For example, looking into the distributed reactance of an inductor we can find that it increases with frequency as a function of $\tan(\theta)$ while a lumped inductor increases linearly with frequency. Given that $\tan(\theta) \approx \theta$ for $\theta << 90°$, the choice of the transmission line length of 45° or shorter is preferable to ensure linear behavior.

Some of the lumped elements-based designs are constrained by the components values that they can achieve. Similarly, the realizability of distributed elements is more critical because of the difficulty in realizing transmission lines with very high and very low characteristic impedances. In addition, depending on the implementation technology in use, microstrip-based distributed circuits enable simple component placement and PCB assembly when compared to stripline circuits which are relatively difficult and expensive to fabricate. However, stripline-based circuits deliver excellent high-frequency performance and good noise immunity between its adjacent circuit elements, unlike CPW and microstrip-based designs.

The fact that the behavior of a TL repeats itself at multiples of 180° in electrical length makes the use of distributed elements undesired when a specific wide stopband performance is required. Spurious parasitic waves in both microstrip- and CPW-based circuits can be an issue in many applications; therefore, distributed elements should be less than a quarter wavelength long at the highest operating frequency of interest.

For broadband applications, wave dispersion is a significant parameter. Microstrip-based distributed circuits are dispersive by nature since a wave traveling through a microstrip line encounters inhomogeneous media composed of air and dielectric. CPW can exhibit less effective inhomogeneity than microstrip-based elements when a tight ground–signal-ground coupling is achieved such that most of the E-fields are traveled through the air medium.

2.4 Design considerations

2.4.1 *Self-resonant frequency*

The resonant frequency is established when the magnitudes of the inductive and the capacitive reactance in a circuit are equal, $X_c = X_L$. SRF of a component relies on the parasitics in the structure. At resonance, the sum of the total reactive impedances of a components equals to zero and only the resistive part of the impedance exist. Therefore, the SRF of a reactive component assuming series LC representation can be obtained by:

$$j\omega L + \frac{1}{j\omega C} = 0 \tag{2.40}$$

and

$$f_{\mathrm{SRF}} = \frac{1}{2\pi\sqrt{LC}} \tag{2.41}$$

where L is the series inductance and C is the series capacitance of the component.

SRF of an inductor is the frequency at which the total inductive reactance and parasitic capacitive reactance are equal. At SRF, an inductor is no longer an energy storage unit; thus, the choice of an inductor component so that its SRF is far as possible from the circuit design frequency is crucial for achieving a realizable design. Hence, an inductor becomes a capacitor above the parallel resonant frequency in the model described in Figure 2.11(b).

2.4.2 Reactance slope

Although the reactance slope is used at resonance to measure the Q-factor of a resonator, it is an important parameter in the design of impedance or admittance matching elements at the terminals of an RF circuit. In fact, the impedance variation due to the terminal reactive elements over frequency can influence the shape of the impedance presented at the terminals and thus can detrimentally impact the matching bandwidth. For example, it is known that the slope of an inductor (inductive reactance) as a function of the increased frequency is positive; however, inductors implemented within an RF module are usually influenced by parasitics from the surrounding EM environment. Designs based on equivalent models of reactive elements extracted in isolation of any EM parasitics and do not take into account the impact of the EM surrounding environment can result in unwanted impedance mismatch losses and different rejection levels as in the case of RF filtering networks.

2.4.3 Inductor orientation

With the several structural configurations that an inductor element may have (either as a planar spiral, 3D helical, or SMT inductors), it is essential to use the correct orientation and values of the inductor during the modeling and optimization of a circuit or an RF module. Using correct or optimized orientation of an inductor helps in reducing unwanted EM coupling propagating in the direction of the inductor's magnetic flux to adjacent components and circuits. Moreover, in order to accurately capture the prober EM environment surrounding an inductor, it is important to use the correct inductor value that corresponds to the design calculations by using the corresponding SMD 3D model or by fully implementing the integrated inductor rather than having a tuning port to adjust the value on a circuit simulator. EM coupling due to the inductor elements can be mitigated by reorienting and optimizing the SMD inductors locations, increasing the separation distance between the inductor elements, adding guard traces grounded with vias, and/or changing the orientation of a printed inductor coil [11]. Between circular and rectangular spiral inductors of equal inductance value, the circular inductor introduces less coupling. This is due to the larger average interface exposure between the two coupled inductors in the case of rectangular spiral inductors.

2.5 Miniaturization of distributed elements

To achieve high levels of miniaturization and low-cost devices, lumped components are usually employed as impedance loading elements to the distributed structures. Miniaturized distributed elements are used in cases where high Q-factor elements are necessary and cannot be realized by using lumped elements alone. Generally from the TL model, one can conclude that a high-impedance trace is inductive in nature and a low impedance line is capacitive. Two basic approaches are usually used to miniaturize TL-based elements as will be discussed next.

2.5.1 High permittivity materials

Microwaves obey the laws of optics. In the presence of a medium, they can be transmitted, reflected, or absorbed depending on the medium characteristics. Waves traveling through a high dielectric material exhibit slower phase velocities than those traveling across a lower dielectric constant material. Therefore, using high dielectric materials enables miniaturization of microwave circuits. In addition, using high dielectric materials is advantageous in achieving low impedance TL elements for reasonably sized circuits and obtaining reduced radiation loss due to the fact that the electric fields are more condensed between the element and its ground plane. A high dielectric material has also some drawbacks represented in increased dispersion, higher parasitic capacitance, and the limitation of implementing high impedance distributed circuits.

2.5.2 Capacitively loaded transmission line

A reduced size, capacitively loaded (C-loaded) TLs equivalent to a quarter-wavelength transformer will be discussed in this section. To effectively obtain a TL that is shorter than a quarter-wavelength, both equivalent inductance and capacitance of the distributed line should be lowered. To ensure equivalency between the shortened and quarter-wave lines, the inductance of the reduced physical length line is compensated by increasing the characteristic impedance of the line, while the capacitance reduction is adjusted by adding a lumped capacitor as shown in Figure 2.19. Figure 2.19(a) shows a conventional two-port quarter-wave transformer with an electrical length of 90°, while Figure 2.19(b) represents its equivalent C-loaded transformer.

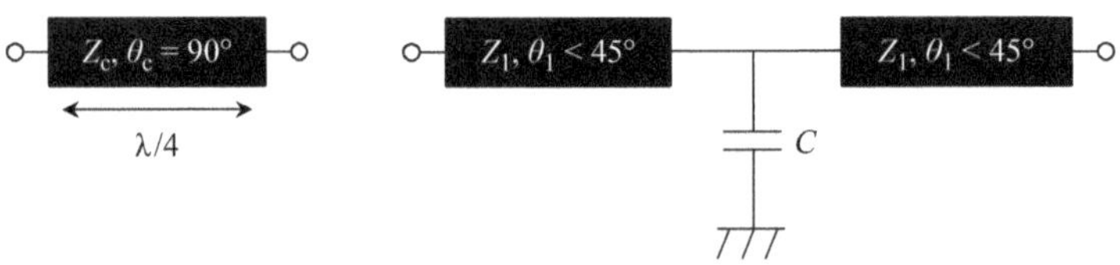

Figure 2.19 (a) The conventional quarter-wave transformer and (b) its equivalent C-loaded transformer

To derive the design equations, consider the ABCD parameters for the conventional quarter-wavelength transmission line as follows:

$$\begin{bmatrix} \cos(90^0) & jZ_c \sin(90^0) \\ jZ_c^{-1} \sin(90^0) & \cos(90^0) \end{bmatrix} = \begin{bmatrix} 0 & jZ_c \\ jZ_c^{-1} & 0 \end{bmatrix} \tag{2.42}$$

The ABCD matrix of the shorted-shunt capacitor shown in Figure 2.19(b) is given in the following form:

$$M_c = \begin{bmatrix} 1 & 0 \\ j\omega C & 1 \end{bmatrix} \tag{2.43}$$

The ABCD matrix for each of the two transmission line transformers that have a characteristic impedance of Z_s and an electrical length of θ_s is given by:

$$M_s = \begin{bmatrix} \cos(\theta_s) & jZ_s \sin(\theta_s) \\ jZ_s^{-1} \sin(\theta_s) & \cos(\theta_s) \end{bmatrix} \tag{2.44}$$

where Z_c, Z_s, θ_s, and ω are the characteristic impedance of the quarter-wavelength line, the characteristic impedance of the shortened line, the electrical length of the shortened line, and the angular frequency, respectively.

Thus, the total ABCD matrix of the whole C-loaded structure is expressed as:

$$\begin{bmatrix} A & B \\ C & D \end{bmatrix}_{\text{Total}} = M_s M_c M_s \tag{2.45}$$

This leads to the following equations:

$$A = \cos^2(\theta_1) - \omega C Z_s \sin(\theta_s)\cos(\theta_s) - \sin^2(\theta_s) \tag{2.46}$$

and

$$B = j2Z_s \sin(\theta_s)\cos(\theta_s) - j\omega C Z_s^2 \sin^2(\theta_s) \tag{2.47}$$

By equating the ABCD parameters, specifically, A in (2.46) with its correspondence of the conventional quarter wave transformer given in (2.42), and applying some trigonometric identities, the following equation is obtained:

$$C = \frac{2\cot(2\theta_s)}{\omega Z_s} \tag{2.48}$$

By substituting (2.48) in (2.47) and using simple trigonometric identities, the transmission line impedance can be expressed by:

$$Z_s = Z_c \cot(\theta_s) \tag{2.49}$$

Then, substituting (2.49) in (2.48) leads to the following shunt capacitor equations:

$$C = \frac{2\tan(\theta_s)}{\omega Z_c \tan(2\theta_s)} \tag{2.50}$$

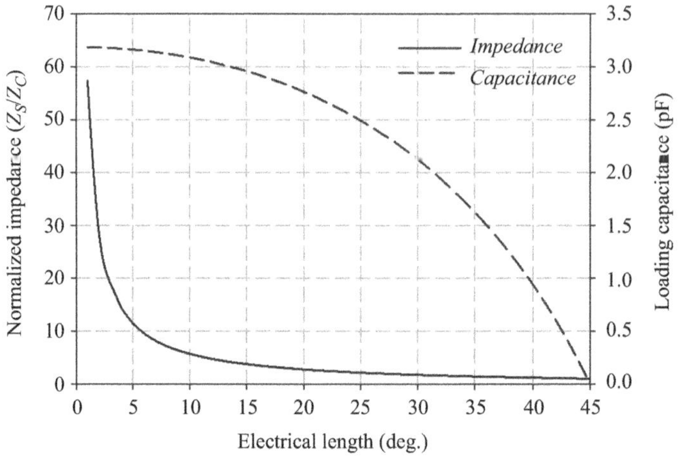

Figure 2.20 Analytical results of Z_s and C of a shortened line as a function of the required physical length at 1 GHz for a conventional 50 Ω quarter-wavelength line

By choosing θ_s such that $2\theta_s < 90°$, a size reduction is achieved. In other words, θ_s is a design parameter and once specified, the impedance Z_s can be calculated from (2.49), and the capacitor value can be calculated from (2.50). This relation is illustrated in Figure 2.20. As can be seen, the electrical length can be made shorter if higher characteristic impedance Z_s is selected and the shunt lumped capacitance C is increased.

The maximum size reduction depends on the line impedance that can be realized without violating the manufacturing design rules for minimum linewidth. The design methodology is fully scalable for different frequencies and impedances, and it is evident from (2.50) that the resonance frequency of the equivalent quarter-wavelength line structure can be shifted by altering the capacitor value C. Thus, by using tunable capacitors (e.g. varactors), a tuning functionality can be added.

2.6 Quality factor calculation

A quality factor is a dimensionless parameter used as a figure of merit to characterize the loss performance of a resonant element besides other resonance parameters such as resonant frequency and coupling coefficient. Performance of several microwave circuits (e.g., filters and couplers) depend significantly on accurate determination of the losses associated with its components. Reactive elements possess some resistive losses which can be described in terms of their Q-factor.

There exist different definitions for calculating the Q-factor of a reactive lumped or a distributed element and resonators. These definitions depend on the element operating frequency whether its Q-factor is calculated at resonance or far from resonance. In terms of power dissipation, the Q-factor is defined for any frequency as:

$$Q = 2\pi \frac{\text{Peak energy stored}}{\text{Energy dissipated per cycle}} = \omega \frac{\text{energy stored}}{\text{average power dissipated}} \tag{2.51}$$

In the case of an individual reactive element operating well below its self-resonant frequency where the reactive component is meaningful, the following Q-factor definition can be used:

$$Q = \frac{\text{Im}(Z_{\text{in}})}{\text{Re}(Z_{\text{in}})} \tag{2.52}$$

where Z_{in} is the input impedance of the reactive element. Q-factor in this context is also referred to as the nodal Q-factor of the reactive element. Assuming a simple series RC or RL model of a capacitor or an inductor element, respectively, the Q-factor can be given by:

$$Q = \frac{|X_C|}{R_s} = \frac{1}{\omega R_s C} \tag{2.53}$$

$$Q = \frac{|X_L|}{R_s} = \frac{\omega L}{R_s} \tag{2.54}$$

where R_s is the effective series resistance of the capacitor or the inductor.

When a reactive element is used as a resonant element or at near the resonance frequency, (2.52) is no longer relevant since the Q-factor yields a value of zero. Thus, the loaded Q-factor can be calculated in terms of the 3-dB bandwidth ($\text{BW}_{3\text{dB}}$) as:

$$Q = \frac{f_r}{\text{BW}_{3\text{dB}}} \tag{2.55}$$

where f_r is the resonance frequency. This method, however, gives inaccurate overestimated Q result in some cases. For example, in an acoustic resonator which comprises two resonances, series and parallel resonances in close proximity to each other, the calculated Q-factor using (2.54) may not be accurate as these resonances influence the narrow resonance bandwidth to be narrower.

For distributed resonators, Dicke [12] formulated the Q-factor definition in terms of the slope of the circuit input reactance as expressed by:

$$Q = \frac{\omega_0}{2R} \left| \frac{dX_{in}}{d\omega} \right|_{\omega_0} \tag{2.56}$$

where ω_0 is the resonant frequency, X and R are the reactance and resistance of the distributed circuit, respectively. This formula may result in erroneous Q values in case of having a reactance response with a multitude of spurious resonances that in turn make the reactance slope at those spurious resonances to largely fluctuate from positive to negative. Another equation [13] that is valid only at the resonance frequency is given by:

$$Q = \frac{\omega_0}{2} \left| \frac{d\phi}{d\omega} \right|_{\omega_0} \tag{2.57}$$

here, Φ is the phase of the input impedance (Z_{in}) of the circuit.

Recently, Feld *et al.* [14] demonstrated a Q-factor formula based on measuring the magnitude and group delay of the reflection coefficient (S_{11}) to provide a resonator Q-factor at below and above the resonant frequency. The derived expression is expressed as:

$$Q = \omega\tau \frac{|S_{11}|}{1 - |S_{11}|^2} \tag{2.58}$$

where τ is the group delay of the reflection coefficient and ω is the angular frequency.

The following discussion describes two basic resonant RLC circuits in terms of their Q-factor. Considering the series and parallel RLC circuits illustrated in Figure 2.21, the Q-factor at each case can be found as follows:

$$Q_S = \frac{1}{\omega R_S C_S} = \frac{\omega L_S}{R_S} \tag{2.59}$$

$$Q_P = \omega R_P C_P = \frac{R_P}{\omega L_P} \tag{2.60}$$

In the series RLC case, with decreasing R_S, the Q-factor (Q_s) increases and the bandwidth decreases, while in the parallel RLC case, increasing R_p increases the Q-factor (Q_p). In a series RLC resonant circuit, the inductive reactance is proportional to the frequency of the applied signal and acts as a low-pass filter (i.e., as the frequency increases, the reactance increases, and the signal attenuation increases). Conversely, the capacitive reactance is inversely proportional to the signal frequency and behaves as a high-pass filter by attenuating the low-frequency contents and allowing the high-frequency contents to pass. The transition (or the intermediate frequency) at which the inductive reactance cancels out the capacitive reactance is

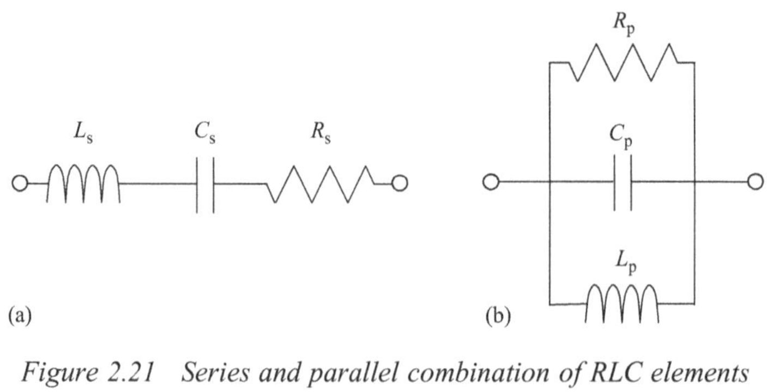

Figure 2.21 Series and parallel combination of RLC elements

the resonant frequency of the circuit. At such frequency maximum current will flow through the resistor element and both the current and voltage will be in phase. A series RLC resonant circuit has an input impedance (Z_{in}) function defined as:

$$
\begin{aligned}
Z_{\text{in}} &= R_S + j\left(\omega L - \frac{1}{\omega C}\right) = R_S\left(1 + j\frac{\omega_o L}{R_S}\left(\frac{\omega}{\omega_o} - \frac{\omega_o}{\omega}\right)\right) \\
&= R_S\left(1 + jQ_S\left(\frac{\omega}{\omega_o} - \frac{\omega_o}{\omega}\right)\right)
\end{aligned}
\tag{2.61}
$$

where ω_{o} is the resonant frequency at which the imaginary part of the input impedance is equal to zero:

$$
\omega_o = \frac{1}{\sqrt{L_S C_S}} \tag{2.62}
$$

As can be observed from (2.61), the real part of the impedance is constant and the imaginary part is a nonlinear function of the signal frequency (ω). Moreover, the input impedance can be related to the circuit Q-factor. Most often, studying the behavior of a resonant circuit at and close to the resonant frequency is of interest to many of the design engineers who design narrowband circuits. In such a case, the input impedance of the circuit can be reduced to:

$$
Z_{in} \approx R_S\left(1 + jQ_S 2\left(\frac{\omega - \omega_o}{\omega_o}\right)\right) \tag{2.63}
$$

Here, the imaginary part of the input impedance function is simplified to a linear (i.e., a straight line) function of frequency that passes through zero ohms at the resonant frequency. This assumption, however, is valid only over a narrow frequency range.

Similarly, for a parallel RLC resonant circuit depicted in Figure 2.21(b), the input admittance function is defined as:

$$
Y_{\text{in}} = \frac{1}{R_P}\left(1 + j\omega_o C_P R_P\left(\frac{\omega}{\omega_o} - \frac{\omega_o}{\omega}\right)\right) = \frac{1}{R_P}\left(1 + jQ_p\left(\frac{\omega}{\omega_o} - \frac{\omega_o}{\omega}\right)\right) \tag{2.64}
$$

And over a narrow range of frequencies can be approximated as:

$$
Y_{in} \approx \frac{1}{R_P}\left(1 + jQ_p 2\left(\frac{\omega - \omega_o}{\omega_o}\right)\right) \tag{2.65}
$$

Assuming a sinusoidal steady-state response of the series RLC circuit illustrated in Figure 2.21(a), the magnitude and the phase response can be related through the following transfer function:

$$
T(j\omega) = \frac{V_o(j\omega)}{V_i(j\omega)} = \frac{1}{\left(1 + jQ_S\left(\frac{\omega}{\omega_o} - \frac{\omega_o}{\omega}\right)\right)} \tag{2.66}
$$

where V_o and V_i are the output and the input voltages.

Figure 2.22(a) and (b) shows the magnitude and phase responses of the transfer function in (2.66) for different circuit Q values. As can be observed from the magnitude response, at the resonant frequency, the input voltage magnitude equals the output voltage magnitude. Also, the higher the circuit Q, the sharper the transition from the passband to the stopband. The behavior over different Q values can be observed on the phase response as well. From the perspective of RF filters, having high Q resonators results in a steeper transition, improved insertion loss, especially, at the passband corners, and improved stopband rejection level as can be concluded from Figure 2.22(a). Below the resonant frequency, the output voltage leads the input voltage and as we move far to the low-frequency side, the output voltage leads the input by 90°. Similarly, above the resonant frequency, the output voltage lags the input voltage. This behavior is depicted in Figure 2.22(b) where the phase response is bounded between −90° and +90°. The sharp transition of the phase response occurs at the resonant frequency.

At resonance, the input impedance of the series RLC circuit reduces to R_s and a lossless circuit will have zero ohm input impedance. In a parallel RLC circuit, at resonance, the input admittance reduces to $1/R_p$, and a lossless circuit will have an infinite input admittance.

To this end, the Q-factor of either lumped or distributed resonators can be expressed as loaded, unloaded, and external Q. The unloaded Q-factor (Q_u) is the Q associated with the reactive elements only without any load. For a resonator Q_u, it is limited by losses due to the conductor and the dielectric and is given by:

$$\frac{1}{Q_u} = \frac{1}{Q_c} + \frac{1}{Q_d} \tag{2.67}$$

where Q_c and Q_d are the conductor and dielectric losses, respectively.

When an external circuit with Q_{ex} influences the resonator this leads to lowering the Q value denoted as the loaded Q-factor (Q_L) and is given by:

$$\frac{1}{Q_L} = \frac{1}{Q_u} + \frac{1}{Q_{ex}} \tag{2.68}$$

Typically, Q_u is the desired parameters for any passive component.

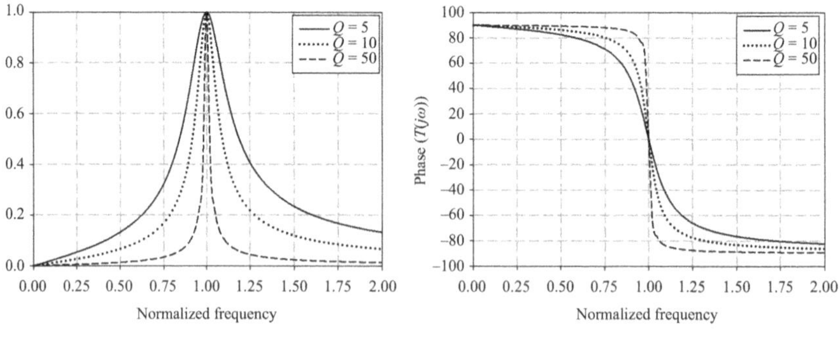

Figure 2.22 (a) Magnitude response of the transfer function of a series RLC resonant circuit as a function of frequency for different circuit Q values, and (b) its phase response

In terms of the coupling coefficient, k, which represents the ratio between the dissipated power in the external circuit to the dissipated power in the resonator, the following relation between the unloaded and the loaded Q can be written as:

$$Q_u = Q_L(1 + k) \tag{2.69}$$

As has been discussed, the Q-factor can be expressed by different formulas depending on whether the measurement of interest is at resonance or far from resonance. To this end, the following discussion will highlight the common methods used to measure the Q-factor of a lossy device. The two measurement methods are prominently used to measure the Q-factor of a device are the transmission-type through S_{21} measurement and the reflection-type using S_{11} measurement [15].

The transmission measurement is a two-port measurement that relies on weakly coupling (decoupling) the device to obtain $1/Q_{ex} \approx 0$, the loaded Q will then equal the unloaded Q which can then be measured from the 3dB BW points as expressed in (2.55).

Considering that the coupling coefficient (k) due to the decoupled input and output of the device are equal, the following relation can be used to calculate k:

$$k = \frac{S_{21}(f_{res})}{1 - S_{21}(f_{res})} \tag{2.70}$$

Substituting (2.70) in (2.69) results in:

$$Q_u = \frac{Q_L}{1 - S_{21}(f_{res})} \tag{2.71}$$

From (2.70 and 2.71), it is noted that at very weak coupling, the coupling coefficient is negligible and the loaded Q equals the unloaded Q.

The reflection-type measurement as indicated from the name is a one port measurement based on measuring the reflection coefficient (S_{11}). By noting the S_{11} response on the Smith chart, a Q-circle is obtained and used for the Q-factor and k calculations of the resonator. While Q-factor can be calculated by identifying three points on the Q-circle (i.e., the center frequency and other two points around it), the coupling coefficient (k) is determined by calculating the diameter of the Q-circle. Here, weakly coupled resonators will result in a small Q circle, while strongly coupled resonators will produce a larger Q circle [16]. The coupling coefficient then can be calculated by:

$$k = \frac{1}{\frac{2}{d} - 1} \tag{2.72}$$

where d is the diameter of the Q-circle.

2.6.1 *Q-factor improvement methods of inductors*

Emanating from the significant impact of the inductors Q-factor on the performance of resonant microwave circuits, there have been extensive studies on finding ways to improve their Q-factor either for those implemented on a chip or on a multilayer laminate. For those implemented on a Si-based chip, the conductivity of the Si substrate and the substrate parasitic capacitances lower the self-resonant frequency of the inductor and degrades its *Q*-factor. Having high *Q*-factor elements improves insertion loss, noise figure, and power amplifier (PA) efficiency which in turn result in lower harmonics and distortion generated by the PA in RF transmit modules [17]. For instance, highly selective passive filters require high-*Q* resonators to achieve the desired steepness level especially in a congested spectrum; low-noise amplifies (LNAs) demand high *Q* matching components in order to achieve a low noise figure; and in general high-*Q* components in RF modules help yielding low overall current consumption, and hence low battery usage given that the module components are properly matched.

There are several methods that exist in order to improve the *Q*-factor of an integrated inductor, especially, when realized on one of the semiconductor substrate technologies. In many instances, it is not viable to improve all inductors in a given module due to size and cost constraints; nevertheless, it is beneficial to identify the most significant inductors in the design at which *Q*-factor improvement techniques can be applied. Typically, inductors located in the series path of an RF chain are the significant ones in terms of loss when compared to inductors placed in a shunt configuration. Also, as the inductance increases, its *Q*-factor decreases at a given frequency because of the parasitic capacitive coupling increase between the coil and the substrate and the coil turns.

The temperature dependency of printed inductors realized on semiconductor substrates has been also studied [18]. The variation of inductance over temperature is found to be not significant; however, an inductor *Q*-factor decreases with increasing temperature at low frequency and increases with increasing temperature at high frequency. This phenomenon can be explained by looking into the loss behavior over temperature of used metallization and substrate. In the case of a conductor loss that can be represented by a series resistance and a series inductance and has a positive temperature coefficient, the conductor resistance increases with the increase in temperature and thus the power loss increases. The substrate loss or substrate resistivity is usually modeled in parallel to the inductance and with a positive temperature coefficient, the substrate resistivity increases with the increase in temperature which implies a decrease in the power loss due to the substrate. As can be concluded, at low frequency, the primary loss mechanism is dominated by the metallization series resistance to the current flowing in the conductor. This is also because at low frequency the substrate loss due to the substrate capacitance is low (presenting a high impedance path to the RF signal). Whereas at high frequency, the substrate shunt capacitance with respect to the inductor provides a low impedance path to the AC current on the conductor. Here, the substrate loss begins to dominate the metal loss.

Given that RFIC transmit modules comprise active devices which contribute to the temperature rise in the module, the knowledge of the inductors' *Q* variation over temperature allows the module designer to account for the worst-case temperature conditions and thereby accurate estimation of the electrical losses.

Although an integrated inductor possesses a simple structure, its optimization on a semiconductive substrates to yield an optimal *Q* value and high inductance per unit area at a given frequency is a complicated task. Some of the commonly used Q-improvement methods compatible with conventional thin-film processes for RF integrated circuits are discussed in Sections 2.6.1.1–2.6.1.6.

2.6.1.1 Layout optimization

Inductor layout optimization [19] by means of varying the conductor width for each turn of the inductor coil is utilized to achieve improved Q-factor. This method is advantageous for inductors implemented on a low loss substrate since it tackles losses mainly introduced by the conductor which has two main contributors such as ohmic and magnetic losses. Given the fact that ohmic losses increase as the width of the inductor coil decreases and also that the magnetically induced loss due to the eddy currents increases as the width of the coil increases, it is possible to conclude that by optimizing the inductor coil width could help in reducing the series resistance loss and thus an improved Q-factor can be obtained.

Higher magnetic fields are typically located at the inner turns of a spiral inductor and therefore, narrow linewidths can be used to reduce magnetic losses. Whereas optimized wide linewidths are used for the outer turns to reduce the ohmic losses and to compensate the increase in ohmic resistance due to the narrower inner turns that tackle the magnetic losses.

Q-factor enhancement can also be obtained by having a large coil area to allow for wider linewidth, wider inside diameter, or increased separation between the turns of the inductor coil [20]. In most of the cases, however, having large area inductors may not possible due to constraints on the available footprint size, both for module or chip designs, and also for low-cost considerations.

2.6.1.2 Patterned ground shield

Shielding an inductor from the substrate losses has been conducted through placing a patterned ground shield (PGS) to reduce the energy loss due to electric coupling between the inductor coil and the Si substrate [21]. The reason for not having a solid ground shield instead of the patterned shield is to reduce the effects of the image eddy currents on the shield which produce an opposing magnetic field; hence the total inductance value and the Q-factor will degrade. In essence, the patterned shield depicted in Figure 2.23 helps in increasing the resistance to the image current induced by the magnetic field on the shield by making it flows perpendicularly to the current paths on the spiral inductor, thus reducing the eddy currents that usually flow parallel and opposite to the inductor conductive path.

2.6.1.3 Thick metal

Thick metal has been used to reduce the ohmic losses as a means to improve the Q-factor of an inductor [22]. The basic idea is to reduce the DC resistance, also referred to as the series resistance. Having thickness that is at least five times the metal skin depth at the operating frequency of interest shall be sufficient to eliminate conductor loss due to the skin depth. DC resistance can be reduced as well by widening the linewidth of the inductor trace; however, the parasitic capacitances and RF resistance increase with the increase in linewidth which can be a limiting

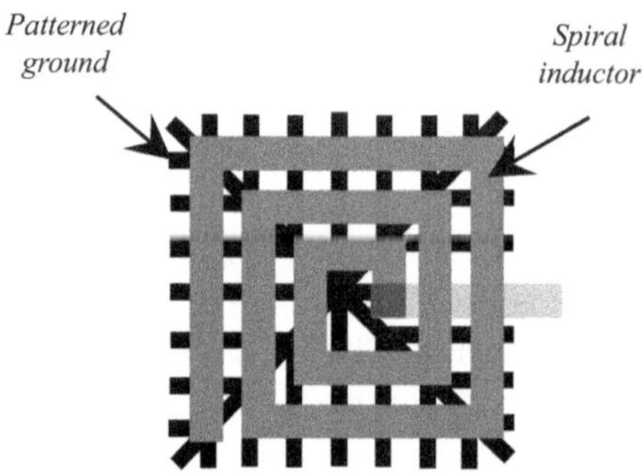

Figure 2.23 A layout of a spiral inductor on a patterned ground shield

factor. Both the conductor linewidth and thickness determine the maximum current a conductor can handle.

2.6.1.4 Substrate modification

Reducing the metallization to substrate loss is an important key to achieving high *Q* on-chip inductors. Micromachining fabrication methods are more popular with Si than any other less lossy substrates. When using inductors on Si substrate, substrate modification can be established by removing the silicon layer underneath the inductor by micromachining methods to provide a sufficient air gap from the lossy substrate. Micromachining can be applied using a front-side etch or a back-side etch process. Generally, the front-side etching process is preferred over the back-side etching because it requires a shorter etch time, doesn't require alignment, and independent of the wafer thickness [1].

High resistive substrates were also utilized such as a silicon on insulator (e.g., SiO_2), or a silicon on sapphire (e.g., Al_2O_3).

2.6.1.5 Magnetic material coating

In an attempt to reduce the conductor loss, several research efforts have been conducted on developing ferromagnetic coating materials, such as a composite of NiFe or CoNbZr, that can be used to enhance the inductance and *Q*-factor of an inductor. The use of magnetic coating material was employed in on-chip inductors to increase the magnetic flux on the inductor coil and therefore, an increase in the self-inductance value. The high resistivity of the magnetic coating material helps to reduce eddy current at high frequency.

2.6.1.6 Differential excitation

The *Q*-factor of an inductor can be substantially improved by using differential or odd-mode excitation rather than single-ended excitation [23]. This type of inductor is suitable for monolithic implementation or when embedded in a multilayer laminate. Differential excitation provides common mode rejection and low susceptibility to noise arising from bias circuits of active devices. A differentially excited inductor exhibits less substrate loss and thus an improved *Q*-factor which can reach up to 50% when compared to its counterpart a single-ended inductor. This is by virtue of the relatively higher equivalent shunt impedance representing the substrate parasitic capacitances seen by the input impedance of the inductor [23]. Moreover, a differentially driven inductor requires less area when compared to a single-ended inductor of equal inductance, leading to a reduction in the effective parasitic capacitances and in turn resulting in an increase in the self-resonance frequency. The increase in self-resonance extends the usable range of frequencies of the inductor. Differential inductors are applicable for use with high performance differentially driven monolithic circuits such as amplifiers, oscillators, and mixers. A drawback of implementing differential circuits is that it requires twice the number of passive and active components when compared to single-ended circuits.

References

[1] Bahl I.J. *Lumped Elements for RF and Microwave Circuits*. Norwood, MA: Artech House; 2003

[2] Daly D.A., Knight S.P., Caulton M., and Ekholdt R. "Lumped elements in microwave integrated circuits." *IEEE Transactions on Microwave Theory and Techniques*. 1967;**15**(12):713–21

[3] Caulton M., Hershenov B., Knight S.P., and DeBrecht R.E. "Status of lumped elements in microwave integrated circuits – present and future." *IEEE Transactions on Microwave Theory and Techniques*. 1971;**19**(7):588–99

[4] DeBrecht R., and Caulton M. "Lumped-elements in microwave integrated circuits in the 1-12 GHz Range." *G-MTT 1970 International Microwave Symposium*; Newport Beach, CA; 1970. pp. 14–18

[5] Rautio J.C. "Synthesis of lumped models from N-port scattering parameter data." *IEEE Transactions on Microwave Theory and Techniques*. 1994;**42**(3):535–37

[6] Angelini A.M. "Transformation of networks containing mutual inductive couplings into equivalent circuit systems with no mutual couplings." *IEEE Transactions on Power Apparatus and Systems*. 1968;**87**(7):1565–78

[7] Goldfarb M.E., and Pucel R.A. "Modeling via hole grounds in microstrip." *IEEE Microwave and Guided Wave Letters*. 1991;**1**(6):135–37

[8] Wight J.S., Chudobiak W.J., and Makios V. "A microstrip and stripline crossover structure." *IEEE Transactions on Microwave Theory and Techniques*. 1976;**24**(5):270

[9] Chiou Y.C., Kuo J.Y., and Lee H.R. "Design of compact symmetric four-port crossover junction." *IEEE Microwave and Wireless Components Letters*. 2009;**19**(9):545–47

[10] Richards P.I. "Resistor-transmission-line circuits." *Proceedings of the IRE*. 1948;**36**(2):217–20

[11] Bahl I.J. "Coupling effects between lumped elements [MMICs]." *IEEE Microwave Magazine*. 2001;**2**(3):73–80

[12] Montgomery C., Dicke R., and Purcell E. *Principles of Microwave circuits*. IEE Electromagnetic Wave Series, **vol. 25**. New York, NY: McGraw-Hill; 1948
[13] Lakin K.M. "Equivalent circuit modeling of stacked crystal filters." *Thirty Fifth Annual Frequency Control Symposium*; Philadelphia, PA; 1981. pp. 257–62
[14] Feld D.A., Parker R., Ruby R., Bradley P., and Dong S. "After 60 years: a new formula for computing quality factor is warranted." *2008 IEEE Ultrasonics Symposium*. Beijing; 2008. pp. 431–36
[15] Kajfez D., *Q Factor Measurements Using MATLAB*. Norwood, MA: Artech House; 2011
[16] Kajfez D., and Hwan E.J. "Q-factor measurement with network analyzer." *IEEE Transactions on Microwave Theory and Techniques*. 1984;**32**(7):666–70
[17] Bahl I.J. "Improved quality factor spiral inductors on GaAs substrates." *IEEE Microwave and Guided Wave Letters*. 1999;**9**(10):398–400
[18] Groves R., Harame D.L. and Jadus D. "Temperature dependence of Q and inductance in spiral inductors fabricated in a silicon-germanium/BiCMOS technology." *IEEE Journal of Solid-State Circuits*. 1997;**32**(9):1455–59
[19] Lopez-Villegas J.M., Samitier J., Cane C., and Losantos P. "Improvement of the quality factor of RF integrated inductors by layout optimization." *IEEE Transactions on Microwave Theory and Techniques*. 2000;**48**(1):76–83
[20] Bahl I.J. "High-performance inductors." *IEEE Transactions on Microwave Theory and Techniques*. 2001;**49**(4):654–64
[21] Yue C.P. and S. S. Wong, "On-chip spiral inductors with patterned ground shields for Si-based RF ICs." *IEEE Journal of Solid-State Circuits*. 1998;**33**(5):743–52
[22] Choi Y.-S., and Yoon J.-B. "Experimental analysis of the effect of metal thickness on the quality factor in integrated spiral inductors for RF ICs." *IEEE Electron Device Letters*. 2004;**25**(2):76–79
[23] Danesh M., and Long J.R. "Differentially driven symmetric microstrip inductors." *IEEE Transactions on Microwave Theory and Techniques*. 2002;**50**(1):332–41

Chapter 3

Basics of microwave network analysis

3.1 Introduction

Solving Maxwell's equations to analyze a given electromagnetic (EM) problem yields an accurate solution with a plethora of spatiotemporal electric and magnetic fields information that may not be needed. It is also evident that quantities resulting from solving Maxwell's equations are interrelated and reconciled to the transmission line (TL) and circuit theory [1]. Practically, port analysis techniques are more prevalent as a simple and effective method in analyzing microwave and mixed radio frequency (RF) digital circuits. Integrating additional components to the original problem has become an easy task by applying port analysis, where network loss, impedance transformation, and other effects are preserved. While most port parameters are valid to represent the device linear behavior due to small signal stimulus, other parameters are also available to cope with the nonlinear characteristics of circuits due to large signal excitation. Some network parameters are convenient to use than others depending on the circuit topology; therefore, conversions between the different port network parameters are also possible [2].

Circuit analysis techniques can be classified into two categories based on their frequency of operation: low frequency based on quasi-static approximations of Maxwell's equations and high frequency (microwave) circuits based on full-wave EM analysis. At low frequencies, circuit physical dimensions are relatively small when compared to the operating wavelength and phase variation at any point in the circuit is considered negligible. There are several techniques used to analyze low-frequency circuits; however, they are not applicable for circuits operating at high (gigahertz) frequencies due to the domination of parasitic effects.

Circuit synthesis techniques greatly benefit from applying port analysis methods to simplify the mathematical representation of the network transfer function and thereby the subsequent steps required to realize the circuit parameters.

3.2 Microwave port analysis

A port in a microwave network is a terminal pair that provides a point of connection from the microwave network reference plane to external circuits. A port is a non-physical interface used to simplify circuit analysis where energy can be applied and extracted. A two terminal is considered as one port if the current entering one terminal equals the current exiting the other (i.e., net terminals current is zero).

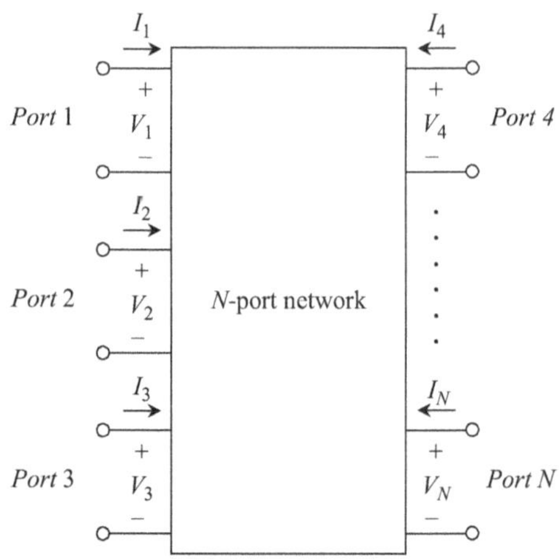

Figure 3.1 An N-port microwave network illustrating the customary description of port voltages and currents

Different port analyses have evolved to accommodate the need for accurate device characterization and the necessity to close the gap between measurement and high fidelity simulation results obtained by computer-aided design (CAD) tools. In addition, port analysis methods are conveniently represented in almost any RF CAD tool and accessible to RF designers. Figure 3.1 shows an arbitrary microwave network comprising N ports. Network parameters are either based on total voltage–current or traveling wave representation at the terminal planes of the network, which will be discussed in Sections 3.2.1 and 3.2.2. Unlike total voltage–current-based parameters, traveling wave-based parameters are dependent upon the source and load impedances of the network [2] and more convenient for characterizing high-frequency circuits.

Given that each excitation mode has distinct characteristic impedance and propagation velocity, network parameters are also referred to as modal parameters to describe the frequency dependent nature of the excitation waves. Note that modal port analyses are evaluated at each frequency of interest. Thus, dispersion characteristics are well preserved.

3.2.1 Voltage–current-based parameters

3.2.1.1 Impedance (*Z*-) parameters

Impedance parameters, also named as Z- or open circuit impedance parameters, which are complex values representing resistance (R) and reactance (X) in ohms as expressed in (3.1). Z-parameters describe the interaction among ports within a microwave network where port voltages are expressed in terms of port currents as depicted in Figure 3.2:

$$Z = R + jX \tag{3.1}$$

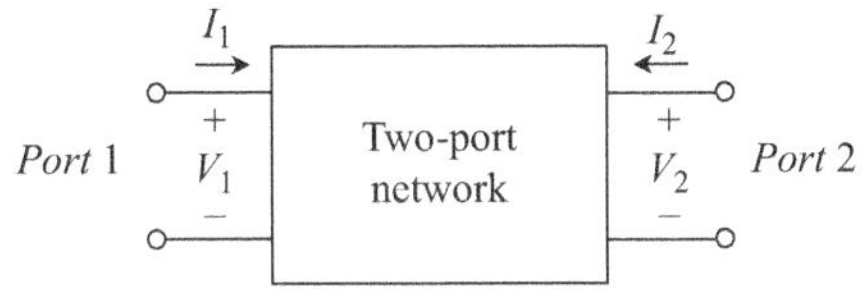

Figure 3.2 A two-port network illustrating the total voltage and current direction at the terminals

For an N-port network, the Z-matrix for the kth mode is given by:

$$\begin{bmatrix} V_1^k \\ V_2^k \\ \vdots \\ V_N^k \end{bmatrix} = \underbrace{\begin{bmatrix} Z_{1,1}^k & Z_{1,2}^k & \cdots & Z_{1,N}^k \\ Z_{2,1}^k & \ddots & & \vdots \\ \vdots & & \ddots & \vdots \\ Z_{N,1}^k & \cdots & \cdots & Z_{N,N}^k \end{bmatrix}}_{Z-\text{Matrix}} \begin{bmatrix} I_1^k \\ I_2^k \\ \vdots \\ I_N^k \end{bmatrix} \tag{3.2}$$

where

$$Z_{i,j}^k = \left. \frac{V_i^k}{I_j^k} \right|_{I_m^k = 0 \,\forall\, m \neq j} \tag{3.3}$$

$$Z_{i,i}^k = \left. \frac{V_i^k}{I_i^k} \right|_{I_m^k = 0 \,\forall\, m \neq i} \tag{3.4}$$

According to (3.3) and (3.4), the calculation of the Z-parameters requires exciting one port while leaving the other port(s) open circuited, hence the name open circuit parameters.

For a purely resistive circuit, the network modal parameters are real constants, whereas a lossless circuit comprising only reactive elements possesses modal network parameters that are imaginary values.

3.2.1.2 Admittance (Y-) parameters

Admittance parameters, also called Y- or short circuit admittance parameters, are available to present the inverse of Z-parameters as in (3.5). Y-parameters are complex values representing conductance (G) in mhos and susceptance (B) in Siemens as illustrated in (3.6) and express port currents in terms of port voltages:

$$[Y] = [Z]^{-1} \tag{3.5}$$

$$Y = G + jB \tag{3.6}$$

While Z-parameters are convenient to use for analyzing elements connected in series, Y-parameters are suitable for elements connected in parallel.

The N port Y-matrix for the kth mode can be written as:

$$\begin{bmatrix} I_1^k \\ I_2^k \\ \vdots \\ I_N^k \end{bmatrix} = \underbrace{\begin{bmatrix} Y_{1,1}^k & Y_{1,2}^k & \cdots & Y_{1,N}^k \\ Y_{2,1}^k & \ddots & & \vdots \\ \vdots & & \ddots & \vdots \\ Y_{N,1}^k & \cdots & \cdots & Y_{N,N}^k \end{bmatrix}}_{Y-\text{Matrix}} \begin{bmatrix} V_1^k \\ V_2^k \\ \vdots \\ V_N^k \end{bmatrix} \tag{3.7}$$

where

$$Y_{i,j}^k = \left.\frac{I_i^k}{V_j^k}\right|_{V_m^k=0\ \forall\ m\neq j} \tag{3.8}$$

$$Y_{i,i}^k = \left.\frac{I_i^k}{V_i^k}\right|_{V_m^k=0\ \forall\ m\neq i} \tag{3.9}$$

In (3.8) and (3.9), the calculation of Y-parameters requires the excitation of one port while leaving the other port(s) as short-circuited.

In many instances, the impedance or admittance parameter is referred to as immittance as defined by Hendrik Bode [3] and is used generally to simplify theoretical analysis. Immittance is a dimensionless quantity as it can represent a general function of impedance or admittance.

To this end, Z- and Y-network parameters rely on relating the total voltages and currents at the ports. Such parameters are convenient for characterizing and measuring low-frequency circuits based on quasi-static approximations of Maxwell's equations, where circuit physical dimensions are relatively small when compared to the operating wavelength. There are a plethora of analysis techniques that can be used for low-frequency circuits; however, they are not applicable for circuits operating at high frequencies (microwaves) due to the domination of parasitic effects. Therefore, at high-frequency circuits, it is more suitable to deal with the incident and reflected wave phasors and matching loads at the network ports.

3.2.1.3 Transmission (ABCD) parameters

ABCD parameters, also known as transmission, chain, or cascade parameters, are beneficial in case of cascaded networks. The ABCD matrix is a 2×2 matrix that can only be used to cascade multiple two-port subcircuits. ABCD parameters relate voltages to currents at the input port of a two-port network to the voltages and currents at the network output port.

The two-port ABCD matrix equations for the kth mode is expressed as:

$$\begin{bmatrix} V_1^k \\ I_1^k \end{bmatrix} = \underbrace{\begin{bmatrix} A^k & B^k \\ C^k & D^k \end{bmatrix}}_{\text{ABCD Matrix}} \begin{bmatrix} V_2^k \\ I_2^k \end{bmatrix} \tag{3.10}$$

where

$$A^k = \left.\frac{V_1^k}{V_2^k}\right|_{I_2^k=0} \tag{3.11}$$

$$B^k = \left.\frac{V_1^k}{I_2^k}\right|_{V_2^k=0} \tag{3.12}$$

$$C^k = \left.\frac{I_1^k}{V_2^k}\right|_{I_2^k=0} \tag{3.13}$$

$$D^k = \left.\frac{I_1^k}{I_2^k}\right|_{V_2^k=0} \tag{3.14}$$

Note in Figure 3.3(a) that the current flows out of the output port (port 2), unlike in *Z*- or *Y*-parameters. This allows the output current of one network to pass through the input of the next network as illustrated in Figure 3.3(b). The matrix product of the two ABCD matrices is equivalent to connecting the individual two-port networks together in a cascade configuration:

$$\begin{bmatrix} V_1^k \\ I_1^k \end{bmatrix} = \underbrace{\begin{bmatrix} A_1^k & B_1^k \\ C_1^k & D_1^k \end{bmatrix}}_{[\mathrm{ABCD}]_1} \underbrace{\begin{bmatrix} A_2^k & B_2^k \\ C_2^k & D_2^k \end{bmatrix}}_{[\mathrm{ABCD}]_2} \begin{bmatrix} V_2^k \\ I_2^k \end{bmatrix} \tag{3.15}$$

In the case of multiple subcircuits in cascade connection, the matrix multiplication of the individual ABCD matrices representing each sub-block must follow the same connection order as in the actual overall circuit since matrix multiplication is not necessarily commutative.

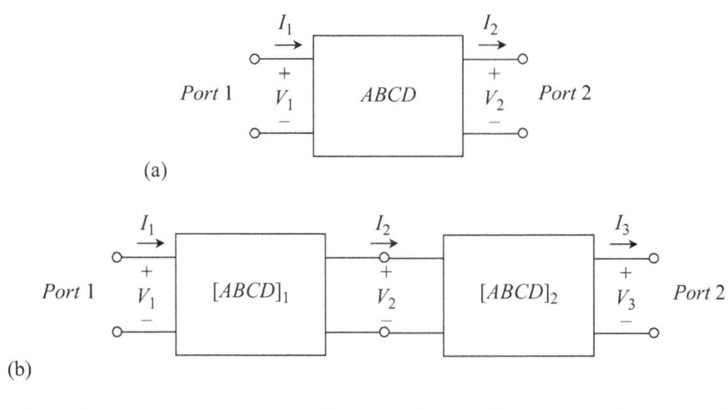

Figure 3.3 (a) A two-port network showing the total voltage and current direction at the terminals, and (b) a cascade connection of two networks

3.2.1.4 Hybrid (*h*-) parameters

Hybrid- or *h*-parameters relate the current at one port of a two-port network in terms of the voltage at the other port. The calculation of the two-port *h*-parameters is based on exciting port 1 while leaving port 2 as a short-circuited, and exciting port 2 while leaving port 1 open circuited. This port *I*–*V* relationship makes *h*-parameters suitable for small signal frequency analysis and modeling of a transistor circuit. Hybrid-parameters are complex numbers representing a mixture of dimensions where h_{11} represents impedance, h_{22} represents admittance, h_{21} and h_{12} are dimensionless transfer quantities describing the forward current gain and the reverse voltage gain, respectively. A reciprocal network that uses *h*-parameters will have $h_{21} = -h_{12}$.

The two-port *h*-matrix for the *k*th mode is described as:

$$\begin{bmatrix} V_1^k \\ I_2^k \end{bmatrix} = \underbrace{\begin{bmatrix} h_{11}^k & h_{12}^k \\ h_{21}^k & h_{22}^k \end{bmatrix}}_{h-\text{matrix}} \begin{bmatrix} I_1^k \\ V_2^k \end{bmatrix} \tag{3.16}$$

where

$$h_{11}^k = \left. \frac{V_1^k}{I_1^k} \right|_{V_2^k=0} \tag{3.17}$$

$$h_{12}^k = \left. \frac{V_1^k}{V_2^k} \right|_{I_1^k=0} \tag{3.18}$$

$$h_{21}^k = \left. \frac{I_2^k}{I_1^k} \right|_{V_2^k=0} \tag{3.19}$$

$$h_{22}^k = \left. \frac{I_2^k}{V_2^k} \right|_{I_1^k=0} \tag{3.20}$$

3.2.2 *Traveling wave-based parameters*

3.2.2.1 Scattering (*S*-) parameters

As discussed in Sections 3.2.1.1–3.2.1.2, analysis based on impedance and admittance parameters requires terminations with open and short loads, respectively. Practically, measurements performed at high frequencies make open and short loads inaccurately interpreted due to the associated parasitic capacitances and inductances. In turn, this hinders the ability of available test equipment to accurately measure the total current and voltage at the device ports. In addition, active devices may not be stable or functional under open and short load conditions. The fundamental principle of S-parameters, however, is premised on using matched load termination which appears as an infinitely long transmission line. This makes S-parameters widely used in analyzing microwave circuits since it is more convenient to describe a given network by means of waves, rather than voltages and currents as illustrated in Figure 3.4. Additionally, S-parameters provide an easier definition of port reference planes where phase references of the traveling wave phasors can be identified.

The N-port S-parameters in a matrix form for the kth wave mode assuming the same characteristic impedance at all ports of a network is given by:

$$\begin{bmatrix} b_1^k \\ b_2^k \\ \vdots \\ b_N^k \end{bmatrix} = \begin{bmatrix} S_{1,1}^k & S_{1,2}^k & \cdots & S_{1,N}^k \\ S_{2,1}^k & \ddots & & \vdots \\ \vdots & & \ddots & \vdots \\ S_{N,1}^k & \cdots & \cdots & S_{N,N}^k \end{bmatrix} \begin{bmatrix} a_1^k \\ a_2^k \\ \vdots \\ a_N^k \end{bmatrix} \tag{3.21}$$

where a_n^k and b_n^k are the incident and reflected modal complex wave amplitudes traveling toward and away from the nth port, respectively. The off-diagonal elements of the S-matrix are computed as:

$$S_{i,j}^k = \left. \frac{b_i^k}{a_j^k} \right|_{a_n^k = 0 \, \forall \, m \neq j} \qquad \text{(transmission coefficients)} \tag{3.22}$$

Equation (3.22) shows that the off-diagonal elements of the S-matrix (S_{ij}^k) can be found by exciting port j with an incident wave a_j^k and calculating the reflected wave at port i while all incident waves at the other ports (a_n^k) are set to zero. This can be established by connecting the other ports by their matched loads so that no reflected waves are returned to the network.

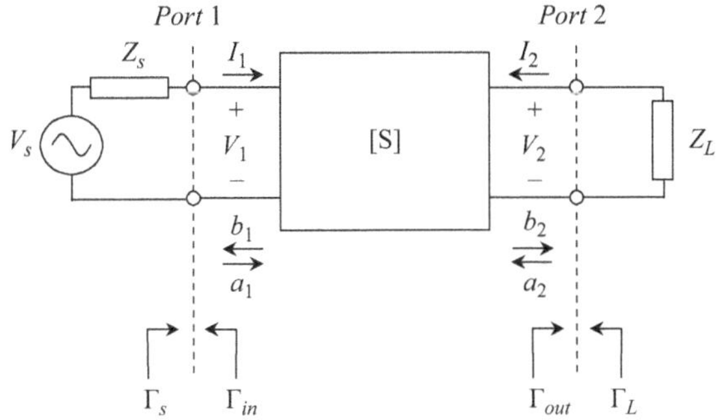

Figure 3.4 Two-port network showing the direction of the incident and reflected waves at the device ports

The diagonal elements of the *S*-matrix describing the matching of each port in the network are defined as:

$$S_{i,i}^k = \left.\frac{b_i^k}{a_i^k}\right|_{a_n^k=0\,\forall\, n\neq i} \quad \text{(reflection coefficients)} \tag{3.23}$$

Considering the two-port network in Figure 3.4, the reflection and the transmission coefficients can be given by:

$$\Gamma_{i,i}^k = \frac{a_i^k}{b_i^k} \tag{3.24}$$

$$T_{j,i}^k = 1 + \Gamma_{i,i}^k \tag{3.25}$$

where the reflection coefficient ($\Gamma_{i,i}^k$) equals $S_{i,i}^k$ at the *i*th port only when all other ports are matched. Similarly, the transmission coefficient ($T_{j,i}^k$) between the ports *j* and *i* equals $S_{j,i}^k$ and $i \neq j$ only when all other ports are terminated with matched loads.

S-parameters are determined under the condition that ports are connected to their matched loads. Having different termination impedances or excitations do not alter the *S*-parameters, which represent the characteristics of the network itself regardless of the load's condition; however, it changes the reflection and transmission coefficients. For example, the input reflection coefficient at the *i*th port of a two-port network is related to the $S_{i,i}$ parameter through the following relation:

$$\Gamma_{i,i} = \frac{b_i}{a_i} = S_{i,i} + \frac{S_{i,j}S_{j,i}\Gamma_{j,j}}{1 - S_{j,j}\Gamma_{j,j}} \tag{3.26}$$

Derivation of this formula is detailed in Section 3.4.

3.2.2.2 Generalized *S*-parameters

In the *S*-parameters discussed in Section 3.2.2.1, the reference impedances (Z_{ref}) of all ports are identical. A generalized scattering matrix can then be defined by removing such restriction and allowing different port impedances. To begin the formularization of the generalized scattering parameters, consider the two-port network shown in Figure 3.1 which illustrates the total voltage and current (V_n and I_{n}) flowing into the *n*th port as described in (3.27) and (3.28) in terms of the incident and reflected voltages (V_n^+, V_n^-) and currents (I_n^+, I_n^-), respectively:

$$V_n^k = V_n^{+,k} + V_n^{-,k} \tag{3.27}$$

$$I_n^k = I_n^{+,k} - I_n^{-,k} \tag{3.28}$$

The average power at the input of the *n*th port is expressed as:

$$P_{\text{avg}} = \frac{1}{2}\text{Re}\{V_n I_n^*\} = \frac{1}{2}\left(\left|a_n^k\right|^2 - \left|b_n^k\right|^2\right) \tag{3.29}$$

From the power wave definition, the incident and reflected port powers in terms of their RMS values of the peak currents and voltages are given by:

$$P_n^{+,k} = \frac{1}{2}\left|a_n^k\right|^2 = \frac{\left|V_n^{+,k}\right|^2}{2Z_{\text{ref}_n}} = \frac{\left|I_n^{+,k}\right|^2}{2}Z_{\text{ref}_n} \tag{3.30}$$

$$P_n^{-,k} = \frac{1}{2}\left|b_n^k\right|^2 = \frac{\left|V_n^{-,k}\right|^2}{2Z_{\text{ref}_n}} = \frac{\left|I_n^{-,k}\right|^2}{2}Z_{\text{ref}_n} \tag{3.31}$$

where Z_{ref_n} is the impedance looking out from the nth port. The square root of the power waves is directly related to the wave amplitudes [4]. Thus, the kth incident and reflected modal power wave amplitudes in terms of the nth port reference impedance can be written as:

$$a_n^k = \frac{V_n^{+,k}}{\sqrt{\text{Re}\{Z_{\text{ref}_n}\}}} \tag{3.32}$$

$$b_n^k = \frac{V_n^{-,k}}{\sqrt{\text{Re}\{Z_{\text{ref}_n}\}}} \tag{3.33}$$

In terms of total modal voltage and current at the nth port, the incident and reflected modal wave amplitudes, can be described as:

$$V_n^k = \sqrt{Z_{\text{ref}_n}}\left(a_n^k + b_n^k\right) \tag{3.34}$$

$$I_n^k = \frac{1}{\sqrt{Z_{\text{ref}_n}}}\left(a_n^k - b_n^k\right) \tag{3.35}$$

Solving (3.34) and (3.35) yields the following algebraic expressions:

$$a_n^k = \frac{V_n^k + Z_{\text{ref}_n} I_n^k}{2\sqrt{\text{Re}\{Z_{\text{ref}_n}\}}} \tag{3.36}$$

$$b_n^k = \frac{V_n^k - Z_{\text{ref}_n}^* I_n^k}{2\sqrt{\text{Re}\{Z_{\text{ref}_n}\}}} \tag{3.37}$$

Equations (3.36) and (3.37) show the transformation between the voltage–current representation and the incident and reflected wave amplitude quantities. It can be observed at this point that the reflection coefficients will remain the same in the generalized S-matrix. However, the generalized transmission coefficients of the S-matrix are changed according to the port impedances and are given by:

$$S_{i,j}^k = \frac{b_i^k}{a_j^k}\frac{\sqrt{Z_{\text{ref}_j}}}{\sqrt{Z_{\text{ref}_i}}}\Bigg|_{a_n^k=0\ \forall\ n\neq j} \tag{3.38}$$

Equation (3.38) shows that an S-matrix determined based on equal port reference impedances can be seamlessly updated to a network possessing different port impedances.

3.2.2.3 Scattering transfer (T-) parameters

Similar to ABCD parameters, T-parameters perform the same function, however, it relates incident and reflected waves instead of voltages and currents. T-parameters are useful in de-embedding cascaded two-port networks. In a matrix form, the T-parameters are expressed as:

$$\begin{bmatrix} a_1^k \\ b_1^k \end{bmatrix} = \begin{bmatrix} T_{1,1}^k & T_{1,2}^k \\ T_{2,1}^k & T_{2,2}^k \end{bmatrix} \begin{bmatrix} b_2^k \\ a_2^k \end{bmatrix} \quad (3.39)$$

3.2.2.4 Hybrid scattering (HS-) parameters

The hybrid S-parameters was first introduced in [5,6] to account for external electromagnetic interference (EMI) sources: both plane-wave (far-field) and near-zone (near-field) sources. In essence, the identified EMI source can be incorporated in the generalized S-matrix as if it was generated by an additional new port with constant voltage sources applied at the network ports. The matrix in (3.40) explains the hybrid S-parameters. In this matrix, a_{N+1}^k and b_{N+1}^k are the modal incident and reflected waves due to the additional port defining an EMI source:

$$\begin{bmatrix} b_1^k \\ b_2^k \\ \vdots \\ b_N^k \\ b_{N+1}^k \end{bmatrix} = \underbrace{\begin{bmatrix} S_{1,1}^k & S_{1,2}^k & \cdots & S_{1,N}^k & \mathrm{HS}_{1,N+1}^k \\ S_{2,1}^k & \ddots & \vdots & \vdots & \mathrm{HS}_{2,N+1}^k \\ \vdots & \cdots & \ddots & \vdots & \vdots \\ S_{N,1}^k & \cdots & \cdots & S_{N,N}^k & \vdots \\ S_{N+1,1}^k & \cdots & \cdots & S_{N+1,N}^k & \mathrm{HS}_{N+1,N+1}^k \end{bmatrix}}_{\text{HS}-\text{Matrix}} \begin{bmatrix} a_1^k \\ a_2^k \\ \vdots \\ a_N^k \\ a_{N+1}^k \end{bmatrix} \quad (3.40)$$

3.2.2.5 X-parameters

Unlike S-parameters that are based on the superposition principle, and thereby can be employed to represent the linear behavior of a device under a small signal excitation, X-parameters are network parameters used to describe the nonlinear behavior of a device under large signal conditions [7]. Moreover, X-parameters are applicable for describing the device linear behavior due to small signal excitation. The formalism of X-parameters is based on the polyharmonic distortion (PHD) model which is a frequency domain modeling technique for nonlinear devices [8]. It also can be thought of as a mathematical extension or generalization of the

S-parameters through using the concept of harmonic superposition to analytically describe the large-signal behavior by linking the scattered frequency components of the output signal with the incident multitone signal. In principle, the frequency components of the output signal represent the combinations of the input tones under spectral linearization around a specific large signal point (i.e., amplitudes and phases of the input tones, bias condition, and loads). *X*-parameters are very helpful in the case of solving cascaded problems by evaluating the individual nonlinear components forming a cascaded system to calculate the overall nonlinear performance.

Figure 3.5(a) illustrates the spectral components generated from a two-port nonlinear device under a small signal condition where the output tone is the linear stimulus response of the same signal frequency presented at the input. Under a small signal condition, there are no new frequencies introduced such as harmonics or intermodulation distortion products. However, Figure 3.5(b) shows the spectral contents that may be produced by large signal excitation. Assuming device excitation is at port 1 in both cases.

The measurements of *X*-parameters are carried out using a nonlinear vector network analyzer (NVNA) by measuring the magnitude and phase of all frequency components of interest including fundamental and harmonic frequencies under large signal excitation. Current instruments allow for sweeping the magnitude and phase of a load impedance using load-pull impedance tuner to observe the variations of the nonlinear device behavior due to load changes. Likewise, sweeps over different bias voltages, input power values, and fundamental frequencies can be carried out to generate a comprehensive set of *X*-parameters that can be used as measurement-based model in a harmonic balance (HB) circuit simulator platform. *X*-parameters, however, are not scalable and do not account for memory effects.

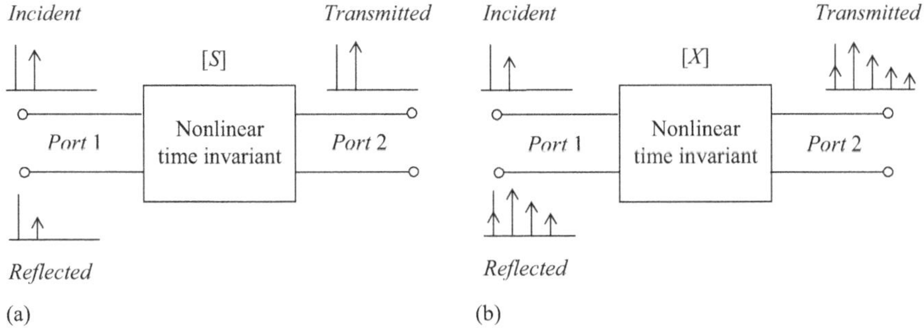

Figure 3.5 Spectral components generated from a nonlinear device due to single tone excitation (a) under small signal condition and (b) under large signal condition

3.3 Microwave network basic properties

Sections 3.3.1–3.3.6 review the fundamental properties of linear time-invariant (LTI) networks. Understanding these properties is very helpful in simplifying the analysis of complex microwave circuits.

3.3.1 Linearity

A linear network must obey the principle of superposition. This entails satisfying two conditions together, additivity and homogeneity (scalability). Referring to Figure 3.2 as an example, a voltage (V_1) yields a current (I_1) at port 1 and similarly, a voltage (V_2) yields a current (I_2) at port 2 and such network is considered linear if the following relations hold:

$$V_1 + V_2 \Rightarrow I_1 + I_2 \quad \text{(additivity)} \tag{3.41}$$

$$\begin{aligned} \alpha V_1 &\Rightarrow \alpha I_1 \\ \beta V_2 &\Rightarrow \beta I_2 \ \text{(homogeneity)} \end{aligned} \tag{3.42}$$

$$\alpha V_1 + \beta V_2 \Rightarrow \alpha I_1 + \beta I_2 \ \text{(superposition : both conditions together)} \tag{3.43}$$

and $\alpha \& \beta$ are constants.

3.3.2 Time invariance

Time invariance is a system property which implies that a network characteristic doesn't change with time and can be illustrated through the following formula:

$$V_1(t - \tau) \Rightarrow I_1(t - \tau) \tag{3.44}$$

This indicates that a delay in the amount of τ in the input signal will cause the same amount of delay in the output signal with no alteration of its response.

3.3.3 Reciprocity

Reciprocal networks are comprised of linear passive components. Reciprocity can be noted by looking at the transfer coefficients of the network parameters. For example, transmission coefficients of a Z-matrix must satisfy:

$$Z^k_{i,j} = Z^k_{j,i} \tag{3.45}$$

while all other ports are terminated by open circuits. Y- and S-matrix of a reciprocal device features the same property, however, with the proper port termination.

3.3.4 Symmetry

A reciprocal network implies that the network is also symmetric along its main diagonal elements. Under the symmetry condition, a Z-matrix of a symmetric network will equal its transpose matrix:

$$[Z] = [Z]^T \tag{3.46}$$

This property applies for all of Z-, Y-, and S-matrices under the reciprocity condition.

3.3.5 Passivity

A passive network is a circuit that doesn't contain active or independent sources so that at any point in time the amount of energy available in the network doesn't exceed the amount of supplied energy. In fact, a network with a positive real part of its port input impedance is considered passive:

$$\mathrm{Re}\{Z_{\mathrm{in}}\} > 0 \tag{3.47}$$

Such circuits comprise resistors, capacitors, inductors, and mutual inductances.

3.3.6 Lossless

Networks with only reactive elements (i.e., no resistances) are considered lossless. When looking at the port input impedance of a lossless network, the real part of the impedance must equal to zero. In general, the real part of any element in the impedance or admittance matrix of a lossless network must equal to zero:

$$\mathrm{Re}\{Z_{\mathrm{in}}\} = 0 \tag{3.48}$$

The S-matrix of a lossless network possesses a special property which is the unitary matrix property satisfying:

$$(S^*)^T S = I \tag{3.49}$$

where $(S^*)^T$ is the conjugate transpose of the S-matrix and I is the identity matrix.

3.4 Signal flow graph

Signal flow graph (SFG) is a graphical technique used to represent a set of algebraic relationships between signals in a linear network. The method is used in various engineering disciplines such as control, power, and microwave systems and aims to simplify the analysis of circuits. Moreover, the method is suitable for solving complex networks with the aid of computer codes using matrix formulations [9]. For a given system, the SFG is not unique and different graphs can be obtained depending on how the system equations are formulated. This section briefly discusses key elements of SFG and an example demonstrating the use of SFG in analyzing a common microwave network problem represented by S-parameters.

The basic decomposition rules depicted in Figure 3.6 are usually utilized to reduce complex SFG into a single equivalent branch between two nodes. Each node in the SFG represents a signal variable and each branch between two nodes represents the signal flow direction accompanied by its transmission factor or the S-parameter that indicates the dependence of one signal variable on another.

Consider the two-port network depicted in Figure 3.4 as an illustrative example which describes the incident and reflected waves at each port and its equivalent SFG representation shown in Figure 3.7. By terminating port 2 to a nonmatched load represented by Γ_L, we would like to obtain an expression to the input reflection coefficient, Γ_{in}, using the SFG rules.

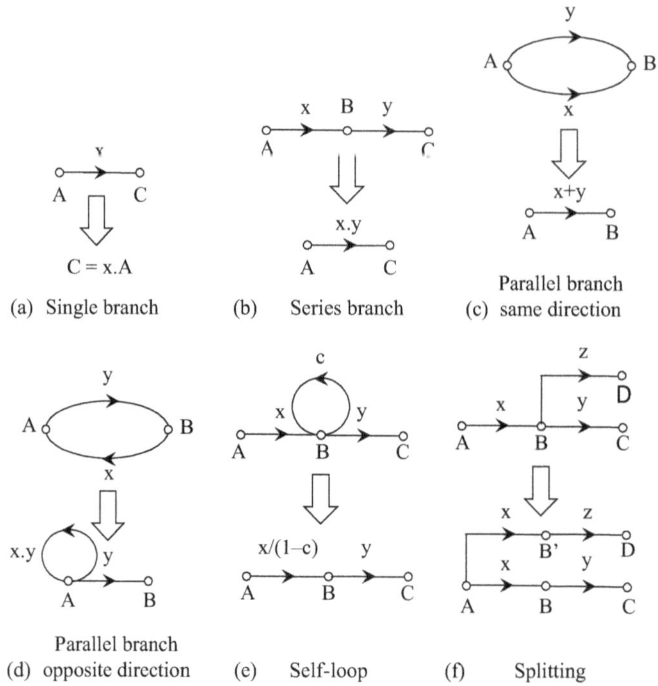

Figure 3.6 Key rules of a signal flow graph

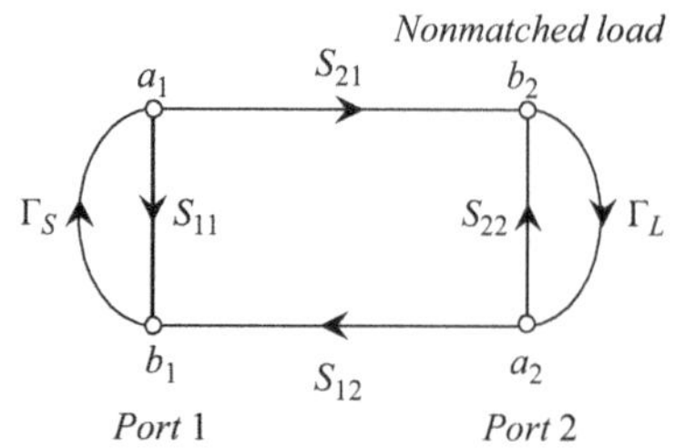

Figure 3.7 Construction of a signal flow graph representation of a two-port network terminated with nonmatched load at port 2

To start, the two-port S-matrix of the network shown in Figure 3.7 can be expanded as:

$$b_1 = S_{1,1}a_1 + S_{1,2}a_2 \tag{3.50}$$

$$b_2 = S_{2,1}a_1 + S_{2,2}a_2 \tag{3.51}$$

Let

$\Gamma_S = \dfrac{a_1}{b_1}$: the source reflection coefficient that is seen from the reference plane at port 1,

$\Gamma_L = \dfrac{a_2}{b_2}$: the load reflection coefficient that is seen from the reference plane at port 2,

$\Gamma_{in} = \dfrac{b_1}{a_1}$: the input reflection coefficient looking at the input of the network at port 1, and

$\Gamma_{out} = \dfrac{b_2}{a_2}$: the output reflection coefficient looking at the output of the network at port 2.

The SGF of the network in Figure 3.7 can be simplified according to the rules illustrated in Figure 3.6 by systematically applying rules d, e, b, and c, respectively. The step-by-step SGF analysis is illustrated in Figure 3.8.

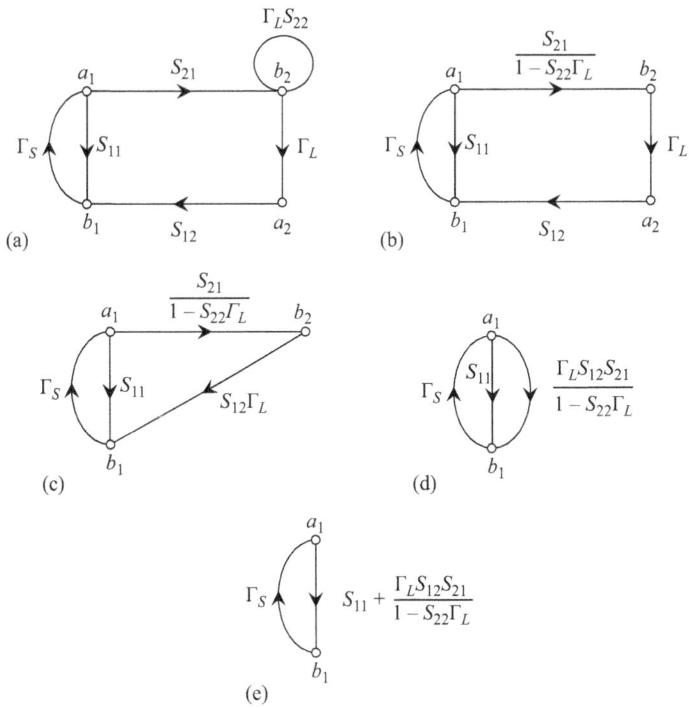

Figure 3.8 Analysis of a two-port network using signal flow graphs

From Figure 3.8(e) and the SFG rule in Figure 3.6(a), it can be concluded that the input reflection coefficient of the network at port 1 with nonmatched load at port 2 can be expressed as:

$$\Gamma_{in} = \frac{b_1}{a_1} = S_{1,1} + \frac{S_{1,2}S_{2,1}\Gamma_L}{1 - S_{2,2}\Gamma_L} \tag{3.52}$$

Note from (3.52) that when $\Gamma_L = 0$ (matched condition at port 2), $S_{1,1} = \frac{b_1}{a_1} = \Gamma_{in}$.

3.5 De-embedding and embedding *S*-parameters

De-embedding and embedding the scattering parameters are important techniques that enable us to mathematically subtract effects of a physical network, or add effects of a virtual network between the measurement equipment reference plane and the device under test (DUT) planes as depicted in Figure 3.9. While de-embedding is used to remove effects from the *S*-parameters, embedding is the opposite process. These effects may include a shift in the network reference plane, resistive loss of interconnects, and parasitic inductance and capacitance.

Typically, RF front-end modules based on semiconductor fabrication technologies are designed in packages with sizes that can range from tens of micrometers to few millimeters. A module with such a small form factor and tens of input/output (I/O) pins requires the use of an evaluation board or a test fixture, which is necessary to allow a practical means of measuring such small-sized modules. This is achievable by having extended traces from the module plane to the

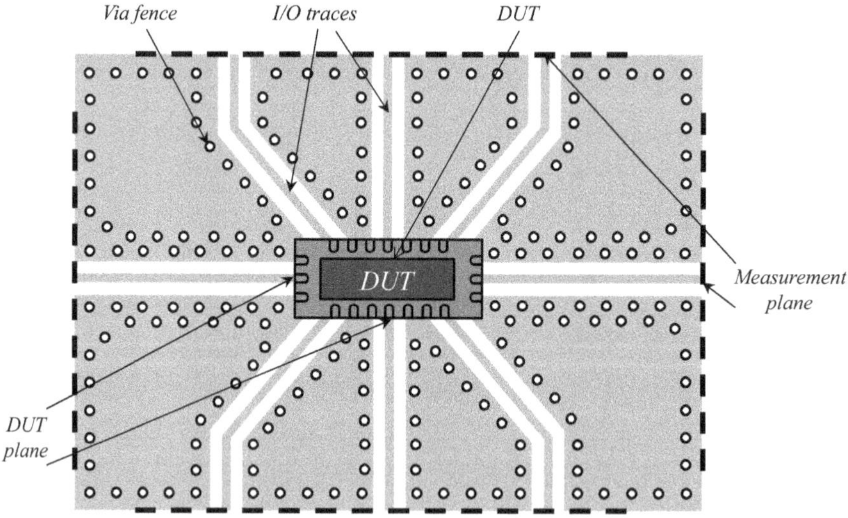

Figure 3.9 A DUT showing a chip mounted on a test fixture which has traces extended to the edge of the board for connection with the test equipment

SMA connectors of the test fixture, where measurement cables of the measuring instruments can be conveniently connected. Note that the extended traces present additional inductances and capacitances added to the module I/O pins that are not part of the actual module under characterization. Therefore, effects due to the physical length and width of these traces must be de-embedded from the measurement data of the full fixture (i.e., test fixture with DUT).

The determination of how much distance the plane should be moved can be measured using a time domain reflectometry (TDR) method. TDR is usually an existing feature in several measuring instruments. Only moving the reference plane may not be sufficient to de-embed especially when the fixture traces and the device I/Os possess different characteristic impedances. The proper method is to characterize the traces of the test fixture without the DUT either through measurement or full-wave electromagnetic simulation so that their full characteristics are captured and then can be properly removed after performing the measurement on the full fixture.

To illustrate the de-embedding process, consider a two-port *S*-parameter measurement setup of a test fixture that includes two traces connecting the DUT at the inner ends and attached to the VNA cables through SMA connectors at the outer ends. A calibrated VNA will provide measured *S*-parameters at the interface of the SMA connectors which represent the measurement plane. The measured *S*-parameters are then converted into *T*-parameters ($T_{\text{Test Fixture}}$) since *T*-parameters are cascadable as expressed by:

$$[T_{\text{Test Fixture}}] = [T_{\text{trace1}}][T_{\text{DUT}}][T_{\text{trace2}}] \tag{3.53}$$

Next, and after removing the DUT from the test fixture, each of the two traces is measured as two-port *S*-parameters and then converting into *T*-parameters (T_{Trace1} and T_{trace2}). As we are interested in the *T*-parameters that correspond to the DUT, the test fixture effects can then be removed by negating/de-embedding the traces from the measured $T_{\text{Test Fixture}}$ matrix by multiplying each side of the matrix by the inverse *T*-parameter matrix of the fixture traces as in (3.54):

$$[T_{\text{DUT}}] = [T_{\text{trace1}}]^{-1}[T_{\text{Test Fixture}}][T_{\text{trace2}}]^{-1} \tag{3.54}$$

Finally, converting back the *T*-parameter of the DUT into *S*-parameters completes the de-embedding process.

3.6 Two-port equivalent circuits

Since most of the discrete passive microwave products are two-port devices, two-port equivalent circuits are often used in their analysis. A linear two-port reciprocal network can be represented by, but not limited to, *T*- or π- equivalent circuit. If the network is lossless, the circuit elements of the *T*- or π-equivalent circuits are purely reactive elements. The two-port network parameters can be generated using measured or simulated data. The elements of the *T*- or π-equivalent circuit are to some extent beneficial in providing a physical meaning to the given parameters as illustrated in Figure 3.10.

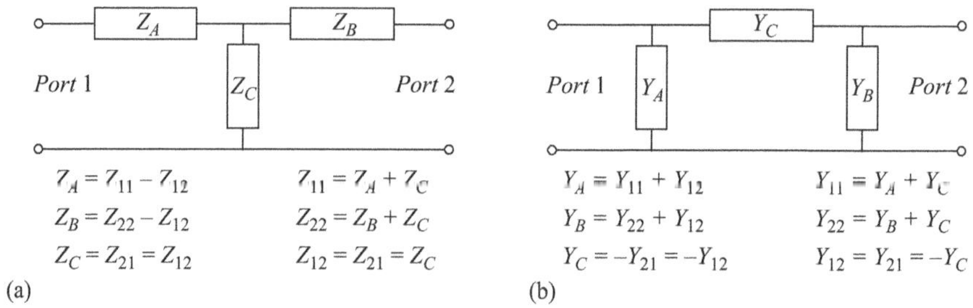

Figure 3.10 Equivalent circuits of a two-port network: (a) T-network and its element values obtained from the Z-parameters and (b) π-network and its element values found from the Y-parameters

Table 3.1 Y–Δ impedance transformations

Y–Δ	Δ–Y
$Z_A\Big\vert_{Y\text{–Network}} = \dfrac{(Z_AZ_B) + (Z_AZ_C) + (Z_BZ_C)}{Z_B}$	$Z_A\Big\vert_{\Delta\text{–Network}} = \dfrac{(Z_aZ_c)}{Z_a + Z_b + Z_c}$
$Z_B\Big\vert_{Y\text{–Network}} = \dfrac{(Z_AZ_B) + (Z_AZ_C) + (Z_BZ_C)}{Z_A}$	$Z_B\Big\vert_{\Delta\text{–Network}} = \dfrac{(Z_bZ_c)}{Z_a + Z_b + Z_c}$
$Z_c\Big\vert_{Y\text{–Network}} = \dfrac{(Z_AZ_B) + (Z_AZ_C) + (Z_BZ_C)}{Z_C}$	$Z_C\Big\vert_{\Delta\text{–Network}} = \dfrac{(Z_aZ_b)}{Z_a + Z_b + Z_c}$

Scrutiny of Figure 3.10 shows that transformation between π- and T-networks is also possible through Y–Δ transformation. As shown, T-networks can be directly extracted from Z-parameters and π-networks can be extracted easily from Y-parameters. The Y–Δ impedance transformation formulas are summarized in Table 3.1.

3.7 Passive and lossless circuit synthesis

Microwave network synthesis has, for many decades, found several applications in the area of microwave impedance matching, filtering, and coupled networks. Synthesis of a circuit is the construction of the circuit topology and finding its element values based on given circuit characteristics or from the knowledge of the transfer function based on analytically oriented circuit theories. Bandwidth, pass-band ripples, and inclusion of circuit parasitic elements can be also specified in the synthesized networks. Synthesis of a circuit does not necessarily yield an optimal design compared to other iterative design-optimization techniques; however, it serves as a good starting point after which, fine-tuning may be required. However, analysis of a circuit is the process of determining its characteristics; hence, it is the inverse of circuit synthesis [10].

There are several methods that can be employed to synthesize a circuit and depending on the problem being solved; synthesis procedures can be employed to form one- and multiport circuits. R. Foster, W. Cauer, O. Brune, and S. Darlington are the pioneers in this field and all have considered the problem in the frequency domain [11]. Their contributions are significant in establishing the governing conditions that led to realizable circuits and providing realization techniques in various canonical forms. Typically, circuit synthesis is associated with the design of filters; however, it can also be generalized to extract equivalent circuits from an arbitrary frequency response. In most cases, synthesis procedures are used to find a compromise and to favor some design specification against others based on design priorities.

Synthesis of an equivalent circuit is not a straightforward procedure owing to the several techniques that can be utilized and the analytical complexities associated with each synthesis method resulting in a nonunique realization of a network. Moreover, different synthesis methods can influence the stability and sensitivity of the realization.

Synthesis of passive circuits allows only the use of R, L, and C components, either as lumped or distributed, to form the desired network elements. Despite the many circuit synthesis approaches, all methods describe the device response in terms of either its power transmission coefficient $T(\omega)$ or its power reflection coefficient $\Gamma(\omega)$ which are completely dependable parameters and for a lossless network is given by $\Gamma(\omega) = 1 - T(\omega)$.

A circuit response can be defined by the singularities (i.e., poles and zeros) of its transfer function or input impedance. In this section, the circuit synthesis based on poles and zeros extraction method will be discussed. A finite circuit comprising a finite number of elements exhibits input immittance function in the form of a rational polynomial of frequency (p). If the numerator and denominator are positive real polynomials, also referred to as Hurwitz polynomials, the circuit is then stable and realizable which means that no poles or zeros exist on the right-half side of the S-plane. For a lossless network that comprises only LC elements, the poles and zeros lie and alternate on the $j\omega$ axis of the S-plane.

Extraction of lumped element circuit models from numerical data such as measurement or simulation data are also available through extracting the pole-zero information from their time- or frequency-domain response by means of optimization or curve fitting techniques [12].

3.7.1 Singly and doubly terminated networks

Considering the two-port networks shown in Figure 3.11, a singly terminated network is a network connected with a resistive load at one of its ports wherein the other port is terminated by an ideal voltage or current source (or with an infinite or a zero impedance, respectively) as illustrated in Figure 3.11(a) and (b). An ideal voltage source must be followed by a series element, while an ideal current source must be connected to a parallel element. Otherwise, the first element in each case will be deemed useless as they will possess the same voltage or same current when using a voltage or a current source, respectively. In fact, singly terminated networks

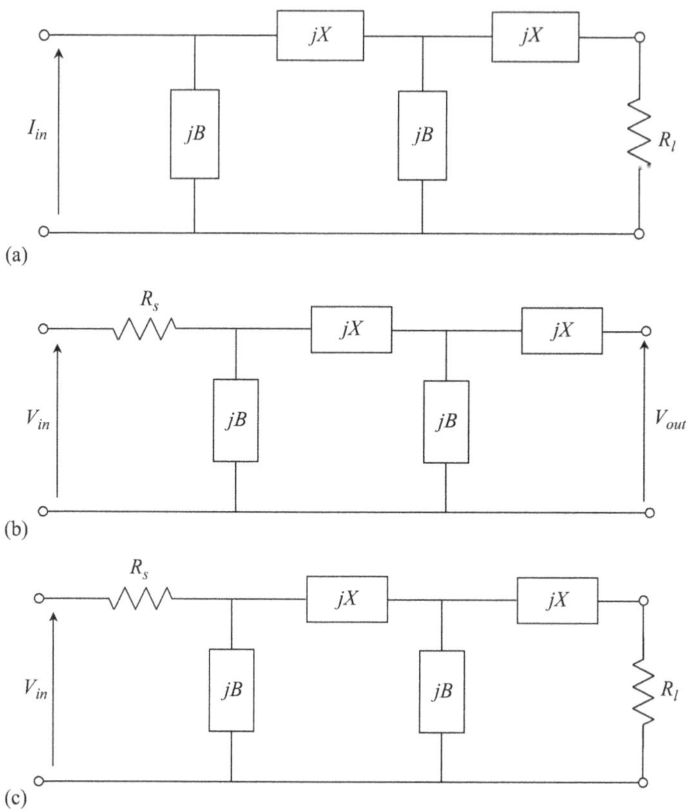

Figure 3.11 Two-port terminated ladder networks: (a) singly with a resistance at the load, (b) singly with a resistance at the source, and (c) doubly terminated

cannot be used as matching networks to match, for example, between a source of 50 ohms and a zero or infinite impedance load and having 0 dB insertion loss in the passband. However, a doubly terminated network has both ports connected to resistive loads to form symmetric or asymmetric networks as depicted in Figure 3.11(c). Here, the network performance parameters such as return loss and insertion loss can be directly calculated, unlike singly terminated networks where the available power from the ideal sources is infinite. In both singly and doubly terminated networks, the resistive load can also be accompanied by a reactance element [10]. Generally, the design procedure of a singly terminated ladder network can serve as a starting point to obtain a doubly terminated network.

Singly terminated networks find applications in broadband matching circuits and in the case of filter synthesis used to form manifold or star-junction multiplexers. Doubly terminated networks are commonly used to design two-port networks such as filters by employing, for example, the Darlington synthesis method as will be discussed in Section 3.7.3.

3.7.2 One-port circuit synthesis

One-port synthesis starts by finding the port immittance function that takes the form of a rational polynomial function. Then, one-port circuit synthesis can be realized by one of four canonical forms as shown in Figure 3.12.

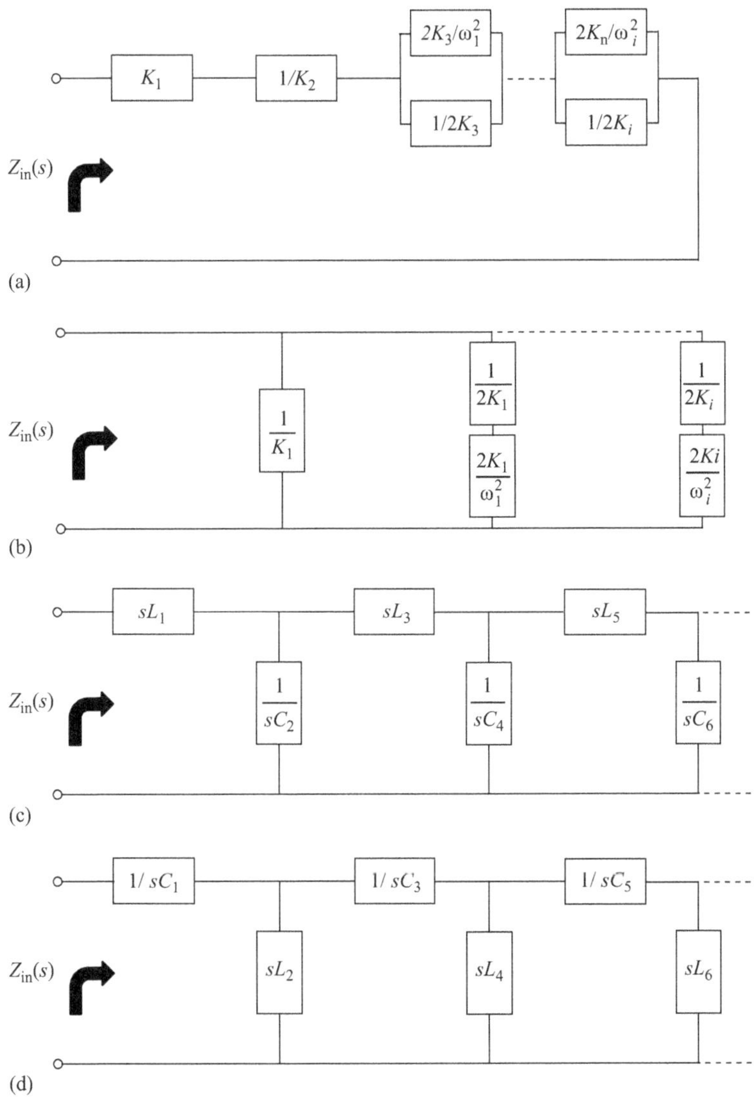

Figure 3.12 Canonical forms of one-port synthesis networks: (a) Foster first form, (b) Foster second form, (c) Cauer first form, and (d) Cauer second form

3.7.2.1 Foster synthesis

Foster synthesis technique was developed in 1924 to conclude that the synthesis of a lossless one-port network can be carried out through pole extraction of the partial fraction expansion of the network input immittance function. This method is known as Foster synthesis where a circuit is realized using a series or parallel branches. Foster synthesis method is available in two forms: the first uses the impedance form and the second uses the admittance form as depicted in Figure 3.12(a) and Figure 3.12(b), respectively. In either case, Foster synthesis approach proceeds in cycles where a pole at $s = 0$ is extracted, if exists, and a pole at $s = \infty$ is extracted, if exists. Then, the subsequent terms are determined by identifying sets of imaginary poles and extracting their residues from the immittance function. The driving point immittance function can take the following general form:

$$Z(s) = k_1 s + \frac{k_2}{s} + \sum_{i=3}^{i=n-1} \frac{2k_i s}{s^2 + \omega_i^2} + Z_n(s) \tag{3.55}$$

$$Y(s) = \tilde{k}_1 s + \frac{\tilde{k}_2}{s} + \sum_{i=3}^{i=n-1} \frac{2\tilde{k}_i s}{s^2 + \omega_i^2} + Y_n(s) \tag{3.56}$$

where $s = \sigma + j\omega$ is the complex frequency variable and in other textbooks is denoted as p. A complex frequency variable is a general term where the real part (σ) refers to amplitude gain or decay of the signal and its imaginary part $(j\omega)$ describe the signal angular frequency. K_i and $i = 1,2,3\ldots$ are the residues of the poles in the partial fraction expansion.

A realizable immittance function must be a positive real function due to the constraints imposed by the causality and stability of the network. A simple realizability test could be examined if the rational immittance function has both the denominator and numerator polynomials as Hurwitz polynomials.

The following is an example concerned with Foster synthesis. Consider a realizable network function represented by its input impedance function as:

$$Z_{\text{in}}(s) = \frac{s^5 + 7s^3 + 10s}{s^4 + 4s^2 + 3} = \frac{s(s^2 + 2)(s^2 + 5)}{(s^2 + 3)(s^2 + 1)} \tag{3.57}$$

By decomposing the expression in (3.57) into a sum of simple polynomial fractions yields:

$$Z_{\text{in}}(s) = s + \frac{2s}{s^2 + 1} + \frac{s}{s^2 + 3} \tag{3.58}$$

Next, finding the zeros and poles of the impedance function gives:

$$\begin{aligned} Z_{1,2} &= \pm j2.236 \\ Z_{3,4} &= \pm j1.414 \\ p_{1,2} &= \pm j1.732 \\ p_3 &= \pm j1.0 \end{aligned} \tag{3.59}$$

Figure 3.13 shows that all poles and zeros are located on the $j\omega$ axis of the S-plane, thus a network with only reactive elements will be obtained.

To extract the network series elements, first, we look at the highest polynomial degree of each of the simple fractions of the impedance function to determine the element type. If s is a first order polynomial, this implies extracting a single reactive element and if s is a second order polynomial, this means extracting a series reactive LC tank. The following is a step-by-step extraction of the circuit elements using Foster synthesis method.

As s tends to ∞, $Z_{\text{in}}(s)$ tends to $1s$, which implies that a series inductor of value = 1 can be extracted from the impedance function:

$$sL = \lim_{s \to \infty} Z_{in}(s) \tag{3.60}$$

$$\lim_{s \to \infty} \frac{Z_{in}(s)}{s} = 1, \quad \text{thus,} \quad L_1 = 1 \tag{3.61}$$

The remaining impedance is:

$$Z_1(s) = Z_{\text{in}}(s) - s = \frac{2s}{s^2 + 1} + \frac{s}{s^2 + 3} \tag{3.62}$$

To extract the next branch (an LC tank), we evaluate the impedance function of $Z_1(s)$ at its imaginary poles. Starting with one of the two fractions as:

$$\text{Let } Z_A(s) = \frac{2s}{s^2 + 1} \tag{3.63}$$

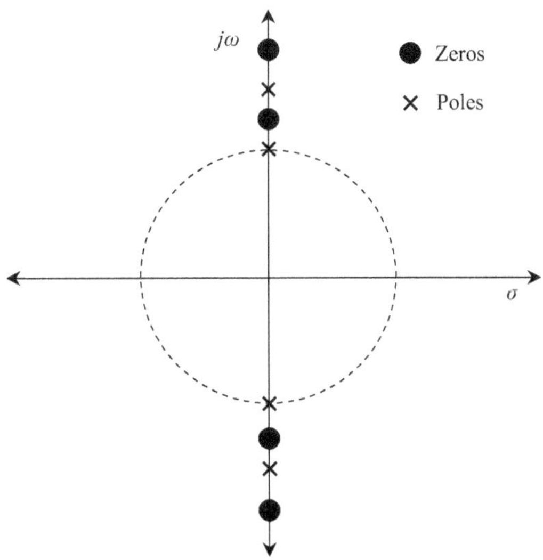

Figure 3.13 Graphical illustration of the pole and zero locations of the immittance function in (3.57) on the S-plane

then

$$Y_A(s) = \frac{s^2+1}{2s} = \frac{s^2}{2s} + \frac{1}{2s} = \underbrace{\frac{s}{2}}_{\text{implies } C=1/2} + \underbrace{\frac{1}{2s}}_{\text{implies } L=2} \tag{3.64}$$

Similarly, to extract the next LC tank:

$$Z_B(s) = \frac{s}{s^2+3} \tag{3.65}$$

and

$$Y_B(s) = \frac{s^2+3}{s} = \underbrace{s}_{\text{implies } C=1} + \underbrace{\frac{3}{s}}_{\text{implies } L=1/3} \tag{3.66}$$

The resultant Foster circuit can then be given as depicted in Figure 3.14.

As another example of Foster realization of an admittance function, consider the rational function given as:

$$Y_{\text{in}}(s) = \frac{s^4+4s^2+3}{s^5+7s^3+10s} = \frac{(s^2+3)(s^2+1)}{s(s^2+2)(s^2+5)} \tag{3.67}$$

By decomposing the expression in (3.67) into a sum of simple polynomial fractions, we obtain:

$$Y_{\text{in}}(s) = \frac{3}{10s} + \frac{1}{6}\frac{s}{s^2+2} + \frac{8}{15}\frac{s}{s^2+5} \tag{3.68}$$

The resultant circuit is illustrated in Figure 3.15 which shows the realized element values.

3.7.2.2 Cauer synthesis

Another realization illustrated in Figure 3.12(c) and (d) is the ladder network which is more prevalent. This method is known as Cauer synthesis developed in 1926 and

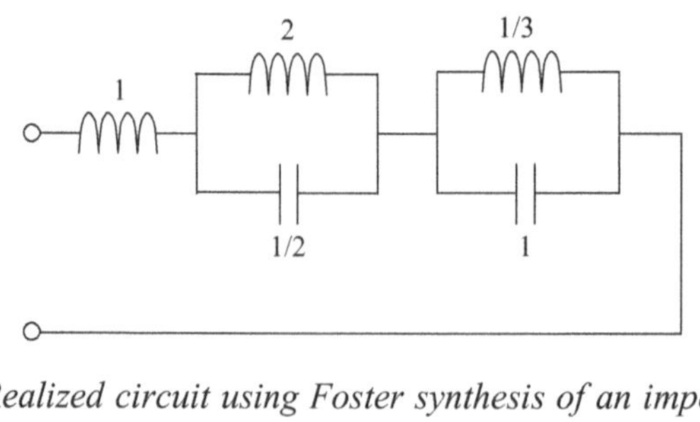

Figure 3.14 Realized circuit using Foster synthesis of an impedance function

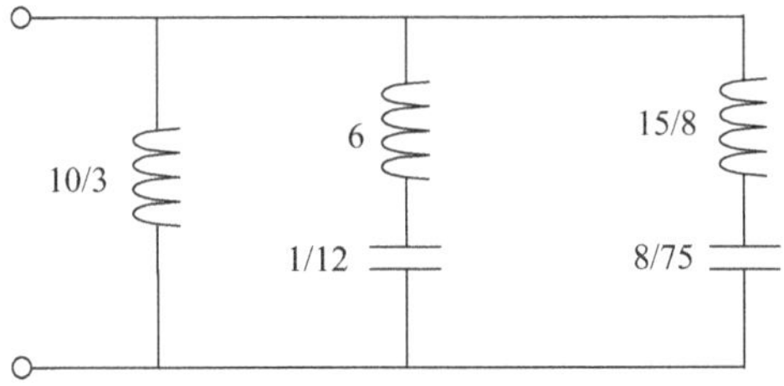

Figure 3.15 Realized circuit using Foster synthesis of an admittance function

based on continued fraction expansion of the network input immittance function. The ladder circuit with a single reactance element per branch can be considered as a low-pass or high-pass filter, and therefore frequency transformations may be applied to obtain other frequency responses with more complex ladder network elements. An important characteristic of a doubly terminated ladder network is the low sensitivity to its component variations when the network terminating resistances are equal.

The following first Cauer canonical form is a generalized form applicable when the degree of the numerator is greater than the degree of the denominator of the impedance function. Otherwise, expansion can be started from the input admittance, $Y(s)$, instead of $Z(s)$:

$$Z(s) = \frac{m(s)}{n(s)} = L_1 s + \cfrac{1}{C_2 s + \cfrac{1}{L_3 s + \cfrac{1}{C_4 s + \ddots}}} \tag{3.69}$$

The second Cauer canonical form has a ladder network with a chain of capacitors in series and inductors in shunt and can be generalized as:

$$Z(s) = \frac{m(s)}{n(s)} = \frac{1}{C_1 s} + \cfrac{1}{\frac{1}{L_2 s} + \cfrac{1}{\frac{1}{C_3 s} + \cfrac{1}{\frac{1}{L_4 s} + \ddots}}} \tag{3.70}$$

Both methods proceed in alternation by first extracting a pole at $s = \infty$ and then finding a pole at $s = 0$ and so on.

With Cauer first form using continued fraction expansion about infinity, the synthesis starts with extracting a pole at $s = \infty$ from the impedance function and the remaining impedance term is inverted to admittance which possesses a pole at $s = \infty$. These cycles continue until the complete extraction of the immittance function. Eventually, the synthesized circuit will comprise of a chain of inductors in the series branches and a chain of capacitors in the shunt branches forming a ladder network. The other form of Cauer synthesis uses continued fraction expansion about a point at origin or a pole at $s = 0$ and follows the same procedure as the expansion about infinity; however, the final ladder circuit has capacitors in the series branches and inductors in the shunt branches.

Consider the following example of a Cauer synthesis for a network represented by its input impedance function:

$$Z_{\text{in}}(s) = \frac{2s^4 + 5s^2 + 1}{s^3 + s} \tag{3.71}$$

Starting with finding the zeros and poles of the impedance function gives:

$$\begin{aligned} Z_{1,2} &= \pm j1.510 \\ Z_{3,4} &= \pm j0.468 \\ p_{1,2} &= \pm j1.0 \\ p_3 &= j0 \end{aligned} \tag{3.72}$$

This immittance function has poles and zeros lie on the $j\omega$ axis of the S-plane as illustrated in Figure 3.16. Furthermore, the coefficients of the numerator and denominator of the input impedance function are all positive real or represent Hurwitz polynomials which imply that no poles or zeros exist in the right half-plane.

The impedance function can be expanded in continued fraction form as:

$$Z_{\text{in}}(s) = 2s + \cfrac{1}{\cfrac{s}{3} + \cfrac{1}{\frac{9s}{2} + \frac{3}{2s}}} \tag{3.73}$$

By evaluating the expression about a point at infinity and considering the arithmetical expression in (3.71), one can extract circuit elements alternating between series and shunt in a ladder network through finding the nonfractional

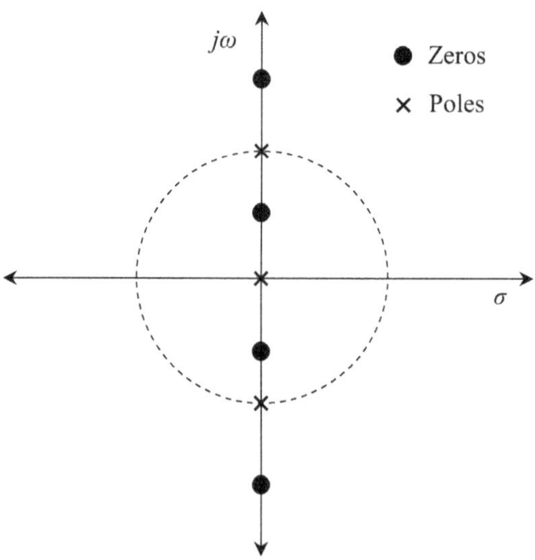

Figure 3.16 Graphical illustration of the pole and zero locations of the immittance function in (3.70) on the S-plane

s coefficient in each row of the continued fraction expansion in (3.73). For example, the first series element of the network can be directly extracted by looking into the nonfractional coefficient of s in the top row of the expression which gives an inductor with a value of 2; while the second shunt element can be interpreted as an admittance coefficient of the nonfractional coefficient of s in the second row of the expression giving a capacitor value of 1/3. Similarly, the rest of the circuit elements can be obtained in a similar fashion.

The synthesized circuit of the impedance function in (3.71) is depicted in Figure 3.17.

Alternatively, s tends to ∞ as $Z_{\text{in}}(s)$ tends to $2s$, which implies a series inductor of value = 2 can be extracted from the impedance function:

$$\lim_{s\to\infty}\frac{Z_{in}(s)}{s} = 2, \ \ \text{thus,}\ L_1 = 2 \tag{3.74}$$

The remaining impedance is:

$$Z_1(s) = Z_{\text{in}}(s) - 2s = \frac{3s^2+1}{s(s^2+1)} \tag{3.75}$$

To extract the next shunt element, we evaluate the admittance function of $Z_1(s)$ at $s = \infty$ as:

$$Y_1(s) = \frac{s(s^2+1)}{3s^2+1} \tag{3.76}$$

and

$$\lim_{s\to\infty}\frac{Y_1(s)}{s} = 1/3, \quad \text{thus} \quad C_2 = 1/3 \tag{3.77}$$

The remaining admittance is then:

$$Y_2(s) = Y_1(s) - \frac{s}{3} = \frac{2}{3}\frac{s}{3s^2+1} \tag{3.78}$$

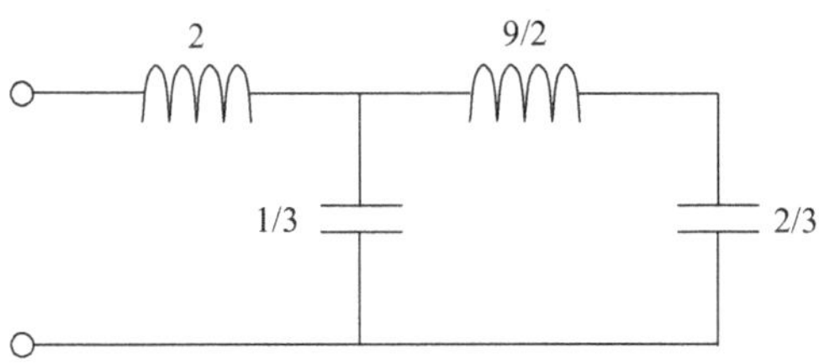

Figure 3.17 Realized circuit using Cauer synthesis of a ladder network from a given input impedance function about a point at infinity

The next series element from the impedance function results in:

$$Z_2(s) = \frac{3}{2}\frac{3s^2+1}{s} \tag{3.79}$$

and

$$\lim_{s\to\infty}\frac{Z_2(s)}{s} = \frac{9}{2}, \quad \text{thus} \quad L_3 = \frac{9}{2} \tag{3.80}$$

The remaining impedance is then given as:

$$Z_3(s) = \frac{3}{2}\frac{3s^2+1}{s} - \frac{9s}{2} = \frac{3}{2s} \tag{3.81}$$

The final shunt element from the admittance function is then calculated as:

$$Y_3(s) = \frac{2s}{3} \tag{3.82}$$

and

$$\lim_{s\to\infty}\frac{Y_3(s)}{s} = \frac{2}{3}, \quad \text{thus} \quad C_3 = \frac{2}{3} \tag{3.83}$$

3.7.3 Two-port circuit synthesis

Similar to a one-port synthesis, a two-port synthesis of a lossless, passive, and reciprocal network can be realized through partial fraction expansion of the transfer function by terminating one of its two ports by a load usually 1-ohm resistor. This method is known as the Darlington synthesis, or the insertion loss method [13] developed in 1939, which forms the foundation of the modern network theory. As was pointed out, the subject of circuit synthesis is mostly associated with the design of filtering networks. The fact that an ideal characteristic of a filter is embodied in the instantaneous transition between the passband to the stopband, such ideal transition makes the network unrealizable. Therefore, several magnitude approximations of filter response such as Butterworth, Bessel, or Chebyshev are available to enable realizable networks by allowing this transition within a certain bandwidth. An ideal filter response requires an infinite degree of the polynomial representing its transfer function which implies an infinite number of circuit elements. Darlington has shown that the two one-port networks discussed in Section 3.7.2 are duals of each other, and that any positive real input immittance function can always be realized by a lossless reciprocal two-port network terminated in a load resistance as depicted in Figure 3.18.

The general synthesis procedure of filters starts by realizing a normalized low-pass prototype that can be implemented using RC, RL, or LC networks. Then, impedance and frequency scaling can be performed on the normalized network to obtain the desired center frequency and source/load impedances. A lossless doubly terminated ladder network comprised of LC components is a highly desirable topology due to the low sensitivity to its component values [14]

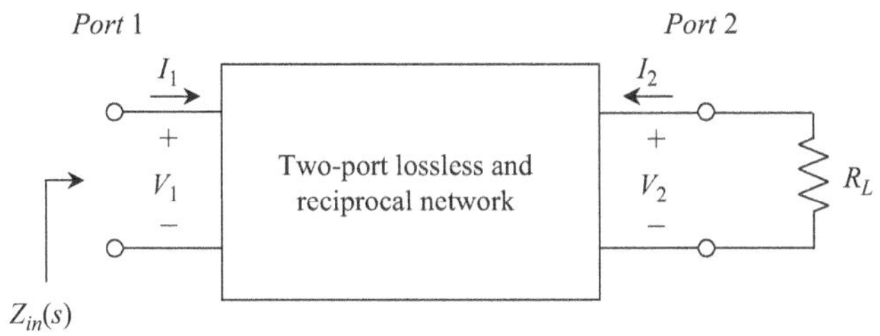

Figure 3.18 Darlington synthesis of a two-port network terminated in a load resistor

and thereby less prone to manufacturing and design tolerances. A reactive ladder network typically achieves higher frequency roll-off than an RC and an RL-based circuits. Moreover, compared to RC and RL networks, the procedures used in LC networks are more generalized. A ladder network is also known as a minimum phase network since energy between its input and output can only take one path which is the minimum phase path (i.e., the phase characteristics are uniquely related to the amplitude characteristics). Nonminimum phase networks usually require inter-resonator coupling between the nonadjacent resonators. In a ladder structure, the amplitude and phase responses are inseparably related by the Hilbert transform [15].

One way to realize the low-pass prototype under a set of specifications is by inferring the circuit input immittance function from the transfer function or the input reflection function which can be formed following one of the filter response approximation methods such as Butterworth or Chebyshev.

A Butterworth filter response has no ripples in the passband or the stopband and for this reason, it is also referred to as a maximally flat response. The amplitude squared forward transfer function of a Butterworth prototype [15] is defined as:

$$|S_{21}(s)|^2 = \frac{1}{1 + (s)^{2N}} \tag{3.84}$$

where s is the normalized frequency of the low-pass prototype of order N as described in [15]. From (3.84), a Butterworth low-pass response has a magnitude at $\omega = 0$ is normalized to one and the magnitude of the transfer function monotonically decreases from this value to $\omega = \infty$.

A Chebyshev filter response of a lossless system has the square of the magnitude of the input reflection function described as:

$$|S_{11}(s)|^2 = 1 - |S_{21}(s)|^2 = \frac{(s)^{2N}}{1 + (s)^{2N}} \tag{3.85}$$

$S_{11}(s)$ can be found with the knowledge of:

$$|S_{11}(s)|^2 = S_{11}(s) \cdot S_{11}^{*}(s) \tag{3.86}$$

Since we are dealing with a rational function comprising real coefficients, the conjugate of a function equals the function of the conjugate variable:

$$|S_{11}(s)|^2 = S_{11}(s) \cdot S_{11}(s)^* \tag{3.87}$$

Given that $s = j\omega$ which is an imaginary frequency variable, its conjugate is equal to the negative of the variable:

$$|S_{11}(s)|^2 = S_{11}(s) \cdot S_{11}(-s) \tag{3.88}$$

Then, the calculation of the input immittance function of the prototype from the reflection coefficient in a $1-\Omega$ system (the circuits are driven from a 1-Ω generator and terminated in a $1-\Omega$ load) by looking at port 1 can be evaluated as:

$$Z_{\text{in}}(s) = \frac{1 + S_{11}(s)}{1 - S_{11}(s)} \tag{3.89}$$

$$Y_{\text{in}}(s) = \frac{1 - S_{11}(s)}{1 + S_{11}(s)} \tag{3.90}$$

Once the driving-point immittance is developed, the elements of a resistively terminated ladder LC network can be synthesized by expanding the network impedance using continued fraction expansion similar to what was discussed under Cauer's synthesis method.

As an example to illustrate the Darlington synthesis method of a two-port network, consider a third order Butterworth filter that has an insertion loss function expressed as:

$$|S_{21}(s)|^2 = \frac{1}{1 + s^6} \tag{3.91}$$

Assuming a lossless network the reflection function can be written as:

$$|S_{11}(s)|^2 = 1 - |S_{21}(s)|^2 = \frac{s^6}{1 + s^6} \tag{3.92}$$

Solving for the roots of the denominator of (3.91) or (3.92) determines the poles of the network. The complex roots occur in conjugate pairs. The poles of the squared magnitude of the reflection function in (3.92) are:

$-0.866 + j0.5$

$-0.866 - j0.5$

$0 + j1.0$

$0 - j1.0$

$0.866 + j0.5$

$0.866 - j0.5$

By selecting the left half-plane roots of the poles to ensure stability, the reflection function can be expressed as:

$$S_{11}(s) = \frac{s^3}{s^3 + 2s^2 + 2s + 1} \tag{3.93}$$

Using (3.89) the input impedance function of a two-port network in a $1-\Omega$ system is then given by:

$$Z_{\text{in}}(s) = \frac{2s^3 + 2s^2 + 2s + 1}{2s^2 + 2s + 1} \tag{3.94}$$

Following the synthesis procedure of Cauer first form discussed in Section 3.7.2.2 and upon using the continued fraction expansion on the input impedance function in (3.94), Figure 3.19 illustrates the circuit representation of the normalized Butterworth low-pass filter prototype. Impedance and frequency transformations can then be carried-out [15] to achieve the desired center frequency, frequency configuration (high-pass, band-pass, or band-stop), and terminating impedances.

To this end, the normalized filter prototype uses a source and load impedance of one ohm and a normalized frequency of one rad/s. Impedance and frequency scaling of the prototype design can be carried out on the circuit component values of the filter network. The element values after scaling its impedance to R_0 can be found as:

$$L_{\text{Imp\_scaled}} = R_0 L_{\text{proto}} \tag{3.95}$$

$$C_{\text{imp\_scaled}} = \frac{C_{\text{proto}}}{R_0} \tag{3.96}$$

$$R_L = R_S = R_0 \tag{3.97}$$

Similarly, to scale the circuit component values to a new cutoff frequency ω_c, ω is replaced by ω/ω_c as:

$$L_{\text{freq\_scaled}} = \frac{L_{\text{proto}}}{\omega_c} \tag{3.98}$$

$$C_{\text{freq\_scaled}} = \frac{C_{\text{proto}}}{\omega_c} \tag{3.99}$$

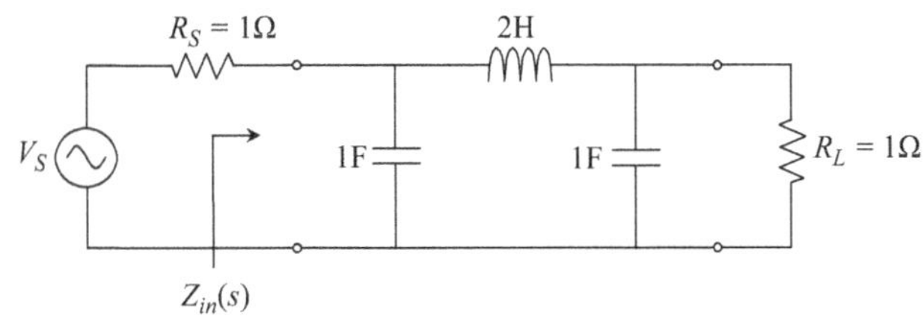

Figure 3.19 Circuit realization of a third order ladder topology in a $1-\Omega$ system

Both transformations can be combined in one step such as:

$$L_{\text{imp\_freq\_scaled}} = \frac{R_0 L_{\text{proto}}}{\omega_c} \tag{3.100}$$

$$C_{\text{imp\_freq\_scaled}} = \frac{C_{\text{proto}}}{R_0 \omega_c} \tag{3.101}$$

Hence, (3.100) and (3.101) describe the LPF to LPF frequency and impedance scaling. Scaling to a different filter response (i.e., HPF, BPF, and SBF) is also possible as discussed in [15].

References

[1] Oliner A. "Historical perspectives on microwave field theory." *IEEE Transactions on Microwave Theory and Techniques*. 1984; **32**(9):1022–45

[2] Frickey D.A. "Conversions between S, Z, Y, H, ABCD, and T parameters which are valid for complex source and load impedances." *IEEE Transactions on Microwave Theory and Techniques*. 1994; **42**(2):205–11

[3] Bode H.W. *Network Analysis and Feedback Amplifier Design*. Princeton, NJ: Van Nostran; 1945.

[4] Kurokawa K. "Power waves and the scattering matrix." *IEEE Transactions on Microwave Theory and Techniques*. 1965;**13**(2):194–202

[5] Bayram Y., and Volakis J.L. "Hybrid S-parameters for transmission line networks with linear/nonlinear load terminations subject to arbitrary excitations." *IEEE Transactions on Microwave Theory and Techniques*. 2007;**55** (5):941–50

[6] Khan Z.A., Bayram Y., and Volakis J.L. "EMI/EMC analysis of printed circuit boards subject to near-zone illuminations." *IEEE Transactions on Electromagnetic Compatibility*. 2009; **52**(2):406–08

[7] Verspecht J., and Root D.E. "Polyharmonic distortion modeling." *IEEE Microwave Magazine*. 2006;**7**(3):44–57

[8] Root D.E., Verspecht J., Horn J., and Marcu M. *X-Parameters: Characterization, Modeling, and Design of Nonlinear RF and Microwave Components*. 1st edn. New York, NY: University Press; 2013

[9] Collin R.E. *Foundation of Microwave Engineering*. 2nd edn. New York, NY: McGraw-Hill; 1992

[10] Cuthbert T.R. *Broadband Direct-coupled and Matching RF Networks*. Arkansas: TRCPEP; 1999.

[11] Darlington S. "A history of network synthesis and filter theory for circuits composed of resistors inductors, and capacitors." *IEEE Transactions on Circuits and Systems*. 1984;**31**(1):3–13

[12] Gustavsen B., and Semlyen A. "Rational approximation of frequency domain responses by vector fitting." *IEEE Transactions on Power Delivery*. 1999;**14** (3):1052–61

[13] Darlington S. "Synthesis of reactance 4-poles with prescribed insertion loss characteristics." *Journal of Mathematics and Physics*. 1939;**18**(1–4):257–353
[14] Sedra A.S., and Brackett P. *Filter Theory and Design: Active and Passive*. Champaign, IL: Matrix Pub; 1977
[15] Hunter I. *Theory and Design of Microwave Filters*. London: The Institution of Engineering and Technology (IET); 2001.

Chapter 4
Impedance matching networks

4.1 Introduction

The concept of electrical impedance matching was initially developed for electric power transmission systems to transfer energy between a source and a load. Most if not all microwave circuits can be thought of in one form or another as matching networks of specific characteristics. To list a few, microwave filters provide impedance matching within their passband frequencies and high impedance mismatch at the out-of-band frequencies. Antennas can be assumed as a matching circuit between the free space characteristic impedance and the antenna input impedance. Also, impedance transformers are means to match between two unmatched stages for achieving maximum power transfer or maximum power efficiency.

This chapter discusses mainly impedance matching circuits that are used to match two different terminations. Among the several impedance matching techniques, the available implementation technology is what usually determines the possible matching technique that can be used. Lumped passive matching elements consisting of reactive components (i.e., inductors and capacitors) have become predominant, mainly, with the advent of highly integrated modules and SoCs. Impedance matching using distributed elements also has broad applications in monolithic microwave integrated circuits (MMICs).

Matching networks in MICs serve two main purposes. The first is to provide maximum power delivery from a radio frequency (RF) source to a load. The second which has an opposite goal to the first is to limit the amount of power delivered to a system where an impedance mismatch circuit is used to cause power reflection and therefore provides a means of system protection from high power sources. In general, the least complex matching network offers a more reliable and less lossy design.

With the different existing matching techniques that a designer can select from, it is feasible to choose a matching network that provides the proper compromise between impedance matching, harmonic attenuation, bandwidth, or network loss.

4.2 Power transfer and power efficiency

Impedance transformation is necessary to maximize power transfer or power efficiency (η) from one stage of a circuit to another. Under a perfect matching condition, the source complex impedance must equal the load complex conjugate impedance resulting in a purely resistive interface between the two stages being matched. This condition implies that the source resistance equals the load resistance and hence only half of the total source power can be transferred to the load. Considering the AC network depicted in Figure 4.1(a), the amount of power delivered to the load is the same as the power dissipated in the source at the maximum power transfer condition. Therefore, the maximum power transfer is only 50% efficient as demonstrated in Figure 4.1(b) and expressed by (4.1):

$$\eta\% = \frac{P_{\text{load}}}{P_{\text{total}}} = \frac{I_{\text{rms}}^2 R_L}{I_{\text{rms}}^2 (R_S + R_L)} = \frac{I_{\text{rms}} V_L}{I_{\text{rms}} V_S} = \frac{I_{\text{rms}} V_L}{2 I_{\text{rms}} V_L} = 50\% \tag{4.1}$$

where P_{load} is the average power dissipated by the load resistance, R_L, and P_{total} is the power delivered by the source. R_S is the real part of the source impedance, Z_S, and I_{rms} is the root-mean-square value of the AC current. From (4.1), one can conclude that having a maximum efficiency and maximum power transfer is not possible since a maximum efficiency implies having a load resistance approaching infinity or a source resistance close to zero. Typically, maximum power efficiency is necessary for AC power distribution systems to establish a low loss transmission; however, this does not satisfy having a maximum power transfer.

For maximum power transfer, therefore, it is important to avoid impedance mismatch losses by properly matching the successive stages in an RF system.

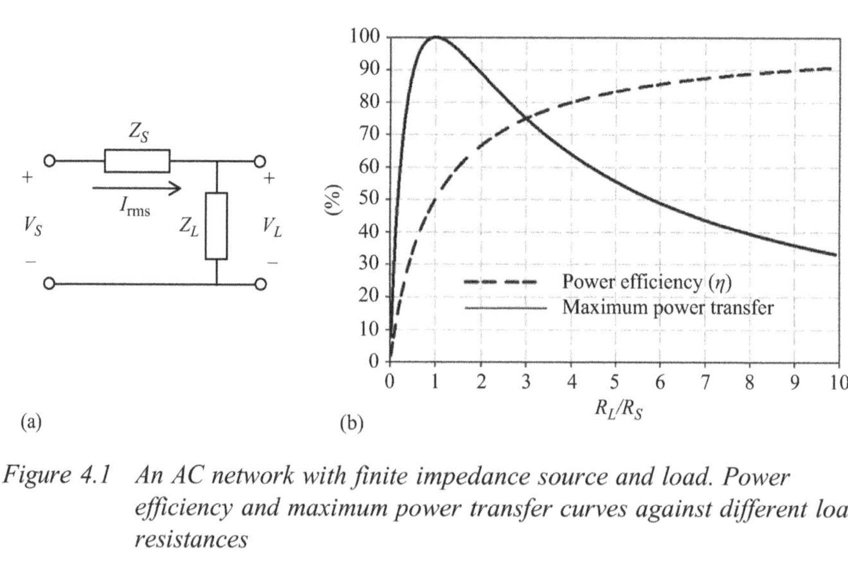

Figure 4.1 An AC network with finite impedance source and load. Power efficiency and maximum power transfer curves against different load resistances

Impedance mismatch losses in RF modules used in wireless mobile devices correspond directly to high current consumption and drain of the device battery.

In terms of the wave reflection coefficient (Γ_{in}) at the matching interface, we would like to keep the input reflection coefficient as low as possible over the operating bandwidth as demonstrated by the following power equation:

$$P_t = P_i - P_r = P_i\left(1 - |\Gamma_{\text{in}}|^2\right) \tag{4.2}$$

where P_t is the transmitted power, P_i is the incident power, and P_r is the reflected power at the matching interface. Impedance matching networks have finite bandwidth over which a load is said to be matched to specific terminal impedance. Challenges associated with impedance matching can be summarized in achieving the required matching bandwidth, having an impedance transformation ratio between the source and the load far from the ratio of one, and if matching has to be accomplished between two complex impedances (e.g., interstage matching).

4.3 Theoretical limitation on lossless matching networks

Perfect impedance matching occurs at a single frequency point, f_o, where the reactive components resonate. However, matching over a band of frequencies is possible with a trade-off. Bode–Fano work established the theoretical limitation on the matching bandwidth that can be achieved from a reactive matching network used to match an arbitrary load to pure resistive impedance [1,2]. In essence, their work showed that the efficiency of transmission and the matching bandwidth are interrelated quantities. Accordingly, wide matched bandwidth will result in poorer return loss when compared to narrow matched bandwidth provided that the same matching network is used in both cases.

The theoretical limitation on the gain-bandwidth of a matching circuit comprising a reactive load is described by Bode–Fano criterion. One of the expressions that define a special case of a parallel RC load is given by:

$$\int_0^{\infty} \ln\frac{1}{|\Gamma(\omega)|}d\omega \le \frac{\pi}{\text{RC}} \tag{4.3}$$

For passive networks, $0 \le |\Gamma| \le 1$, where $|\Gamma| = 0$ indicates perfect transmission and $|\Gamma| = 1$ means total reflection. Typically matching is required over a limited bandwidth ($\omega_1 \le \omega \le \omega_2$). Thus the integral equation in (4.3) can be limited to:

$$\int_{\omega_1}^{\omega_2} \ln\frac{1}{|\Gamma(\omega)|}d\omega \le \frac{\pi}{\text{RC}} \tag{4.4}$$

From the integral equation definition and from (4.4), the area under $\ln(1/|\Gamma(\omega)|)$, within ω_1 and ω_2, must not exceed π/RC. Hence, the larger the $\ln(1/|\Gamma(\omega)|)$ term, the better the power transmission. Γ is a function of frequency, and for the sake of

Table 4.1 Bode gain-bandwidth limitations for different load configurations

Load configuration	Bode integral
Series *R-C*	$\int_0^\infty \ln\frac{1}{\|\Gamma(\omega)\|}d\omega \le \pi\omega_o^2\mathrm{RC}$
Parallel *R-L*	$\int_0^\infty \ln\frac{1}{\|\Gamma(\omega)\|}d\omega \le \frac{\pi\omega_o^2 L}{R}$
Series *R-L*	$\int_0^\infty \ln\frac{1}{\|\Gamma(\omega)\|}d\omega \le \frac{\pi R}{L}$

simplicity, Γ can be assumed constant over the matching bandwidth. With this assumption solving (4.4) yields to:

$$|\Gamma| = e^{\frac{-\pi}{(\omega_2-\omega_1)\mathrm{RC}}} \tag{4.5}$$

Equation (4.5) entails that there has to be a trade-off between the matching bandwidth and the matching level represented by $|\Gamma|$. Table 4.1 lists the gain-bandwidth constraints for different loads presented to a matching network.

Although there are several forms of matching networks, those based on lumped elements are most suitable for RFIC modules, since circuit dimensions in RFICs are assumed to be electrically small. In general, the objective of a matching network is to negate the reactive component of a complex load and transform the real part so that maximum power is delivered into the real part of the load impedance (at which, a conjugate match is obtained). Among the several existing matching analysis techniques that fall between graphical methods and numerical methods, in this chapter, the focus is made to demonstrate those based on numerical formulas as they are more applicable for computer scripting and thereby more efficient for frequent use.

4.4 Single reactive element matching

Figure 4.2 shows the impedance movement due to adding a single reactive element. Adding a parallel inductor or capacitor to a load will move the impedance along the constant conductance circle in a counterclockwise or clockwise direction, respectively. Similarly, a series inductor or capacitor to a load will move the impedance along the constant resistance circle in a clockwise or counterclockwise direction, respectively [3]. This single reactive element can be considered as a simple means of matching capable of providing a perfect match when the impedance real part of the load equals to the impedance real part of the source, and one of the impedances comprises a reactive part. In this special case, adding a reactive element of the same magnitude, however, with opposite phase will provide a perfect match at a single frequency point because perfect cancelation occurs at one frequency point and diverges as we move away from this point.

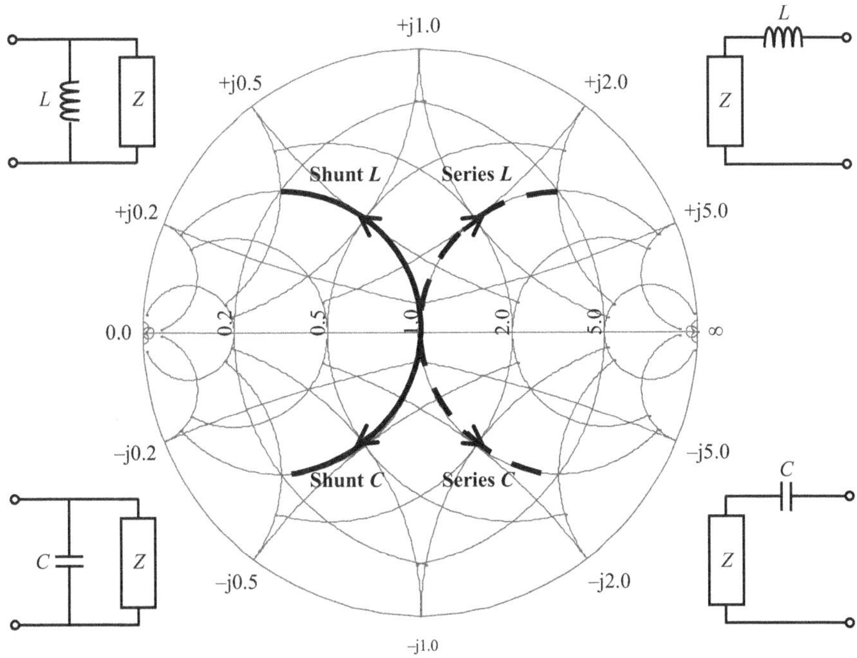

Figure 4.2 Impedance motion on Smith chart due to adding a single reactive element, either in series or parallel to a load. In this plot, the real part impedance of the load and the source assumed to be equal (i.e., 1 ± jx)

4.5 Reactive *L*-section matching networks

In an *L*-section matching network, there are two possibilities: one configuration starts with a parallel reactance from the load, and another begins with a series reactance as depicted in Figure 4.3. The general approach is to transform the load impedance into a complex conjugate of the source impedance or vice versa using a reactive network. The selection of these two configurations depends on the source resistance (i.e., the real part of the impedance) being matched. The impedance motion on the Smith chart demonstrates that an *L*-match network must start with a shunt reactance from the Z_L side in case of a load comprising a larger resistance than the source resistance (i.e., Z_L inside the $1+jx$ circle), while configurations that start with a series reactive element from the Z_L side is used in case of a source resistance larger than the load resistance (i.e., Z_L outside the $1+jx$ circle). Note that the two reactive elements in an *L*-section can match any positive-real complex impedance. Analytical expressions of both cases are illustrated in [4] and summarized in Sections 4.5.1 and 4.5.2. Depending on how each of the networks in Figure 4.3 uses an inductor or a capacitor in the series or the parallel branch of the

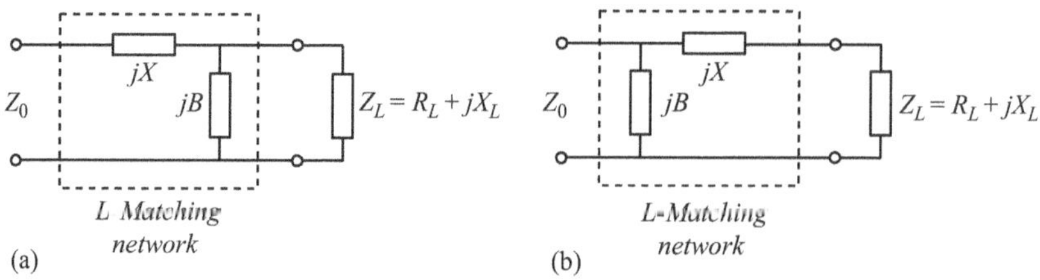

Figure 4.3 Two main configurations of an L-match reactive network: (a) case I: $R_L > Re\{Z_0\}$ and (b) case II: $R_L < Re\{Z_0\}$

matching network, the resulting network will realize either a low-pass filter (LPF) or a high-pass filter (HPF) impedance characteristics. Between the two solutions of each configuration, the matching bandwidth over the passband is relatively the same, however, the selectivity of the insertion loss response and out-of-band characteristics are different since one network represents an LPF response and the other represents an HPF response.

4.5.1 Case I: $R_L > \mathrm{Re}\{Z_0\}$

Under an impedance-matched condition, the total input impedance of the network in Figure 4.3(a) satisfies the following expression:

$$Z_0 = Z_L // \left(\frac{1}{jB}\right) + jX \tag{4.6}$$

Upon arranging the equation in (4.6) and equating the real and imaginary parts, the X and B unknowns can be found as:

$$B_{R_L > Z_0} = \frac{X_L \pm \sqrt{R_L/Z_0}\sqrt{R_L^2 - R_L Z_0 + X_L^2}}{R_L^2 + X_L^2} \tag{4.7}$$

$$X_{R_L > Z_0} = \frac{1}{B} + \frac{X_L Z_0}{R_L} - \frac{Z_0}{\mathrm{B}R_L} \tag{4.8}$$

Table 4.2 shows the locations on the Smith chart where matchable impedances can be found for a particular L-matching network satisfying $R_L > Re\{Z_0\}$. The matchable areas on the Smith chart show Yin-Yang like shapes, a Chinese drawing, that aid in easily finding the matching regions for a specific L-matching topology.

4.5.2 Case II: $R_L < \mathrm{Re}\{Z_0\}$

Similar to the analysis in Section 4.5.1, the configuration in Figure 4.3(b) leads to the following expressions:

$$Z_0 = (Z_L + jX) // \left(\frac{1}{jB}\right) \tag{4.9}$$

Table 4.2 Yin-Yang like shapes defining the matching areas of the L-matching topology illustrated in Figure 4.3(a)

L-section lossless impedance matching networks $R_L > Re\{Z_0\}$	**Areas of matchable impedances on a YZ Smith Chart**
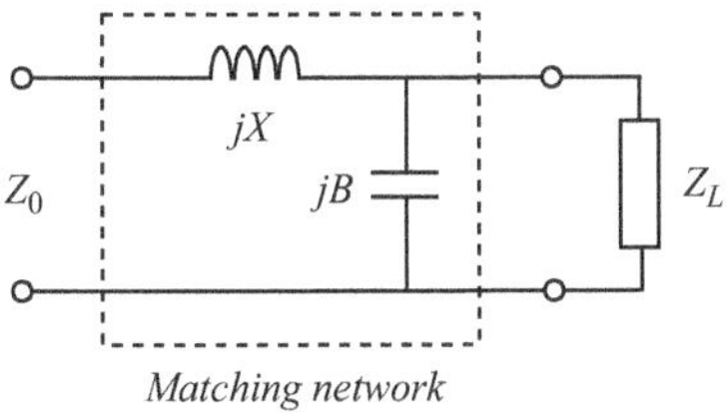	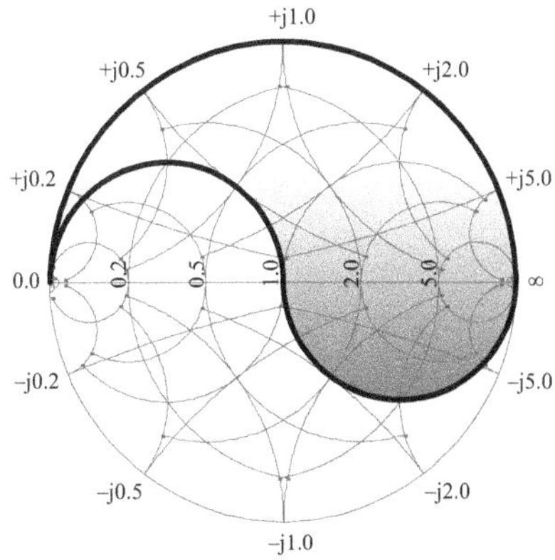
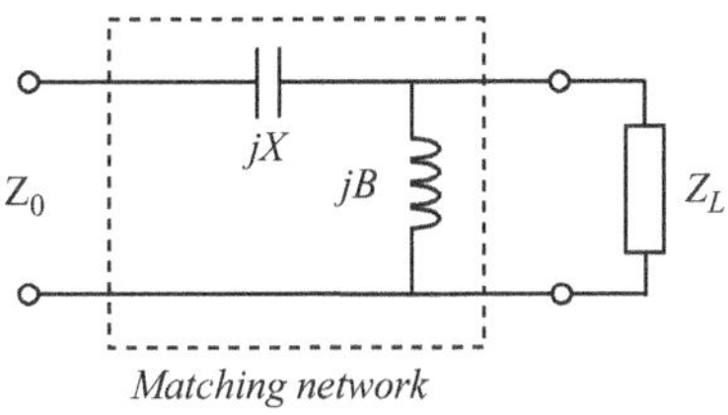	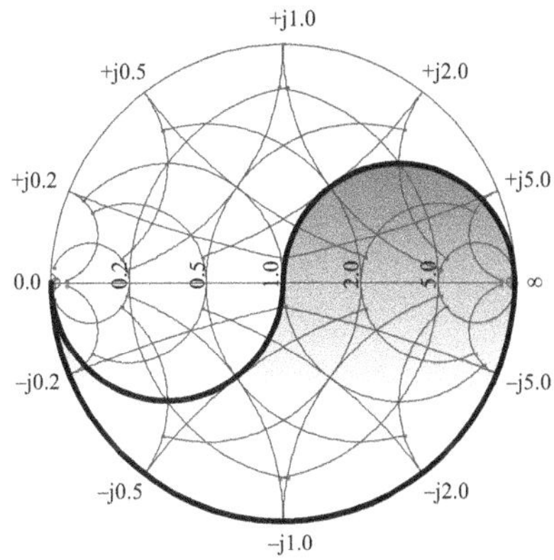
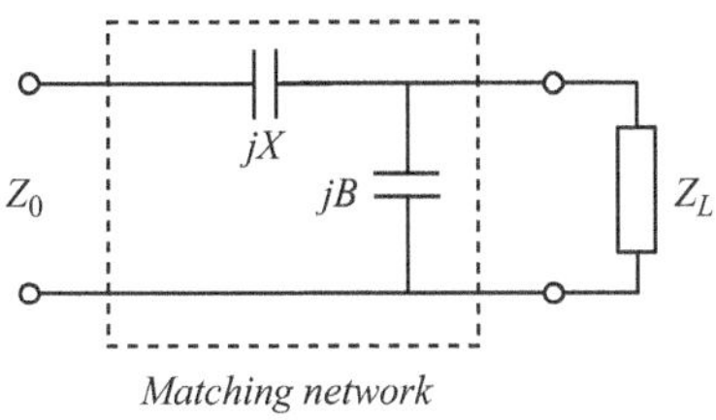	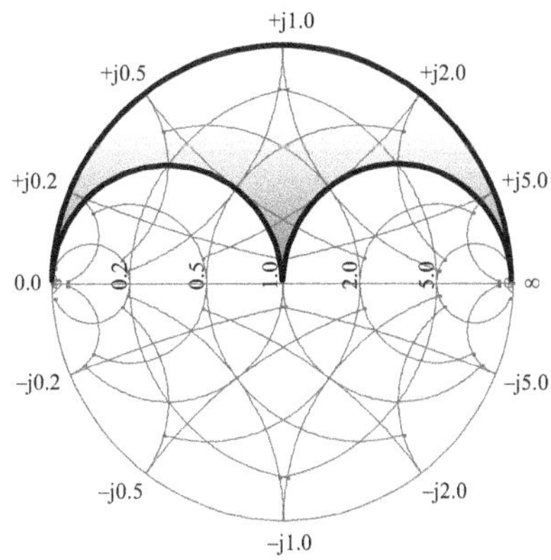

(*Continues*)

Table 4.2 (Continued)

***L*-section lossless impedance matching networks $R_L > Re\{Z_0\}$**	**Areas of matchable impedances on a YZ Smith Chart**
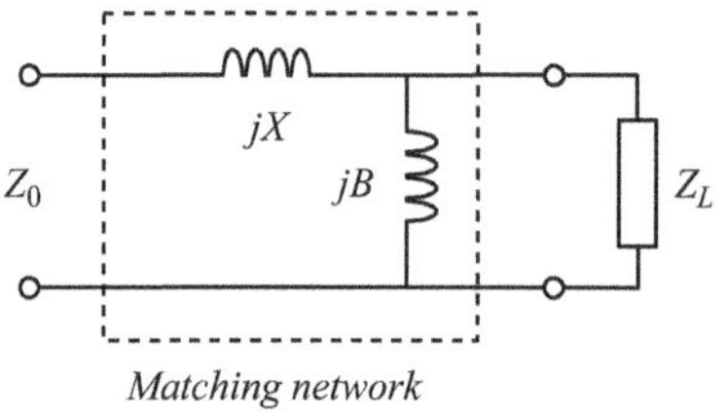	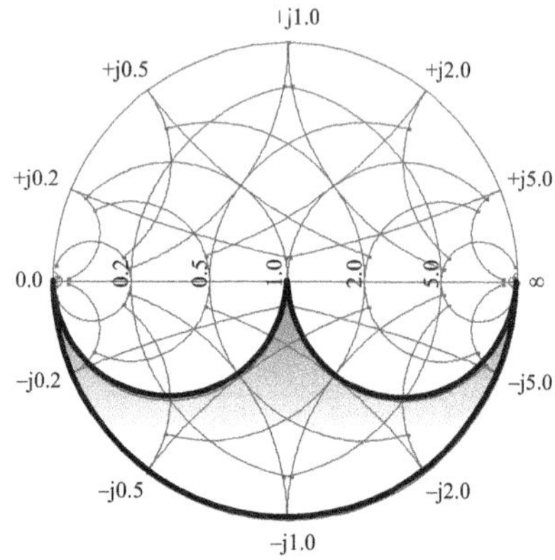

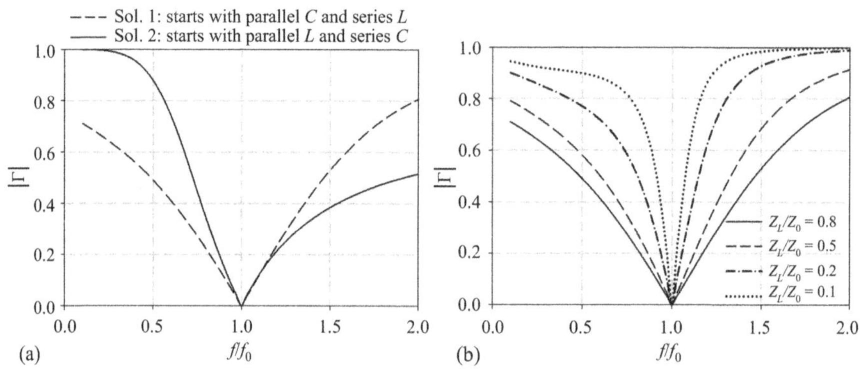

Figure 4.4 Matching bandwidth variation as a function of frequency for (a) two possible L-match solutions for a 40 + j100 ohms load at $f_0 = 1.0$ GHz and a 50 ohms source impedance and (b) over different real impedance ratios using the network resulted from the first solution as an example

Re-arranging (4.9) and solving for X and B, yields:

$$B_{R_L<Z_0} = \pm\frac{1}{Z_0}\sqrt{\frac{Z_0 - R_L}{R_L}} \tag{4.10}$$

$$X_{R_L<Z_0} = \pm\sqrt{R_L(Z_0 - R_L)} - X_L \tag{4.11}$$

Figure 4.4(a) shows the matching bandwidth of both possible solutions of an L-matching network for a $40 + j100$ ohms load and a 50 ohms source. In this case, solution 1 seems to favor the low-frequency side of the perfect matching point, while solution 2 has a relatively better match on the high-frequency side. Figure 4.4(b) illustrates the matching bandwidth over the different source and load real impedance ratios. It is evident from this result that the lower the impedance transformation ratio, the narrower the bandwidth which limits the matching performance of the matching network. An important aspect of the L-networks in Figure 4.3 is that their Q-factors do not depend on the matching components and can be determined only from the source and load impedances. In general, an L-matching network is suitable for narrowband matching as the value of the network Q-factor is determined mainly by the transformation ratio.

Table 4.3 shows the Yin-Yang like shapes specifying the locations on the Smith chart where matchable impedances can be found for a particular L-matching network satisfying $R_L < Re\{Z_0\}$.

Considering the circuit illustrated in Figure 4.5 and upon dividing it into two subnetworks one includes the parallel Z_L and jB elements and another comprises the series jX and Z_0 elements, a matched impedance condition implies that the impedance seen looking at the interface between these two subnetworks should result in complex conjugate input impedances at the matching frequency point. Here, the Q-factor of the series and parallel subnetworks are equal. This leads to the relationship in (4.12) showing that the Q-factor is dependent on the ratio of the source and load impedances. A detailed derivation of this relation is discussed in Section 4.8:

$$Q = \sqrt{\frac{R_S}{R_L} - 1} \tag{4.12}$$

From the design equation in (4.12), one can infer that the higher the network Q-factor, the higher the impedance transformation which indicates higher insertion loss and narrower matching BW. This can also be explained through the Bode–Fano's case example in (4.5) which implies that increasing R (or the source/load impedance ratio) must be accompanied with decreasing $\Delta\omega$ or $1/\Gamma_{min}$ under the same matching condition. In other words, it is difficult for a high Q-factor network to support broad BW matching.

Improving the BW is possible by cascading L-networks in series wherein each L-section is designed to a virtual impedance (R_{int}) comprising a value that falls between the source and the load impedances. This allows achieving a lower Q-factor network and thus an improved bandwidth. In a sense, the transmitted power of the matching network rolls off at a rate determined by the loaded Q-factor of the circuit. Adding more L-networks results in smoother impedance transformations, lower network Q-factor, and thereby broader BW. In general, the selection of a particular L-matching network relies on several factors such as the required matching bandwidth, implementation technology, tunability, and the tolerable loss in performance. Typically, selecting the simplest possible matching networks that can achieve the desired performance is a highly desirable as adding more elements to improve further the performance beyond this point becomes negligible and close to the theoretically optimal performance [5].

Table 4.3 Yin-Yang like shapes defining the matching areas of the L-matching topology illustrated in Figure 4.3(b)

***L*-section lossless impedance matching networks $R_L < Re\{Z_0\}$**	**Areas of matchable impedances on a YZ Smith Chart**
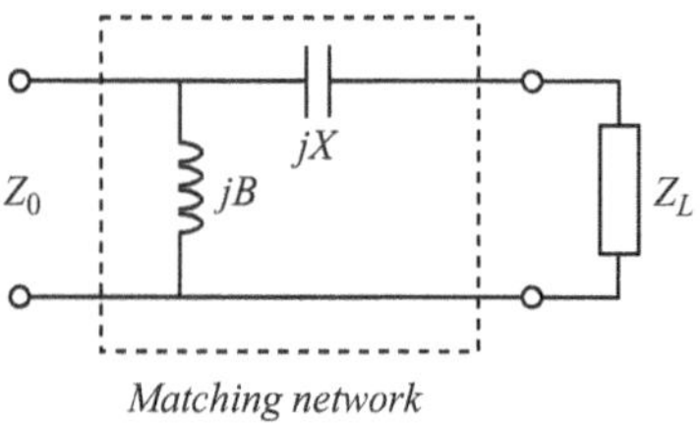	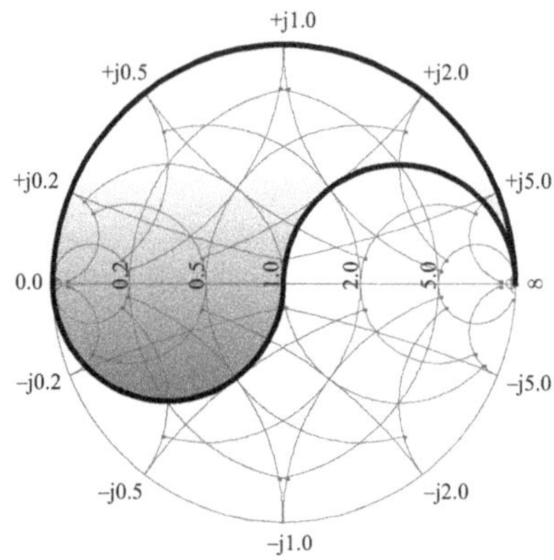
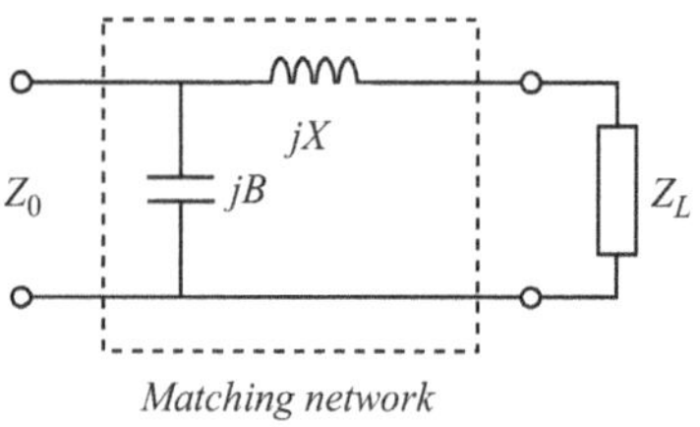	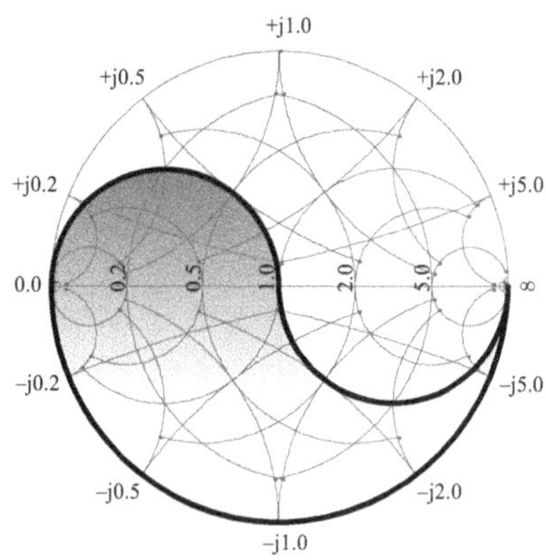
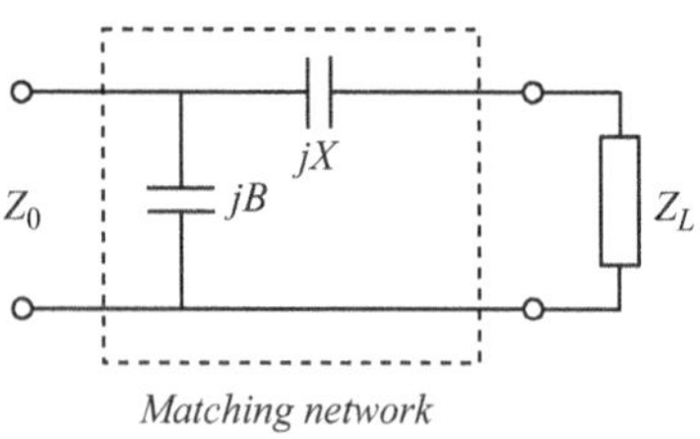	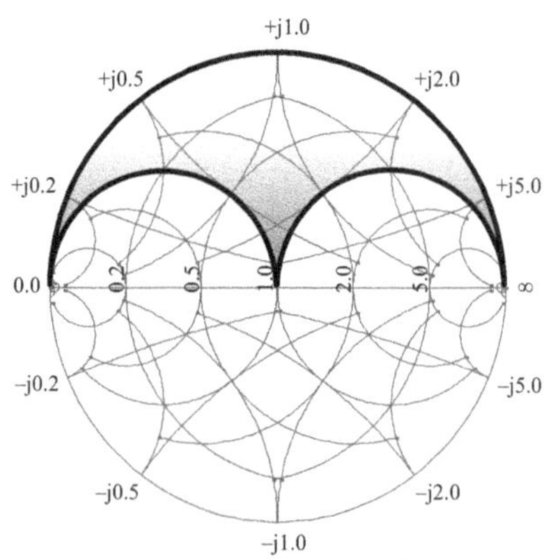

(*Continues*)

Table 4.3 (Continued)

***L*-section lossless impedance matching networks $R_L < Re\{Z_0\}$**	**Areas of matchable impedances on a YZ Smith Chart**
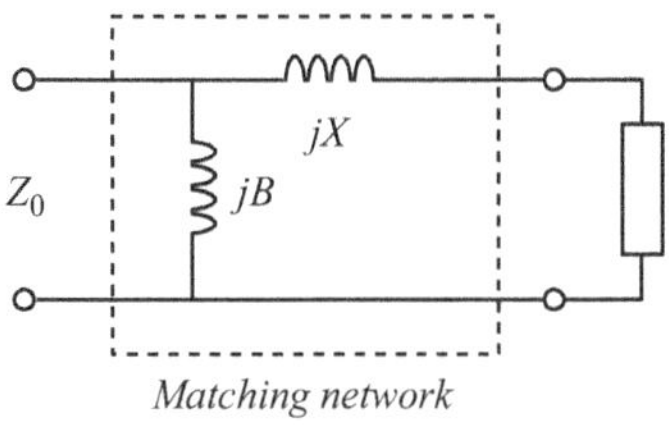	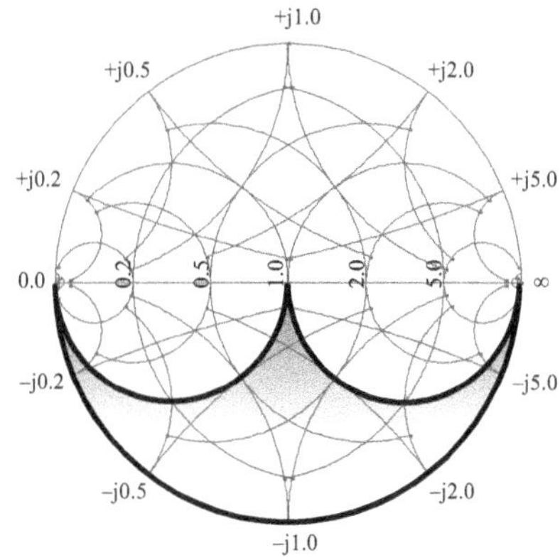

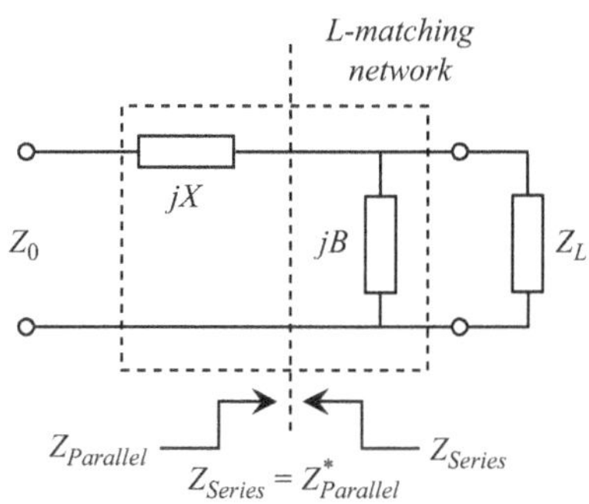

Figure 4.5 An L-section divided into two subnetworks series and parallel branches for the $R_L > R_S$ case

Maximum achievable BW occurs when the intermediate impedance equals the geometric mean of the source and load impedances [6]. This can be concluded for the particular circuit in Figure 4.6 by taking the derivative of the following Q_t equation with respect to R_{int} equation which are described as:

$$
\begin{aligned}
Q_a &= \sqrt{\frac{R_{\text{int}}}{R_S} - 1}, \quad R_{\text{int}} > R_S \\
Q_b &= \sqrt{\frac{R_L}{R_{\text{int}}} - 1}, \quad R_{\text{int}} < R_L \qquad (4.13)\\
Q_t &= (Q_a + Q_b)/2
\end{aligned}
$$

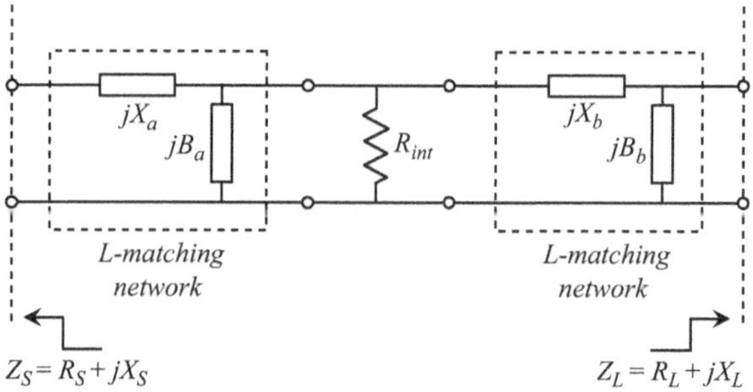

Figure 4.6 Matching bandwidth improvement by cascading two L-matching networks in series

thus

$$R_{\text{int}} = \sqrt{R_S R_L} \tag{4.14}$$

4.6 *T*- and *π*-matching networks

In an *L*-network, impedance transformation is done directly in one step where the network *Q*-factor is constrained by the source to load impedance ratio. Hence, a designer has no control over determining the network *Q*-factor or the matching bandwidth. A three reactive elements network, however, comprises two *Q* values that correspond to the two *L*-sections forming a *π*- or *T*-network. Having a third reactive element provides added flexibility in terms of controlling the network *Q* and therefore the matching bandwidth through the transformation of the load or source impedances into an intermediate resistance defined by the circuit designer.

The *π*- and *T*-networks allow matching of any load impedance condition, unlike an *L*-section network that is capable of only matching loads less or higher, but not both, than the source real part of the impedance. Similar to the two cascaded *L*-matching networks, in three reactive elements configuration the impedance is transformed first into fictitious intermediate impedance, R_{int}, using one *L*-network and then another *L*-network is used to transform the R_{int} into the target impedance. Figure 4.7 shows a *T*-matching network that is formed by two back-to-back *L*-sections where R_{int} must be greater than R_{s} and R_{L}. The opposite holds for a *π*-network depicted in Figure 4.8 whereas R_{int} must be smaller than R_S and R_L. The choice of these topologies is determined by whether DC continuity is required and/or whether low-pass or high-pass impedance characteristic is needed in the design.

Sections 4.6.1–4.6.2 will illustrate that a three-reactance configuration makes the *Q*-factor of a matching network and its transformation ratio (R_S/R_L) independent of each other since the network *Q* is defined as the ratio between the intermediate resistor and the load/source resistances.

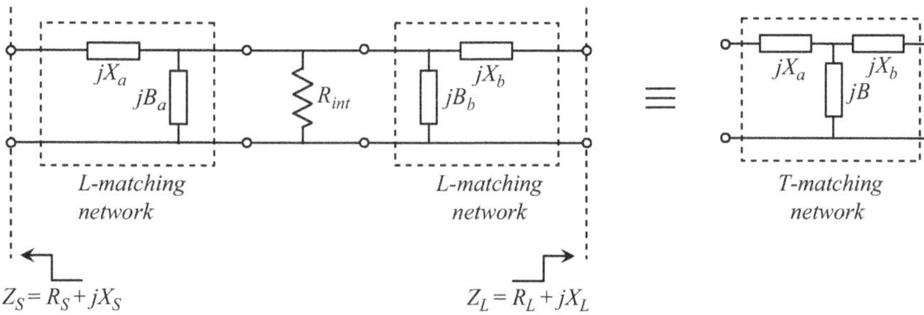

Figure 4.7 A T-network formed by cascading two back-to-back L-networks and an intermediate resistor

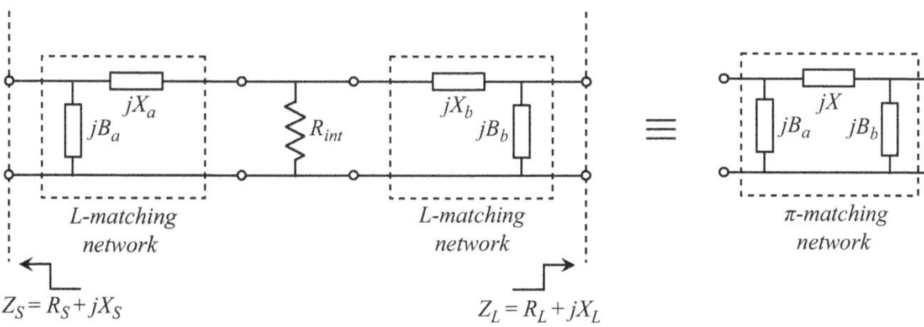

Figure 4.8 A π-network formed by two back-to-back L-networks and an intermediate resistor

4.6.1 T-matching Circuit

Considering the circuit equivalency of a *T*-network illustrated in Figure 4.7 where $R_{\text{int}} > R_S$ and $R_{\text{int}} > R_L$, the total Q of the network formed by the Q-factor of the two back-to-back *L*-sections is derived as:

$$
\begin{aligned}
Q_a &= \sqrt{\frac{R_{\text{int}}}{R_S} - 1}, \quad R_{\text{int}} > R_S \\
Q_b &= \sqrt{\frac{R_{\text{int}}}{R_L} - 1}, \quad R_{\text{int}} > R_L \\
Q_t &= (Q_a + Q_b)/2
\end{aligned}
\tag{4.15}
$$

4.6.2 π-matching circuit

The π impedance transformation circuit shown in Figure 4.8 is dual to the *T*-matching network in Figure 4.7. Similar to the design formulas of the *T*-matching network, the total Q formulas for the π-network are expressed as:

$$Q_a = \sqrt{\frac{R_S}{R_{\text{int}}} - 1}, \quad R_{\text{int}} < R_S$$
$$Q_b = \sqrt{\frac{R_L}{R_{\text{int}}} - 1}, \quad R_{\text{int}} < R_L \tag{4.16}$$
$$Q_t = (Q_a + Q_b)/2$$

The two back-to-back L-networks combined yields a network with a higher Q-factor than that of a single L-network; however, this is an undesirable characteristic especially for broadband matched circuits. Here, a series cascade of L-networks instead of back-to-back configuration is a viable option. Additionally, low Q matching networks provide less loss and are less sensitive to the component variation than that of a high Q network.

4.6.3 Inverters

An impedance or admittance inverter diagrammed in Figure 4.9 is a two-port network that is lossless and reciprocal. The primary function of an inverter is to inversely relate the load immittance to the input immittance of the inverter with a gain factor representing the characteristic immittance of the inverter. Impedance inverters are more convenient to use with circuits connected in series, and admittance inverters are suitable for shunt circuits. The relationship between the input and output of an immittance inverter is expressed as

$$Z_{in} = \frac{K^2}{Z_L}$$
$$Y_{in} = \frac{J^2}{Y_L} \tag{4.17}$$

where K is the characteristic impedance and equivalently $J = 1/K$ is the characteristic admittance of the inverter. As can be observed from (4.17), the inversion is normalized by the square of the characteristic immittance. In a sense, the operation of an impedance inverter is similar to that of a quarter-wavelength transformer of characteristic impedance K.

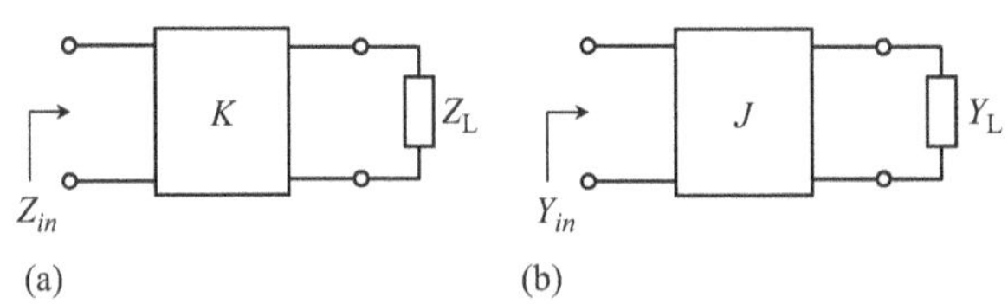

Figure 4.9 A circuit block diagram showing an inverter terminated in a load: (a) an impedance inverter (K-inverter) and (b) an admittance inverter (J-inverter)

Immittance inverters are frequently used in the synthesis of narrowband microwave filters to simplify the synthesis procedure, particularly, when studying different topologies or when the values of the components of the filter are difficult to realize. For instance, one of the commonly used microwave filter topologies is the ladder topology where a quarter wavelength transformer is typically used to transform a series connected element or resonator into a shunt element, and vice versa.

An impedance inverter can be defined by its transfer parameters as:

$$T = \begin{bmatrix} 0 & jK \\ j/K & 0 \end{bmatrix} \tag{4.18}$$

Or in terms of the scattering parameters as:

$$S = \begin{bmatrix} 0 & -j \\ -j & 0 \end{bmatrix} \tag{4.19}$$

Immittance inverters can be realized in different forms such as transmission line stubs, coupled transmission lines, and lumped elements formed as a symmetrical *T*- or Π-network, where the reactance shunt elements possess an opposite sign of the series reactance elements and both have the same absolute value of reactance in ohms at a single design frequency. The combination of the lumped and distributed elements can also be utilized to form an inverter. Table 4.4 lists different impedance inverter configurations using *T*- and Π-equivalent lumped element networks and a quarter-wavelength transmission line.

Inverter circuits consisting of negative component values can be absorbed into the adjacent components of the network being transformed, as the case in microwave filter synthesis, to result in a positive-valued realizable element. Having negative component values is in fact one of the limiting factors of inverters. Besides, inverters are not suitable for use with broadband networks due to their inherent narrow bandwidth, specifically, when high impedance inversion is required.

4.7 Impedance matching using transformers

To match a specific load resistance to a source resistance, the transformer depicted in Figure 4.10 can be used to convert a source power from one specific voltage and current level to another voltage and current level, given that the power input to the transformer equals the power output from the transformer except for a small amount of transformation loss [7]. Hence, the load impedance is transformed as a square of the voltage-transformation ratio determined by the number of turns in the primary and secondary coils of the transformer as expressed by:

$$n = \left(\frac{N_S}{N_P}\right)^2 = \left(\frac{V_S}{V_P}\right)^2 = \frac{Z_S}{Z_P} \tag{4.20}$$

where *n* is the turns ratio.

Table 4.4 Equivalent lumped element networks of immittance inverters formed as T- and π-networks

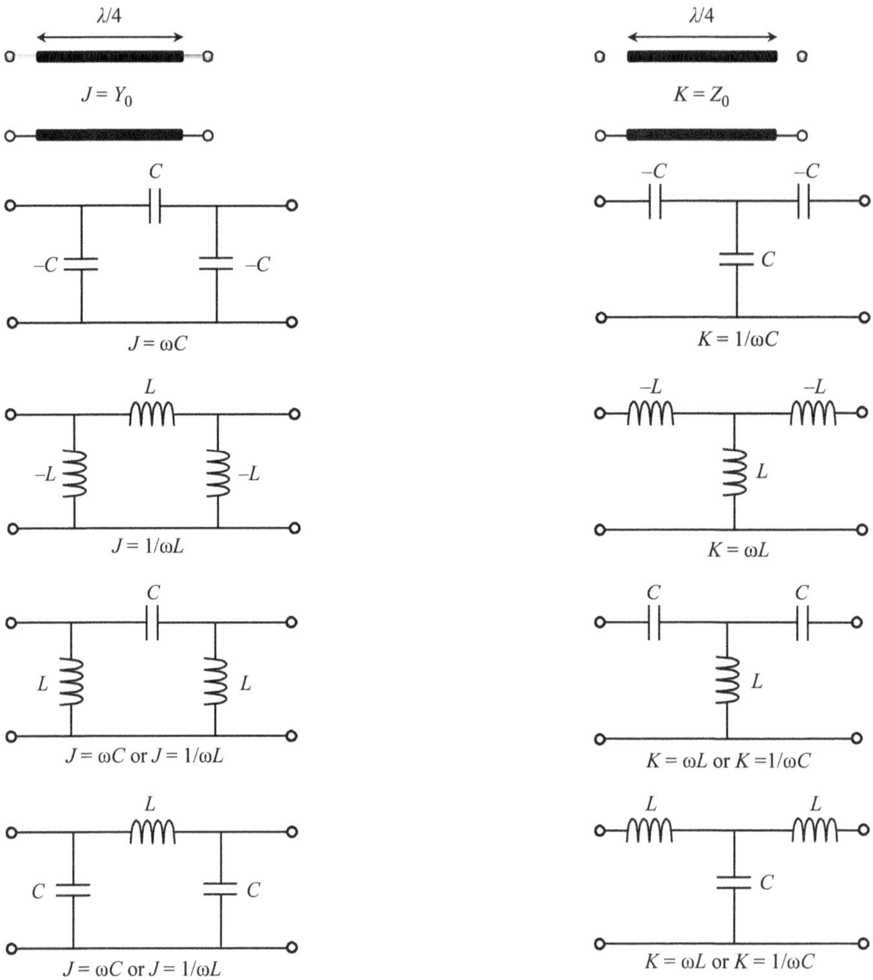

The advantage of using transformers is that they provide DC isolation between the circuits and broader bandwidths if compared to *L*-matching networks.

In the synthesis techniques of RF circuits, it is common to switch a shunt element into its equivalent series element plus a transmission line that is one-quarter wavelength long with the appropriate impedance transformation (e.g., through a transformer). Similarly, a series element can be switched into its equivalent shunt element plus a quarter-wavelength line. In microelectronic circuits which mostly comprise lumped elements, the same concept can be employed using Π or T lumped equivalent quarter-wavelength transformer. Using the two-port junction illustrated in

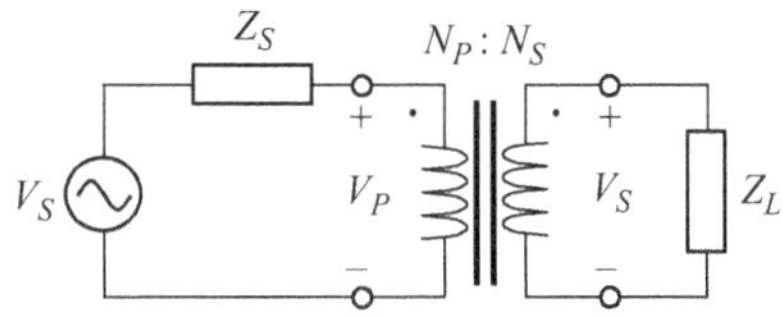

Figure 4.10 Circuit schematic of impedance matching using a DC isolated transformer

Table 4.5 Impedance parameters of the circuit in Figure 4.10 and its dual assuming an impedance transformer with n:1 transformation ratio

Circuit in Figure 4.10	Dual of Figure 4.10
$Z_{11} = Z_S + n^2 Z_L$	$Z_1 = Z_{11} - Z_{12}{}^2/Z_{22}$
$Z_{22} = Z_2$	$Z_2 = Z_{22}$
$Z_{12} = \pm\ nZ_L$	$n = Z_{12}/Z_{22}$

Figure 4.10, the impedance parameters of its dual equivalent circuit (i.e., Z_S becomes a shunt element and Z_L becomes a series element) are listed in Table 4.5.

4.8 Tapped capacitor resonator matching network

Similar in principle to matching using inductors as in transformers, impedance matching with capacitors is also possible using the diagram in Figure 4.11.

The concept behind this configuration is that at a given design frequency, the parallel C_2R_2 branch can be converted into series RC branch. Then, the resultant tapped capacitor with the inductor (L) in the circuit can form an L-matching circuit. The design process starts with the parallel RC to series RC conversion as illustrated in Figure 4.12. We assume both branches have similar Q-factor and therefore this conversion is valid over a narrow frequency range.

The equivalency between parallel and series branches, in general, can be established at a single frequency point assuming both branches have the same complex impedance, thus:

$$\begin{aligned} \mathrm{Re}\left(\frac{1}{Y_P}\right) &= \mathrm{Re}(Z_S) = R_S \\ \mathrm{Im}\left(\frac{1}{Y_P}\right) &= \mathrm{Im}(Z_S) = \omega C_S \end{aligned} \tag{4.21}$$

This implies that:

$$Q_P = Q_S \tag{4.22}$$

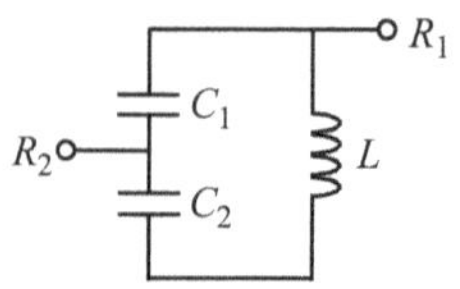

Figure 4.11 Circuit schematic of a tapped capacitor impedance transformer

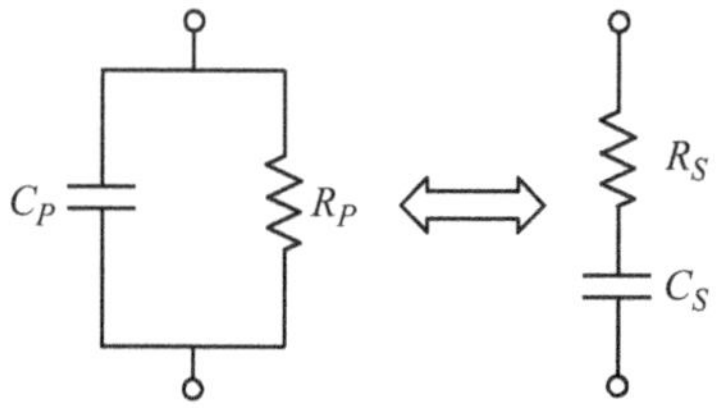

Figure 4.12 Parallel to series RC conversion

The impedance of the parallel RC circuit is:

$$Z_p = \frac{R_P}{1 + j\omega R_P C_P} \tag{4.23}$$

The real part of Z_p is:

$$\mathrm{Re}(Z_p) = \frac{R_P}{1 + (\omega R_P C_P)^2} \tag{4.24}$$

The Q-factor of the parallel RC branch in Figure 4.12 can then be given by:

$$Q_P = \frac{R_P}{X_P} = \omega C_P R_P \tag{4.25}$$

By rewriting R_S and C_S in terms of Q_P using (4.22) and (4.23), we obtain:

$$\mathrm{Re}(Z_S) = R_S = \frac{R_P}{1 + Q_P^2} \tag{4.26}$$

$$\mathrm{Im}(Z_S) = X_S = \frac{1}{\omega C_S} = \frac{X_P Q_P^2}{1 + Q_P^2} \tag{4.27}$$

Hence for high Q-factor value, (4.24) can be approximated as:

$$R_P \cong Q_P^2 R_S \tag{4.28}$$

and similarly, the reactive parts become equal:

$$X_S \cong X_P \tag{4.29}$$

Given that $Q_P^2 = \frac{R_P^2}{X_P^2}$, then

$$R_S R_P \cong X^2 \tag{4.30}$$

Table 4.6 summarizes the formula for the two-way conversion between the series and parallel RC circuits. To this end, the tapped capacitor circuit in Figure 4.11 can now be converted to an equivalent L-network by transforming the parallel C_2R_2 into series C_{2S} R_{2S} and assuming high Q_P ($Q_P > 10$). This yields the equivalent circuit illustrated in Figure 4.13.

C_1 and C_{2S} are in series, and from (4.29), C_{2S} equals to C_2. The total capacitance is then given by:

$$C_T = \frac{C_1 C_2}{C_1 + C_2} \tag{4.31}$$

The circuit in Figure 4.13 represents an L-matching circuit. By applying L-match between R_{2S} and R_1 (here R_1 must be greater than R_{2S} since it is in parallel with the reactive element of the L-matching circuit) or by moving one step farther and transforming the series R_{2S} and C_T into parallel branch $R_{2p}C_T$, we obtain the circuit shown in Figure 4.14.

Table 4.6 Series-to-parallel and parallel-to-series circuit conversion

Element of complex impedance	**Series to parallel**	**Parallel to series**
Q	$Q_S = X/R$	$Q_P = R/X$
R	$R(1 + Q_S^2)$	$R/(1 + Q_P^2)$
X	$X(1 + Q_S^2)/Q_S^2$	$XQ_P^2/(1 + Q_P^2)$

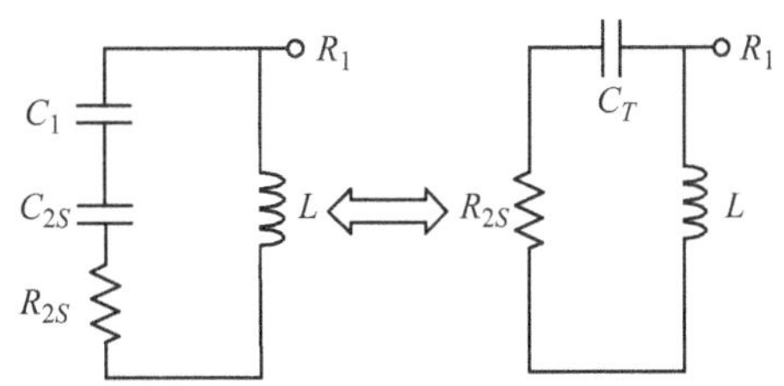

Figure 4.13 Conversion of tapped capacitor network to an L-network

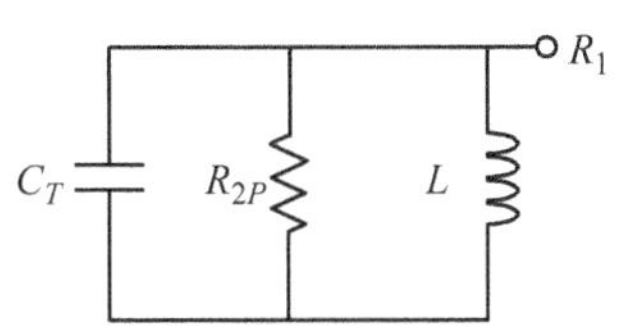

Figure 4.14 Series-to-parallel RC conversion of the L-match network in Figure 4.13

From (4.29), the reactive part (C_T) will remain the same, and the new parallel resistance (R_{2p}) is calculated as follows. Applying (4.28) to C_T and R_{2S}, we obtain $R_{2P} = \frac{1}{R_{2S}(\omega C_T)^2} = \frac{X_{2P}^2}{R_{2S}}$ and $R_{2S} = \frac{1}{R_2(\omega C_2)^2} = \frac{X_P^2}{R_2}$. Therefore, the resultant R_{2P} in Figure 4.14 in terms of R_2 can be expressed as:

$$R_1 = R_{2P} = \frac{X_{2P}^2}{X_P^2} R_2 = \left(\frac{C_2}{C_T}\right)^2 R_2 = \left(\frac{C_1 + C_2}{C_1}\right)^2 R_2 = n^2 R_2 \tag{4.32}$$

where

$$n = \frac{C_1 + C_2}{C_1} \tag{4.33}$$

In summary, the configuration of a tapped capacitor impedance transformer is applicable for matching resistive or real part of the impedance where n is the transformation ratio and R_1 must be greater than R_2 according to Figure 4.11.

4.9 Transmission lines-based matching networks

Transmission line-based matching networks provide relatively wider bandwidths when compared to lumped element matching circuits. Moreover, transmission line matching circuits are suitable for RFIC designs operating at high microwave frequencies at which the component size is comparable to the wavelength, or when it is difficult to control the parasitics of lumped elements. Depending on the choice of the implementation technology, transmission line matching elements is beneficial in realizing a very small capacitor or inductor values that are usually difficult to obtain using lumped elements. Sections 4.9.1–4.9.5 discuss common transmission line-based matching elements.

4.9.1 Quarter wavelength transformer

Lumped or distributed elements or a combination of the two can be used to realize a quarter wavelength ($\lambda/4$) transformer to match a given real load impedance to a real source impedance. The characteristic impedance of a quarter wavelength transformer realized as a $\lambda/4$ section of a transmission line at the matching frequency is the geometric mean of Z_0 and Z_L:

$$Z_Q = \sqrt{Z_0 Z_L} \tag{4.34}$$

A lumped-element equivalent circuit of the quarter wavelength transformer comprising LC elements in T or π configuration is discussed in Table 4.4 and Table 2.1 of Chapter 2. In brief, an equivalency of a $\lambda/4$ transformer can be realized as three $\lambda/8$ transformers in a T and PI configuration where a lumped element inductor is realized as $\lambda/8$ short-circuited stub and a lumped element capacitor is realized as $\lambda/8$ open-circuited stub.

4.9.2 Radial stub

A radial stub [8] is an open-circuited transmission line section that is usually realized in microstrip or stripline technology. When used as a quarter-wavelength open-circuited shunt stub at a given operating frequency, it establishes a point of low impedance or a short circuit. A low impedance rectangular open-circuited stub exhibits a wide line width where higher order modes can be easily excited. Alternatively, a radial stub is advantageous in providing a cleaner response in terms of higher order modes. Moreover, the performance of a radial stub has been found to work better than a low-impedance rectangular stub especially when a broadband short and accurate localization of a zero-point impedance is needed. The continuously varying profile of the radial stub results in reduced effects of discontinuity by avoiding abrupt changes in the stub width. Radial stubs are often associated with high-frequency applications to realize a low impedance match as in bias lines for high-frequency amplifiers.

4.9.3 Multisection transformer

Since quarter-wavelength impedance transformers exhibit limited bandwidth, a broadband matching is possible by cascading multisection transformer and varying their characteristic impedance to provide matching at multiple frequency points over the desired band. The theory of small reflections demonstrates this concept [4]. Considering a simple case of a multisection transformer comprising commensurate sections as shown in Figure 4.15, the overall reflection coefficient in terms of the partial reflections and transmissions on the discontinuities of the matching transformer is expressed as:

$$\Gamma_{\text{total}}(\theta) \simeq \Gamma_0 + \Gamma_1 e^{-2j\theta} + \Gamma_2 e^{-4j\theta} + ... + \Gamma_N e^{-2jN\theta} \tag{4.35}$$

The formula in (4.35) is an approximate expression which assumes that all Z_N change monotonically along the transformer wave propagation direction. This implies that the relation is accurate in the case of small discontinuities between the successive sections. Specific passband characteristics of the multisection matching transformer can be obtained by applying one of the common amplitude approximation methods such as Binomial or Chebyshev frequency characteristics.

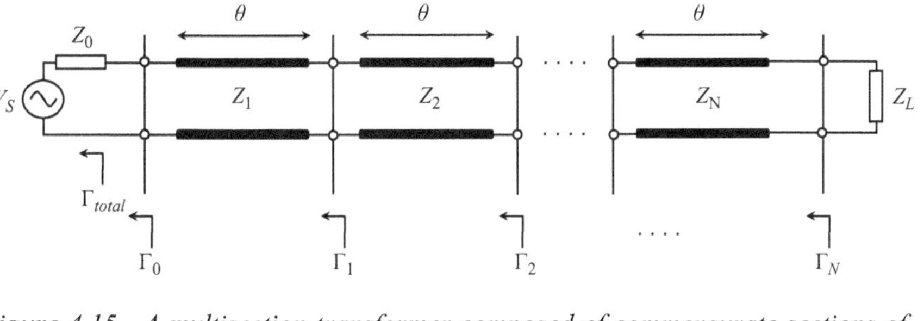

Figure 4.15 A multisection transformer composed of commensurate sections of transmission lines

The binomial approximation is used to achieve a flat passband response at the design frequencies; however, this is accomplished at the expense of a poorer bandwidth when compared to the same number of sections of a Chebyshev transformer by allowing a level of passband ripple.

4.9.4 Tapered transmission lines

Tapered line transformer is another form of broadband matching that exhibits continuously varying characteristic impedance along the wave propagation direction. Tapering the impedance of a transmission line can be established by reducing the step changes in the characteristic impedance of the successive sections by increasing the number of segments between the source and load impedances. Similar to multisection transformers, different passband characteristics can be obtained using, for example, exponential, Klopfenstein, or triangular amplitude approximation methods. Also the bandwidth of a tapered line can be extended by increasing the number of sections. Typically, the length of a taper section should be between 0.5λ and $1.5\ \lambda$ to achieve the required matching. In both tapered and multisection transmission line matching, it is not possible to obtain zero reflection coefficients at all frequencies within the matching bandwidth. Section 4.9.5 discusses in details one case of the tapered line transformers that incorporates the Fourier series as an approximation method to control the line impedance profile.

4.9.5 Nonuniform transmission lines

The advantage of using a nonuniform transmission line (NTL) over a tapered or uniform transmission line (UTL) is the ability to miniaturize the line length along its wave propagation direction. This is feasible in the case of transverse electromagnetic (TEM) or quasi-TEM modes of operation as outlined in [9]. The analogy between the uniform and nonuniform line is established by modulating the UTL impedance profile via one of the series expansion methods such as the Fourier series expansion where governing equations are employed to vary the width or impedance of several cascaded small segments of the UTL to obtain an equivalent compact NTL. NTL is advantageous in a sense that the miniaturization mechanism is achieved without the need for additional reactive loading elements or high dielectric materials. Moreover, NTLs do not comprise sharp discontinuities and thereby produce minimal parasitics.

Figure 4.16 illustrates a typical UTL that has a length d_u, a characteristic impedance Z_0, and a propagation constant β_0 and its equivalent NTL of length d_n, a varying characteristic impedance $Z(z)$, and a propagation constant $\beta(z)$. The design of an NTL begins by evaluating the *ABCD* matrix of a UTL [9] that will be used consequently as an optimization goal for the nonuniform line as:

$$\begin{bmatrix} A_0 & B_0 \\ C_0 & D_0 \end{bmatrix} = \begin{bmatrix} \cos(\theta_0) & jZ_0 \sin(\theta_0) \\ jZ_0^{-1} \sin(\theta_0) & \cos(\theta_0) \end{bmatrix} \tag{4.36}$$

where $\theta_0 = \beta_0 d_u$ is the electrical length of the UTL. The NTL is designed so that the cascaded ABCD parameters at a given frequency f are equal to the UTL parameters.

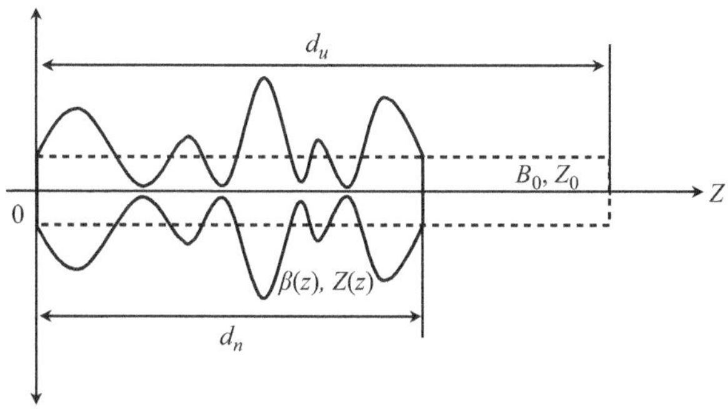

Figure 4.16 A uniform transmission line (dashed) and its equivalent miniaturized nonuniform line (solid) following a Fourier series profile along the propagation axis Z

By setting the targeted length d_n to be smaller than d_u and subdividing the line into k electrically commensurate short sections, each with a length of Δz according to:

$$\Delta z = \frac{d_u}{k} \ll \lambda_g = \frac{c}{f} \tag{4.37}$$

where c is the speed of light and λ_g is the guided wavelength. The total ABCD matrix of the NTL is obtained by multiplying the ABCD matrices of each of the individual segments as follows:

$$\begin{bmatrix} A & B \\ C & D \end{bmatrix}_{Z(z)} = \begin{bmatrix} A_1 & B_1 \\ C_1 & D_1 \end{bmatrix} \cdots \begin{bmatrix} A_i & B_i \\ C_i & D_i \end{bmatrix} \cdots \begin{bmatrix} A_K & B_K \\ C_K & D_K \end{bmatrix} \tag{4.38}$$

The ABCD parameters of the ith segment are:

$$A_i = D_i = \cos(\Delta\theta) \tag{4.39}$$

$$B_i = Z^2((i - 0.5)\Delta z)C_i = jZ((i - 0.5)\Delta z)\sin(\Delta\theta) \tag{4.40}$$

and

$$\Delta\theta = \frac{2\pi}{\lambda}\Delta z = \frac{2\pi}{c} f\sqrt{\varepsilon_{\text{eff}}}\Delta z \tag{4.41}$$

The normalized characteristic impedance of each segment is then determined through the following truncated Fourier series expansion as:

$$\ln(\bar{Z}(z)) = \sum_{n=0}^{N} C_n \cos\left(\frac{2\pi nz}{d}\right) \tag{4.42}$$

where $\bar{Z}(z) = Z(z)/Z_o$, C_n's ($n = 1, 2, \ldots, N$) are the Fourier coefficients and N is the number of coefficients. N equals to ten is usually sufficient to obtain the desired electrical and physical properties. It is worth pointing out that symmetric or asymmetric line profiles are attainable by excluding or including the Fourier series orthogonal sine terms, respectively. The ABCD matrix of the compact NTL can be found by having its parameters optimized through the C_n's coefficients to yield the closest possible parameters to those of the UTL at a specific design frequency.

The shorter the NTL, the more extreme the widths of the line segments become. A very wide line width along the direction of propagation may result in non-TEM lines that may complicate the general behavior of the intended impedance transformer. This sets a limit to the amount of miniaturization that can be achieved. Typically a 30 percent reduction on the UTL can is achieved following this method without compromising the performance.

4.10 Power gains

Power amplifiers when placed in a module are connected to the source and load impedances which are not necessarily matched to the input and output impedances of the amplifier. The performance of the matching network is critical for a power amplifier as it directly impacts the amount of power delivered to the output, gain, and the produced noise. Therefore, several power gain definitions exist and mainly associated with the design and analysis of RF amplifiers. Power gains describe different mismatch levels between the amplifier source and load impedances. Figure 4.17 shows a block diagram of a general two-port RF amplifier circuit that consists of input and output matching networks. The reflection coefficients seen at the different device interfaces are also depicted in Figure 4.17 to build upon them and other related power gain definitions as will be discussed in this section.

Symbols illustrated in Figure 4.17 are defined as follows:

- V_S and Z_S: Represent the source voltage and source internal impedance.
- Z_L: Represent the load impedance.
- IMC and OMC: Input and output matching circuit, respectively.
- [*S*]: the amplifier network scattering parameters.
- Γ_a: the reflection coefficient at the input portion of the input matching circuit.

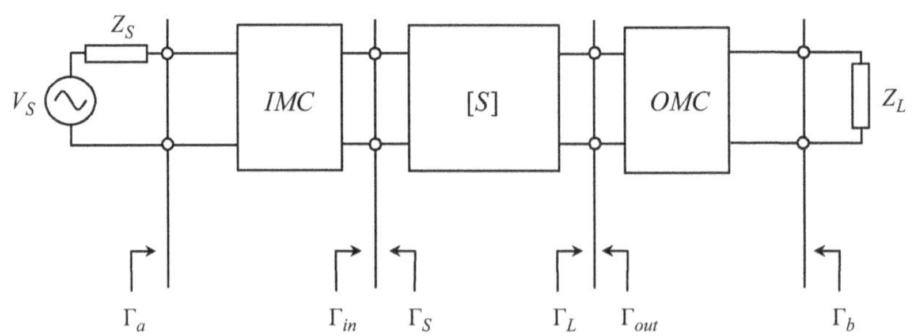

Figure 4.17 A two-port block diagram of a unilateral gain amplifier with its matching networks at the input and output sides

- Γ_b: the reflection coefficient at the output portion of the output matching circuit.
- Γ_S: the source reflection coefficient.
- Γ_L: the load reflection coefficient.
- Γ_{in}: the input reflection coefficient looking at the input of the transistor.
- Γ_{out}: the output reflection coefficient looking at the output of the transistor.

In the next set of gain equations, P_{avn} denotes the power delivered by the network to a conjugately matched load, P_{avs} denotes the maximum available power from the source to a conjugately matched network, P_L is the power delivered to the load by the amplifier network and not necessarily conjugately matched to the output of the network, and P_{in} is the power input to the network and source not necessarily conjugately matched to the input of the amplifier network. These four time-average power quantities can be expressed in terms of the network S-parameters and the input and output reflection coefficients as:

$$P_{\text{in}} = \frac{|V_S|^2}{8Z_S} \frac{\left(1 - |\Gamma_{\text{in}}|^2\right)|1 - \Gamma_S|^2}{|1 - \Gamma_{\text{in}}\Gamma_S|^2} \tag{4.43}$$

$$P_{\text{avs}} = P_{\text{in}}\big|_{\Gamma_{\text{in}}=\Gamma_S^*} = \frac{|V_S|^2}{8Z_S} \frac{|1 - \Gamma_S|^2}{1 - |\Gamma_S|^2} \tag{4.44}$$

$$P_L = \frac{|V_S|^2}{8Z_S} |S_{21}|^2 \frac{\left(1 - |\Gamma_L|^2\right)|1 - \Gamma_S|^2}{|1 - \Gamma_L S_{22}|^2 |1 - \Gamma_S \Gamma_{\text{in}}|^2} \tag{4.45}$$

$$P_{\text{avn}} = \frac{|V_S|^2}{8Z_S} |S_{21}|^2 \frac{|1 - \Gamma_S|^2}{|1 - \Gamma_S S_{11}|^2 \left(1 - |\Gamma_{\text{out}}|^2\right)} \tag{4.46}$$

Several power gain expressions can now be defined using the time-average power quantities in (4.43–4.46) as follows.

Transducer power gain (G_T) is the ratio of the power delivered to the load to the power available from the source. From (4.47), it is observed that no matching is assumed at the source and load terminals of the amplifier and therefore the transducer power gain depends on two independent parameters which are the source and load reflection coefficients:

$$G_T = \frac{P_L}{P_{\text{avs}}} = \frac{1 - |\Gamma_S|^2}{|1 - \Gamma_S \Gamma_{\text{in}}|^2} |S_{21}|^2 \frac{1 - |\Gamma_L|^2}{|1 - S_{22}\Gamma_L|^2} \tag{4.47}$$

Operating power gain (G_P) also known as power gain is the ratio of the power delivered to the load to the power delivered from the source. If the source is matched, then the transducer power gain becomes equal to the operating power gain which only depends on the load reflection coefficient:

$$G_P = G_T \frac{P_{\text{avs}}}{P_{\text{in}}} = \frac{P_L}{P_{\text{in}}} = \frac{|S_{21}|^2 (1 - |\Gamma_L|^2)}{|1 - S_{22}\Gamma_L|^2 (1 - |\Gamma_{\text{in}}|^2)} \tag{4.48}$$

Available power gain (G_A) is the ratio of the power available from the amplifier to the power available from the source. On the opposite of the operating power gain, the available power gain assumes load matching and depends only on the source reflection coefficient. The available gain is also called insertion gain. This power expression is typically used in the analysis of low noise amplifiers (LNAs):

$$G_A = \frac{P_{\text{avn}}}{P_{\text{avs}}} = \frac{1 - |\Gamma_S|^2}{|1 - S_{11}\Gamma_s|^2} |S_{21}|^2 \frac{1}{1 - |\Gamma_{\text{out}}|^2} \tag{4.49}$$

Maximum available gain ($G_{\max}$) is the ratio of the power available from the amplifier to the power delivered from the source. Maximum gain assumes simultaneous conjugate matched input and output ports of the amplifier:

$$G_{\max} = \frac{P_{\text{avn}}}{P_{\text{in}}} = \frac{1}{1 - |S_{11}|^2} |S_{21}|^2 \frac{1}{1 - |S_{22}|^2} \tag{4.50}$$

Typically, the output matching network of a PA can be considered as a step up transformer of the RF voltage since the output of the amplifier requires low-impedance to ensure high signal gain, and the subsequent circuits (e.g., filters or RF switches) connected in series to the output of the amplifier possesses relatively higher real impedance.

4.11 Source- and load-pull analysis

Since most of the power devices are far from 50 ohms, impedance source- and load-pull is a key procedure associated with the design and characterization of nonlinear devices (e.g., amplifiers) to efficiently determine the desired load impedance according to a targeted performance [10]. Source/load-pull analysis is beneficial in the case of analyzing circuits under large signal conditions where analyses based on linear approximations are not applicable. The analysis comprises a set of contours on the Smith chart as depicted in Figure 4.18 used to determine the maximum achievable gain or maximum constant output power for multiple design criteria. Different load and/or source impedance presented to the nonlinear device help in understanding changes in the device performance and sensitivity to varying the load.

For instance, the analysis is important in deciding between two desired performance parameters: maximum gain and low noise factor as in LNA-based designs. The optimal input return loss needed to achieve a maximum gain can be achieved only when the input-matching network terminates the device with a conjugate of S_{11}. However, in many cases, this is different from the gamma optimum that is needed for establishing a low noise figure.

Measurements of load- and source-pull are carried-out by using impedance tuners between the respective input and the output sides of the nonlinear device under characterization. Moreover, load/source-pull analysis is also beneficial in monitoring the impact of deviating from the optimum load/source impedance on the ruggedness performance of the device.

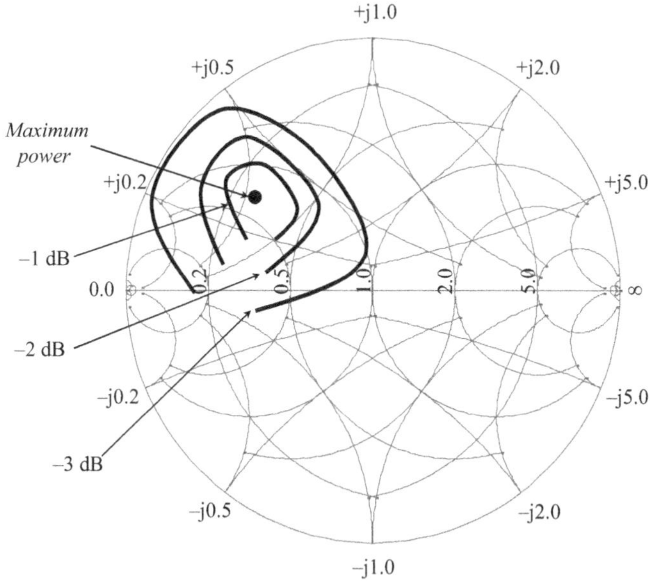

Figure 4.18 A graphical illustration of load-pull contours on Smith chart

4.12 Design considerations

Given that one may end up with several options of matching topologies, some important considerations should be taken into account when selecting the suitable matching circuit. To list some:

- Large inductors and capacitor values usually accompanied with low self-resonant frequency. Some design specifications impose restrictions on the maximum current and voltage levels allowed in the design that is related in one way or another to the selection of the component's values. Additionally, larger inductor values in the series branch of a matching network correspond to higher loss when compared to small inductor values. Hence, at high frequencies, skin depth effect and interwinding capacitance decrease the Q-factor of an inductor.
- The slope of a load reactance is a frequency dependent parameter, and in most of the design scenarios, it is important to ensure that the load reactance slope aligns and works in synergy with the impedance behavior of the matching network, especially, over a broadband frequency. For instance, careful design of a matching network can significantly improve harmonic attenuations in an RF transmitting module.
- As discussed in Section 4.3, there is always a trade-off between loss and achievable bandwidth in any matching network. In designs where loss can be tolerated, one can achieve a broader impedance matching with a lower Q circuit (e.g., adding a series resistance to a high reactance load or low resistance load).

- The underlying principle of Bode–Fano criterion implies that broader bandwidth and flat impedance curves can be established by cascading more matching elements to achieve smaller impedance ratios.
- Impedance matching of a balanced circuit can be performed by assuming a virtual ground plane between the terminals of the balanced circuit so that the impedance matching problem can be solved in a similar fashion to those of a single-ended circuit by taking into account that the input and output impedances of the unbalanced terminals are half of the balanced ones.
- An all resistive network can also be used to match between two real impedances; however, such networks possess high loss in the signal transmit path.

References

[1] Fano R.M. "Theoretical limitations on the broadband matching of arbitrary impedances." *Journal of the Franklin Institute*. 1950; **249**(1):57–83

[2] Bode H.W. *Network Analysis and Feedback Amplifier Design*. New York, NY: D. Van Nostrand Co; 1945

[3] Gonzalez G. *Microwave Transistor Amplifiers: Analysis and Design*. 2nd edn. Upper Saddle River, NJ: Prentice-Hall, Inc.; 1996

[4] Pozar D.M. *Microwave Engineering*. 4th edn. Hoboken, NJ: John Wiley; 2012.

[5] Besser L., and Gilmore R. *Practical RF Circuit Design for Modern Wireless Systems*. Norwood, MA: Artech House; 2003.

[6] Matthaei G., Young I., and Jones E. Microwave Filters, *Impedance Matching Networks and Coupling Structures*. Dedham, MA: Artech House; 1980

[7] Bloch E., and Socher E. "Beyond the Smith chart: a universal graphical tool for impedance matching using transformers." *IEEE Microwave Magazine*. 2014;**15**(7):100–09

[8] Vinding J.P. "Radial line stubs as elements in strip line circuits." *NEREM Record*. 1967. pp. 108–09

[9] Khalaj-Amirhosseini M. "Closed form solutions for nonuniform transmission lines." *Progress in Electromagnetics Research B*. 2008;**2**:243–58

[10] Takayama Y. "A new load–pull characterization method for microwave power transistors." *1976 IEEE-MTT-S International Microwave Symposium*; Cherry Hill, NJ, Jun 1976. pp. 218–20

Chapter 5
Electromagnetic field couplings

5.1 Introduction

With the continuous shrinkage of the minimum feature sizes of semiconductor integrated circuits and their packages along with the increase in the operating frequency have caused parasitic electromagnetic (EM) coupling to become a serious design concern. Realizing the significant role of EM coupling in influencing the system- and device-level radio frequency (RF) performance entails a good understanding of the coupling mechanism occurring in a certain EM environment. EM coupling between the circuit elements can be considered as a desirable or undesirable phenomenon depending upon the intended application, the intrinsic EM structure, and components forming the RF circuit. EM coupling is necessary for the design of coupled microwave structures such as coupled line filters, couplers, microwave baluns, and several other distributed circuits. In other cases, however, EM coupling is not beneficial and considered as a source of EM interference (EMI) that can drastically impact the electrical performance of the circuit. For example, undesired EM couplings in transceiver modules may cause cross-isolation issues between transmit and receive RF paths that may lead to a total system paralysis.

The EM coupling mechanism in RF modules can be simply attributed to the fact that adding conductive traces, planes, or components in proximity to a signal trace can cause field lines to couple from the signal conductor into the surrounding conductors instead of the intended return path. The wave characteristics that might couple from an aggressor circuit onto an adjacent victim circuit are completely independent of the wave that might already be present on the victim circuit. In return, this makes it possible to identify the source of the EM coupling. In distributed structures, waves travel in backward and forward directions on transmission lines (TLs) and thus a transverse electromagnetic (TEM) coupled lines can support four independent modes of propagation. A traveling wave in the forward and backward directions on a TL will have similar characteristic impedance and wave velocity. Therefore, the number of independent modes can be reduced into two. In the case of asymmetric coupled lines, the two modes of propagation are known as π- and c-modes and are used to fully analyze the coupling behavior between the lines. Similarly, in the case of symmetric coupled lines, which can be considered as a special case of the more generalized asymmetric coupled line analysis, two propagation modes, namely, even- and odd-modes are used.

With today's congested RF spectrum and the increasing complexity of RF modules, spectrum management becomes important to coordinate, plan, and manage through engineering and administrative procedures in a way that enables electronic and electrical systems to perform their functions in a shared EM environment without causing unacceptable interference. As a result, several electromagnetic compatibility (EMC) and signal integrity (SI) disciplines and standards have been put into effect for design engineers to use and to design better countermeasures against EMI emissions. Usually, SI deals with noise levels in micro-Volts (μV) and micro-Amperes (μA), while EMC deals with milli-Volts (mV) and milli-Amperes (mA) noise levels. In a broader sense, EMI impacts the efficient utilization of bandwidth and wireless system throughput. These detrimental effects result in poorer overall system performance which can lead to longer product time-to-market and even product cancelation if the RF system is not characterized carefully at the design phase. EMI typically occurs if all the following three components exist together:

- An aggressor—EM element that is a source of the EMI;
- A victim—system or EM element that is susceptible to the EMI; and
- A coupling path—conducted or radiated path in which EMI is transferred.

Based on the three components mentioned above, EMI mitigation methods can be carried out by either one or more of the following:

- Reducing the interference at the source by changing the operating frequency, reducing the signal power level, changing the circuit orientation, or location.
- Increasing the immunity of the victim by changing the operating frequency, changing the circuit orientation, or location.
- Attenuating the coupling path through increasing the spatial separation between the victim and aggressor circuits, adding guard shields between the impacted components, or altering and improving the RF grounding.

Typically reducing the interference at the source or at the victim circuits is difficult if not impossible in RF modules since changing the operating frequency is not usually an option and changing the device location is limited by the given real-estate. However, EMI mitigation is often tackled through configuring the coupling path. As will be discussed in this chapter, it is important to identify the dominant coupling type (e.g., capacitive or inductive, near-field or far-field, and conducted or radiated) in order to find the proper EMI shielding technique.

5.2 Electromagnetic coupling types

The EM coupling from one conductor to another in an RF module, also so-called crosstalk, can be attributed to different radiated and conducted coupling types: inductive, capacitive, a combination of both inductive and capacitive (i.e., mixed coupling), and common path coupling.

5.2.1 Inductive coupling

Inductive or magnetic-field coupling between circuit elements occurs when magnetic (H-) fields transfer from one element to another by means of electromagnetic induction. This is because the current movement in a conductor will produce a magnetic field that can couple into nearby circuit components and induce voltages and currents on them. Example of circuits where the inductive coupling is beneficial include transformers, wireless chargers, and electric motors. EM induction is governed by Maxwell–Faraday's equation, which states that the rate of change of the magnetic flux (B) through a closed path conductor is proportional to the voltage induced on that conductor:

$$\nabla \times E = -\frac{\partial B}{dt} \tag{5.1}$$

where E is the electric field representing the induced voltage.

For a given device, signal traces with high current and low voltage exhibit low impedance characteristics that potentially result in magnetic coupling to adjacent conductors as depicted in Figure 5.1.

5.2.2 Capacitive coupling

Capacitive or electric-field coupling occurs when energy in the form of electric (E-) fields couples from one circuit element to another. Signal traces carrying low current and high voltage possess high impedance characteristics that allow electric field lines to couple to adjacent traces and circuit components that have relatively lower electric potential as illustrated in Figure 5.2. Maxwell–Gauss' electric field equation governs the electric capacitance as given by the following differential form:

$$\nabla \times E = \frac{\rho}{\varepsilon_0} \tag{5.2}$$

where ε_0 is the permittivity of free space, and ρ is the volume charge density.

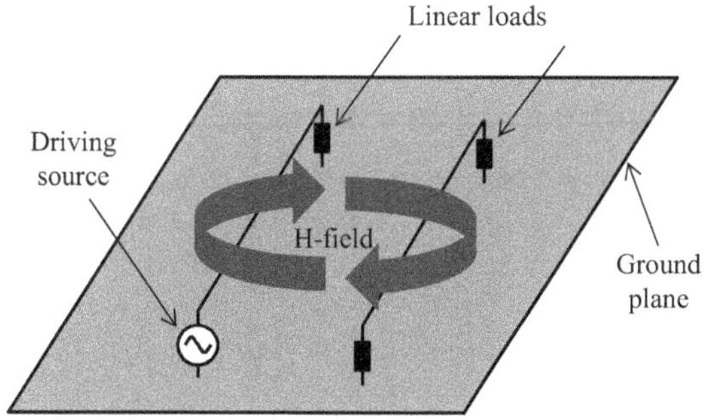

Figure 5.1 Inductive coupling between two traces showing the direction of magnetic coupling path from a trace with a high current to another with a relatively lower current

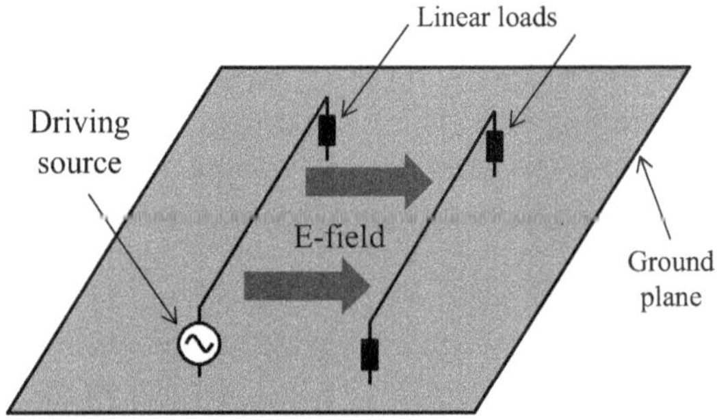

Figure 5.2 Capacitive coupling between two adjacent traces showing the direction of electric field lines from a trace with high electric potential to another with a relatively lower electric potential

Capacitive coupling usually exists between neighboring traces, unlike inductive coupling that can extend farther in the distance to impact distant circuit elements. Having signal traces in close proximity to each other increases the mutual capacitance levels which can be sufficient to couple significant energy from one element to another.

5.2.3 Mixed coupling

Mixed coupling is a combination of electric and magnetic fields coupling. Finding the proper shielding method for this type of coupling is generally a challenging task. In resonant coupling structures, mixed coupling is beneficial and prominently utilized in several microwave circuits design. For instance, mixed coupling is utilized in the design of microwave bandpass filters for creating a transmission zero to attenuate co-existing bands in the filter's stopband region or to improve the filter's selectivity. Moreover, mixed coupling can also be used to reject specific nonlinear components generated from a power amplifier (PA), often called harmonic traps, and incorporated at the PA output matching network.

5.2.4 Common path coupling

The ground plane of a system serves as a reference potential for all the terminals of circuits sharing the same system. Therefore, any two points that are physically separated on an ideal ground plane must be equipotential to ensure a seamless path for current to return to the source. Practically, ground conductors are not ideal and may have finite low impedance causing a potential difference on the surface of the ground plane where a return current is flowing.

Common impedance or path coupling occurs when two or more signal traces share a nonideal ground conductor as illustrated in Figure 5.3. For sensitive analog circuits, having a small amount of impedance in the shared ground path can result in interfering voltages and return currents that can significantly impact other circuitry sharing the same ground plane. This detrimental effect can be seen on different occasions such as increased noise level in the RF receiver analog circuits and

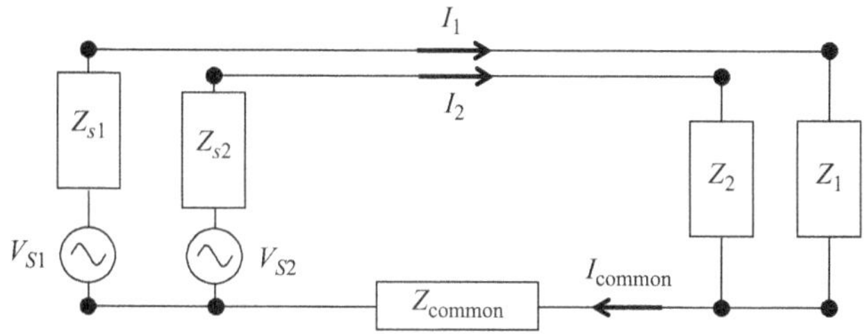

Figure 5.3 Common impedance coupling due to sharing a nonperfect ground path

degraded rejection floor of highly selective filters. Usually, high-frequency return currents take the path of least inductance, while low-frequency currents take the path of least resistance.

5.3 Coupled transmission lines

A coupled TL in a homogeneous medium (e.g., stripline) can support a pure TEM mode which is the lowest order mode of wave propagation along the coupled lines. However, in the presence of inhomogeneous dielectric medium (e.g., microstrip), waves propagate in unequal phase velocities because of the different media surrounding the coupled lines that preclude having a pure TEM mode. Instead, microstrip-based circuits support quasi-TEM that tends to have degraded performance when compared to pure TEM circuits. For example, the TEM nature of wave propagation of directive couplers presents a directive backward-coupling when the lines are tightly coupled; while with different mode velocities co-directional coupling (i.e., forward-coupling) is also possible resulting in reduced directivity. In a coupled TL, the maximum coupling occurs when the electrical length of the lines are a quarter wavelength long.

Figure 5.4 illustrates an LC distributed model of a pair of lossless coupled transmission lines. Exact and approximate analyses of pure and quasi-TEM mode, respectively, have been widely discussed in the literature [1]. Important analysis of coupled transmission lines is highlighted in the subsequent Sections 5.3.2 and 5.3.3.

5.3.1 Single transmission line

As a review of the TL theory, consider the equivalent distributed-element model of a lossy TL depicted in Figure 5.5 that takes the form of *L*-line segments or symmetric PI-line segments cascaded in series. Having a TL with relatively large electrical length requires that the line to be segmented into *N* sections where *N* is a sufficient number that can provide a good representation on the TL. In Figure 5.5, a lossy TL of total length *d* can be represented as a cascade of *N* copies of *L*-type TL model, each formed by resistance, inductance, conductance, and capacitance (RLGC) lumped elements. The series *L* and *R* elements account for the line length and its metal skin effect, while *G* and *C* elements describe the dielectric losses and capacitance between the line and the ground plane or the current return path.

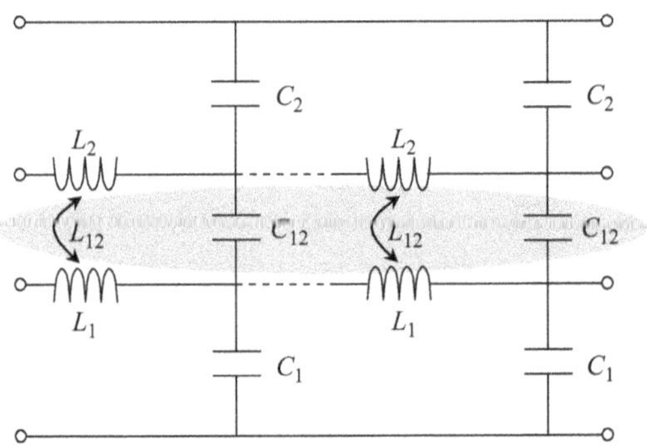

Figure 5.4 A lossless lumped equivalent circuit of a coupled transmission line. Coupling region is highlighted in gray color

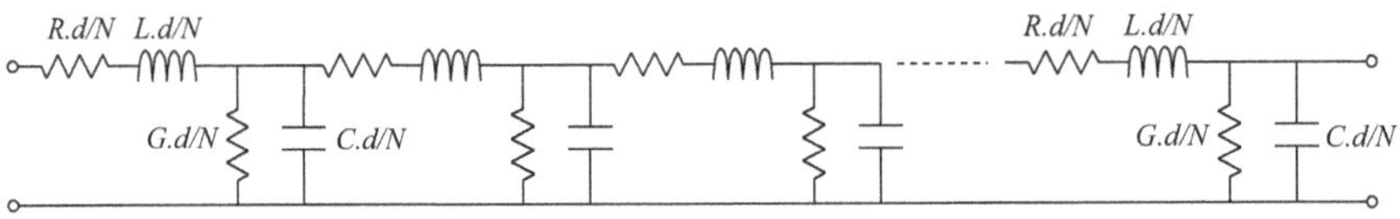

Figure 5.5 A lossy transmission line model represented by N-number of RLGC line segments

Generally, the total loss in a TL is attributed to four types of losses: conductor loss, dielectric loss, radiation loss, and leakage loss. The conductor and dielectric losses are more significant than the others. Each of the N sections represents a small fraction of the lowest operating wavelength, and accordingly, it may require many sections to approximate a TL response. The calculated response using this model usually contains spurious oscillations if less number of segments are used; however, this challenge has been resolved in [2,3].

The transmission line lumped parameters are summarized as below:

- R is the resistance per unit length,
- L is the inductance per unit length,
- G is the insulation conductance per unit length,
- C is the capacitance per unit length,
- d is the total length of the line, and
- N is the number of series segments.

Upon solving the Telegrapher's time domain differential equations by assuming a sinusoidal steady-state condition, the voltage across a TL and the current flowing through it can be expressed as:

$$\begin{aligned} \frac{d^2V(\mathrm{z})}{dz^2} - \gamma^2 V(\mathrm{z}) = 0 \\ \frac{d^2I(\mathrm{z})}{dz^2} - \gamma^2 \mathrm{I}(\mathrm{z}) = 0 \end{aligned} \tag{5.3}$$

and

$$\gamma = \alpha + j\beta = \sqrt{(R + j\omega L)(G + j\omega C)} \tag{5.4}$$

The propagation constant (γ) comprises the attenuation constant (α) and the phase constant (β). The characteristic impedance (Z_0) of the TL in terms of the TL lumped parameters can be defined as:

$$Z_0 = \sqrt{\frac{R + j\omega L}{G + j\omega C}} \tag{5.5}$$

The propagation time delay (t_d) in seconds can be simply calculated as:

$$t_d = \frac{d}{v_p} = d \times \frac{\beta}{\omega} = d \times \sqrt{LC} \tag{5.6}$$

where v_p is the phase velocity, and d is the TL length.

And the group velocity of a TL can be calculated as:

$$v_g = \frac{\partial \omega}{\partial \beta} \tag{5.7}$$

The group velocity is used to describe how fast a modulated signal or an information signal propagates through the propagation medium. For a lossless and dispersion less line, the group and phase velocities are equal.

A frequency dependent lumped element model of a lossy TL given in terms of RLGC parameters can be described using the line propagation constant, operating frequency, and the characteristic impedance as shown in the following set of equations:

$$\begin{aligned} R &= \mathrm{Re}\{\gamma Z_0\} \\ G &= \mathrm{Re}\left\{\frac{\gamma}{Z_0}\right\} \\ L &= \frac{\mathrm{Im}\{\gamma Z_0\}}{\omega} \\ C &= \frac{\mathrm{Im}\left\{\frac{\gamma}{Z_0}\right\}}{\omega} \end{aligned} \tag{5.8}$$

At some instances where the ratio of R/L is large, a simple RC model formed from a series resistor and a shunt capacitor can be used as a special case of the lossy TL model. In this case, the voltage across the TL and the current flowing through it can be expressed as:

$$\begin{aligned} \frac{dI(\mathrm{z})}{dz} &= -C\frac{dV(\mathrm{z})}{dt} \\ \frac{dV(\mathrm{z})}{dz} &= -RI(\mathrm{z}) \end{aligned} \tag{5.9}$$

Combining the equations in (5.9) produces the following diffusion equation:

$$\frac{d^2V(z)}{dz^2} = \mathrm{RC}\frac{dV(z)}{dt} \tag{5.10}$$

This type of TL model is commonly used in microelectronic circuits, where electrically short TLs are used in a capacitive laminate environment. The RC TL model is also beneficial in the analysis of circuits based on low-*K* dielectric materials, which are commonly used in MICs and high-speed digital interconnects to reduce RC time delay and capacitive couplings. It is worth mentioning that radiated emissions from a high-dielectric substrate are very low when compared to a low dielectric material. In some applications, significantly low dielectric materials are preferred such as in antennas.

5.3.2 Symmetric coupled transmission lines

Symmetric transmission line circuits can be readily analyzed using the method of even- and odd-modes or, the method of common and differential modes as will be discussed in Sections 5.3.2.1 and 5.3.2.2. Similar to the RLGC model of a single TL, a distributed lumped model of symmetric coupled TL can be represented in terms of the even- and odd-modes of the line propagation constant and characteristic impedance. Here, each RLGC parameter will be a 2×2 matrix to account for the mutual effects between the coupled lines.

5.3.2.1 Common and differential mode impedance

Differential impedance is the impedance between two symmetric coupled lines when they are driven with voltages that are equal in magnitude, but opposite in phase (i.e., differentially driven) as illustrated in Figure 5.6(a). When the same excitation drives the two lines, the impedance between the two lines is called common mode impedance as shown in Figure 5.6(b). Differential and common mode impedance analysis are predominately used in solving differential amplifier problems, high-speed circuits, and circuits transmitting logic/clock signals. This type of analysis is concerned with the signal integrity of systems that may be influenced by common mode noise effects such as power supply noise.

In the differential mode, the voltage doubles and the current remains the same, whereas, for the common mode, the currents doubles and the voltage remains the same. This is explained by the following mode impedance relationship:

$$Z_{\mathrm{common}} = 0.5Z_0, \quad Z_{\mathrm{differential}} = 2Z_0 \tag{5.11}$$

5.3.2.2 Even- and odd-mode impedance

Even- and odd-mode analysis is a commonly used approach in analyzing symmetric coupled transmission lines. Similar to the differential and common mode impedance, the odd-mode impedance is defined as the impedance of either of the two lines when they are excited with signals of the same amplitude and opposite polarity. And the even-mode impedance is the impedance of either of the two lines when they are driven with signals of the same amplitude and the same polarity.

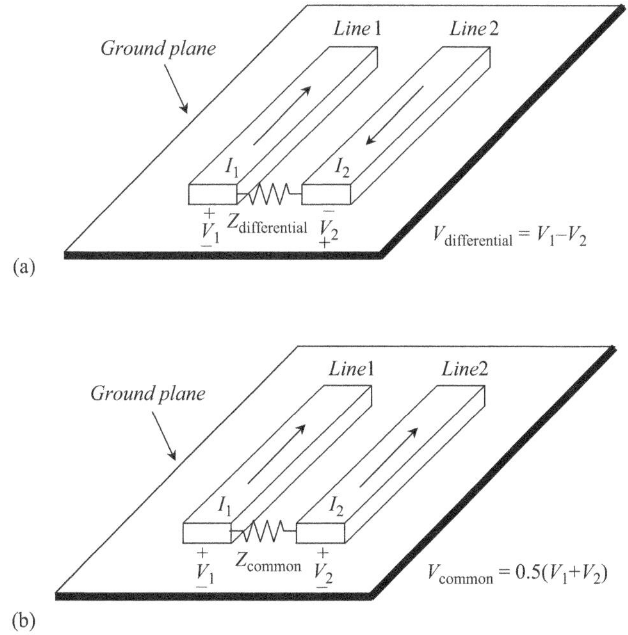

Figure 5.6 Two signal traces driven by (a) differential and (b) common stimulus

From the principle of superposition, the circuit behavior of the coupled lines shown in Figure 5.7(a) can be obtained through the algebraic sum of the two independent excitation modes depicted in Figure 5.7(b) and (c). This definition effectively makes the differential mode impedance equal to twice the odd-mode impedance and the common mode impedance to be equal to half the even-mode impedance. Thus, mode conversion between the two definitions can be related as in (5.12) and (5.13):

$$Z_{common} = 0.5Z_{even}, \quad \gamma_{common} = \gamma_{even} \tag{5.12}$$

$$Z_{differential} = 2Z_{odd}, \quad \gamma_{differential} = \gamma_{odd} \tag{5.13}$$

In the even-mode, the electric field lines are evenly symmetric along the plane-of-symmetry (POS) due to the equal amplitude and same polarity of the two propagating signals on the coupled lines. The lines, in this case, appear as decoupled traces as depicted in Figure 5.8(a) where the mutual capacitance, C_{12}, is effectively open-circuited (*OC*). In the odd-mode, the equal amplitude and opposite polarity of the two propagating signals make the electric field lines perpendicular along the POS. This implies that the POS is effectively acting as a ground plane to the field from both lines similar in principle to the image theory.

Practically, with the aid of an EM field solver, one can obtain the even-mode impedance of one of the coupled lines by placing a magnetic-wall at the POS as

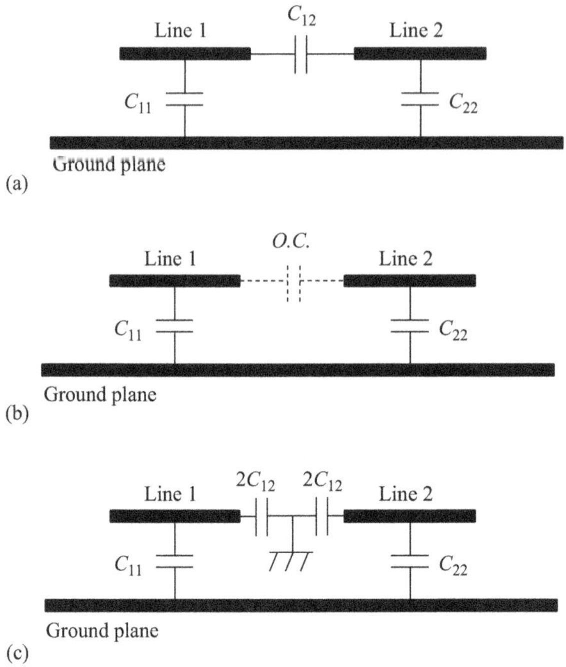

Figure 5.7 Cross-sectional views of (a) two edge-coupled lines above a ground plane, and its equivalent capacitance network when excited by (b) even-mode and (c) odd-mode

illustrated in Figure 5.8(a), while the odd-mode impedance can be found in a similar fashion but by replacing the POS with an electric-wall as depicted in Figure 5.8(b). The effect of these walls is to alter the characteristic impedance and phase velocity of the line close to the wall.

Coupled TLs, especially, electrically short lines are capacitively coupled. In this case, coupled TLs can be analyzed by finding the equivalent capacitances and the propagation constants accompanying each mode as discussed below. Examining Figure 5.7, the equivalent capacitance of the even- and odd-mode line can be written as:

$$C_{\text{even}} = C_{11} = C_{22} \tag{5.14}$$

and

$$C_{\text{odd}} = C_{11} + 2C_{12} = C_{22} + 2C_{21} \tag{5.15}$$

Substituting (5.14) in (5.15), the mutual capacitance in terms of even- and odd-mode capacitances can be obtained as follows:

$$C_{12} = \frac{C_{\text{odd}} - C_{\text{even}}}{2} \tag{5.16}$$

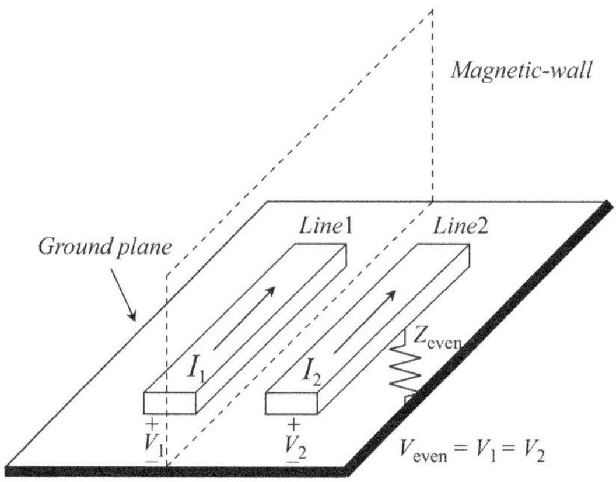

(a)

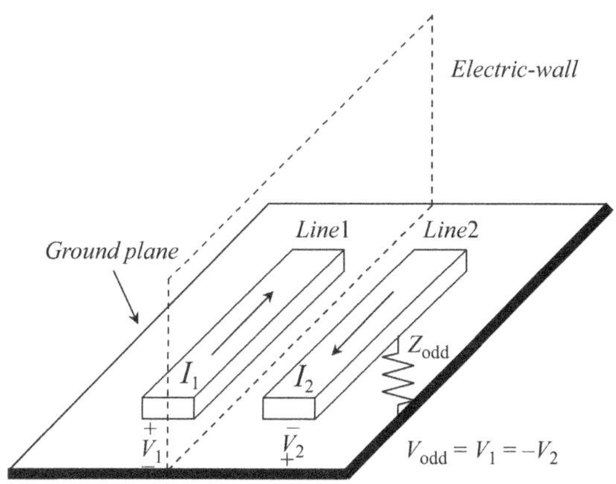

(b)

Figure 5.8 Coupled transmission line with two modes of propagation: (a) even-mode and (b) odd-mode

Phase velocities of the even- and odd-modes, v_{odd} and v_{even}, respectively, are equal in the case of TLs embedded in a homogeneous dielectric medium (e.g., striplines) and are given by:

$$v_{odd} = v_{even} = \frac{c}{\sqrt{\varepsilon_r}} \tag{5.17}$$

where c is the wave speed in vacuum and ε_r is the effective dielectric constant of the medium.

In the case of inhomogeneous dielectric medium (e.g., microstrip lines), the propagation velocities of either line vary depending on the even- and odd-mode effective dielectric constants as given by:

$$v_{odd} = \frac{c}{\sqrt{\varepsilon_{r,odd}}}, \quad \varepsilon_{r,odd} = \frac{C_{odd}}{C_{0,odd}} \tag{5.18}$$

$$v_{\text{even}} = \frac{c}{\sqrt{\varepsilon_{r,\text{even}}}}, \quad \varepsilon_{r,\text{even}} = \frac{C_{\text{even}}}{C_{0,\text{even}}} \tag{5.19}$$

The effective dielectric constants for each excitation mode can be calculated by taking the ratio of the mode capacitance due to the actual dielectric material surrounding the coupled lines to the mode capacitance assuming unity dielectric constant (i.e., $C_{0,\text{odd}}$ and $C_{0,\text{even}}$).

By knowing the propagation velocities and capacitances of the even- and odd-mode excitations, the mode impedances can be calculated as follows:

$$Z_{\text{odd}} = \frac{1}{v_{\text{odd}} C_{\text{odd}}} = \frac{\omega}{\beta_{\text{odd}} C_{\text{odd}}} = \sqrt{\frac{L_{\text{odd}}}{C_{\text{odd}}}} \tag{5.20}$$

$$Z_{\text{even}} = \frac{1}{v_{\text{even}} C_{\text{even}}} = \frac{\omega}{\beta_{\text{even}} C_{\text{even}}} = \sqrt{\frac{L_{\text{even}}}{C_{\text{even}}}} \tag{5.21}$$

From (5.20 and 5.21) note that the knowledge of the mode inductance is not necessarily needed in this analysis.

The characteristic impedance of a TEM TL is equal to the geometric mean of the even- and odd-mode impedances as:

$$Z_0 = \sqrt{Z_{\text{even}} Z_{\text{odd}}} \tag{5.22}$$

For uncoupled TLs where the mutual inductance (L_{12}) and the mutual capacitance (C_{12}) do not exist, the even- and odd-mode impedances become equal to the TL characteristic impedance:

$$Z_0 = Z_{\text{odd}} = Z_{\text{even}} \tag{5.23}$$

The frequency dependent lumped element model of a lossy coupled TL given in terms of the RLGC parameters can be described using the modes propagation constant, operating frequency, and the modes characteristic impedance as:

$$\begin{aligned}
R_{11} &= \text{Re}\{\gamma_{\text{even}} Z_{\text{even}} + \gamma_{\text{odd}} Z_{\text{odd}}\} \\
R_{12} &= \text{Re}\{\gamma_{\text{even}} Z_{\text{even}} - \gamma_{\text{odd}} Z_{\text{odd}}\} \\
G_{11} &= \text{Re}\{\gamma_{\text{even}}/Z_{\text{even}} + \gamma_{\text{odd}}/Z_{\text{odd}}\} \\
G_{12} &= \text{Re}\{\gamma_{\text{even}}/Z_{\text{even}} - \gamma_{\text{odd}}/Z_{\text{odd}}\} \\
L_{11} &= \text{Im}\{\gamma_{\text{even}} Z_{\text{even}} + \gamma_{\text{odd}} Z_{\text{odd}}\}/\omega \\
L_{12} &= \text{Im}\{\gamma_{\text{even}} Z_{\text{even}} - \gamma_{\text{odd}} Z_{\text{odd}}\}/\omega \\
C_{11} &= \text{Im}\{\gamma_{\text{even}}/Z_{\text{even}} + \gamma_{\text{odd}}/Z_{\text{odd}}\}/\omega \\
C_{12} &= \text{Im}\{\gamma_{\text{even}}/Z_{\text{even}} - \gamma_{\text{odd}}/Z_{\text{odd}}\}/\omega
\end{aligned} \tag{5.24}$$

One should take into account the simultaneous effects of both forward and backward coupling in the analysis of coupled lines so that the determination of the coupling mechanism can be identified and mitigation methods can be found. It was reported in the literature that backward coupling (from port 1 to port 3 in

Figure 5.9) arises from the difference in the characteristic impedance between the odd- and even-mode of the coupled lines, while forward coupling (from port 1 to port 4) occurs due to the difference between the odd- and even-mode phase velocities [4]. In fact, tightly coupled lines (i.e., closely spaced lines) possess appreciably different line impedances for the odd- and even-modes and thus high backward coupling occur. Conversely, forward coupling accompanies weakly coupled lines.

Figure 5.10 illustrates the even- and odd-mode impedance behavior as a function of the spacing between the coupled lines. Both mode impedances monotonically tend toward the single TL characteristic impedance as the spacing between the lines increases. As can be observed, the even-mode impedance is always higher than the odd-mode impedance. The fact that the odd-mode impedance is always less than the even-mode impedance implies that the odd-mode waves are always traveling faster than the even-mode waves, and as a result, a co-directional forward coupling occurs.

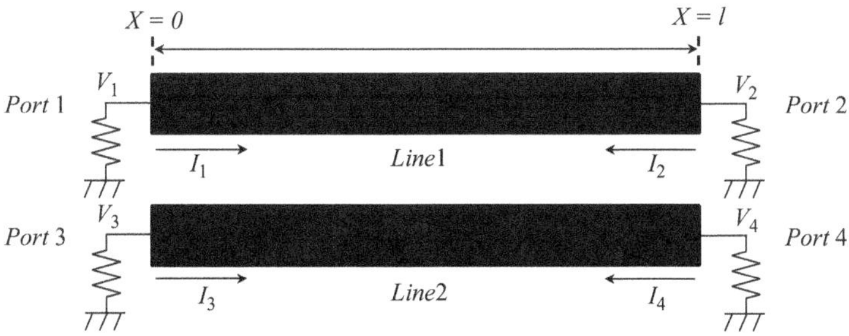

Figure 5.9 A four-port network showing symmetric coupled lines

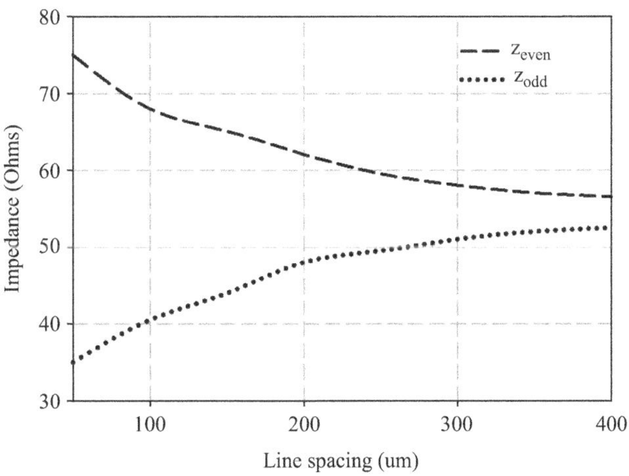

Figure 5.10 Variations of the even- and odd-mode impedances as a function of the spacing between two coupled lines for characteristic line impedance of 54 ohms

One can also look at (5.14 and 5.15) along with (5.20 and 5.21) to verify that the impedance of the even-mode is always higher than the odd-mode. For a wide separation between the lines, the odd-mode loss decreases strongly due to the reduction of current density at the inner strip edges whereas the even-mode loss increases slightly as an opposite effect.

In terms of S-parameters, the four-port network S-matrix of a coupled TL is expressed as:

$$[S] = \begin{bmatrix} S_{11} & S_{12} & S_{13} & S_{14} \\ S_{21} & S_{22} & S_{23} & S_{24} \\ S_{31} & S_{32} & S_{33} & S_{34} \\ S_{41} & S_{42} & S_{43} & S_{44} \end{bmatrix} \tag{5.25}$$

Given that the coupled line network is reciprocal, symmetric, and assuming a lossless network, the following elements of the S-matrix are equal:

$$\begin{gathered} S_{11} = S_{33},\ S_{22} = S_{44} \\ S_{21} = S_{12},\ S_{43} = S_{34} \\ S_{31} = S_{13}, S_{41} = S_{14},\ S_{32} = S_{23},\ S_{42} = S_{24} \\ S_{34} = S_{12},\ S_{14} = S_{23} \end{gathered} \tag{5.26}$$

The elements of the S-matrix can then be reduced to six elements instead of sixteen, and the updated S-matrix can be then expressed as:

$$[S] = \begin{bmatrix} S_{11} & S_{12} & S_{13} & S_{14} \\ S_{12} & S_{22} & S_{14} & S_{24} \\ S_{13} & S_{14} & S_{11} & S_{12} \\ S_{14} & S_{24} & S_{12} & S_{22} \end{bmatrix} \tag{5.27}$$

The four-port S-matrix in (5.27) can be further simplified and rewritten in a compact fashion by partitioning it into 2×2 block as in (5.28) with two sub-matrices: S_{line} contains the S-parameters of either one of the coupled lines and S_{cpl} that includes the intercoupling parameters between the two lines:

$$\begin{bmatrix} b_1 \\ b_2 \\ b_3 \\ b_4 \end{bmatrix} = \begin{bmatrix} [S_{\text{line}}] & [S_{\text{cpl}}] \\ [S_{\text{cpl}}] & [S_{\text{line}}] \end{bmatrix} \begin{bmatrix} a_1 \\ a_2 \\ a_3 \\ a_4 \end{bmatrix} \tag{5.28}$$

where

$$[S_{\text{line}}] = \begin{bmatrix} S_{11} & S_{12} \\ S_{12} & S_{22} \end{bmatrix} \tag{5.29}$$

and

$$[S_{\text{cpl}}] = \begin{bmatrix} S_{13} & S_{14} \\ S_{14} & S_{24} \end{bmatrix} \tag{5.30}$$

In the case of even-mode wave propagation, ports 1 and 3 are excited by equal magnitude sources that are in phase. The incident and reflected wave amplitudes are then related to the even-mode waves as $a_1 = a_3 = a_{1,\text{even}}$, $b_1 = b_3 = b_{1,\text{even}}$, $a_2 = a_4 = a_{2,\text{even}}$, and $b_2 = b_4 = b_{2,\text{even}}$:

$$\begin{bmatrix} b_{1,\text{even}} \\ b_{2,\text{even}} \\ b_{1,\text{even}} \\ b_{2,\text{even}} \end{bmatrix} = \begin{bmatrix} [S_{\text{line}}] & [S_{\text{cpl}}] \\ [S_{\text{cpl}}] & [S_{\text{line}}] \end{bmatrix} \begin{bmatrix} a_{1,\text{even}} \\ a_{2,\text{even}} \\ a_{1,\text{even}} \\ a_{2,\text{even}} \end{bmatrix} \tag{5.31}$$

In the case of odd-mode wave propagation, ports 1 and 3 are excited by equal magnitude but out of phase sources. The incident and reflected wave amplitudes can then be related to the odd-mode waves as $a_1 = -a_3 = a_{1,\text{odd}}$, $b_1 = -b_3 = b_{1,\text{odd}}$, $a_2 = -a_4 = a_{2,\text{odd}}$, and $b_2 = -b_4 = b_{2,\text{odd}}$:

$$\begin{bmatrix} b_{1,\text{odd}} \\ b_{2,\text{odd}} \\ -b_{1,\text{odd}} \\ -b_{2,\text{odd}} \end{bmatrix} = \begin{bmatrix} [S_{\text{line}}] & [S_{\text{cpl}}] \\ [S_{\text{cpl}}] & [S_{\text{line}}] \end{bmatrix} \begin{bmatrix} a_{1,\text{odd}} \\ a_{2,\text{odd}} \\ -a_{1,\text{odd}} \\ -a_{2,\text{odd}} \end{bmatrix} \tag{5.32}$$

Both (5.31) and (5.32) can be simplified even further and rewritten as:

$$\begin{bmatrix} b_{1,\text{even}} \\ b_{2,\text{even}} \end{bmatrix} = \underbrace{[[S_{\text{line}}] + [S_{\text{cpl}}]]}_{[S_{\text{even}}]} \begin{bmatrix} a_{1,\text{even}} \\ a_{2,\text{even}} \end{bmatrix} \tag{5.33}$$

and

$$\begin{bmatrix} b_{1,\text{odd}} \\ b_{2,\text{odd}} \end{bmatrix} = \underbrace{[[S_{\text{line}}] - [S_{\text{cpl}}]]}_{[S_{\text{odd}}]} \begin{bmatrix} a_{1,\text{odd}} \\ a_{2,\text{odd}} \end{bmatrix} \tag{5.34}$$

Once the even- and odd-mode *S*-parameters of either of the coupled lines are known, the 2×2 submatrices $[S_{\text{line}}]$ and $[S_{\text{cpl}}]$ can be found as follows:

$$[S_{\text{line}}] = \frac{[S_{\text{even}}] + [S_{\text{odd}}]}{2} = \begin{bmatrix} \dfrac{S_{11,\text{even}} + S_{11,\text{odd}}}{2} & \dfrac{S_{12,\text{even}} + S_{12,\text{odd}}}{2} \\ \dfrac{S_{12,\text{even}} + S_{12,\text{odd}}}{2} & \dfrac{S_{22,\text{even}} + S_{22,\text{odd}}}{2} \end{bmatrix} \tag{5.35}$$

and

$$[S_{\text{cpl}}] = \frac{[S_{\text{even}}] - [S_{\text{odd}}]}{2} = \begin{bmatrix} \dfrac{S_{11,\text{even}} - S_{11,\text{odd}}}{2} & \dfrac{S_{12,\text{even}} - S_{12,\text{odd}}}{2} \\ \dfrac{S_{12,\text{even}} - S_{12,\text{odd}}}{2} & \dfrac{S_{22,\text{even}} - S_{22,\text{odd}}}{2} \end{bmatrix}, \tag{5.36}$$

Finally, the four-port S-matrix elements can be determined as follows:

$$\begin{gathered} \left(\begin{aligned} S_{11} &= \frac{S_{11,\text{even}} + S_{11,\text{odd}}}{2} \\ S_{22} &= \frac{S_{22,\text{even}} + S_{22,\text{odd}}}{2} \end{aligned} \right) \\ \left(S_{21} = \frac{S_{21,\text{even}} + S_{21,\text{odd}}}{2} \right) \\ \left(\begin{aligned} S_{41} &= \frac{S_{21,\text{even}} - S_{21,\text{odd}}}{2} \\ S_{42} &= \frac{S_{22,\text{even}} - S_{22,\text{odd}}}{2} \\ S_{31} &= \frac{S_{11,\text{even}} - S_{11,\text{odd}}}{2} \end{aligned} \right) \end{gathered} \tag{5.37}$$

5.3.3 *Asymmetrical coupled transmission lines*

Analysis and characterization methods of symmetric coupled lines have been investigated extensively in the literature by employing the even- and odd-mode approach. However, the demand for more generalized coupled TL analysis that supports and include asymmetric multicoupled lines on single and multilayer substrates has become immensely important to cope with the complexity of modern microwave circuits and modules. Although several symmetric microwave circuits can benefit from the even- and odd-mode analysis approach, it is unlikely that symmetry can be attained between interconnecting traces in MICs and many other important circuits that require asymmetrical coupled lines. For example, the use of asymmetrical coupled lines is essential in the realization of coupled line bandpass filters exhibiting advanced electrical characteristics such as improved stopband performance and also directional couplers with improved bandwidth.

Even- and odd-modes of propagations cannot be utilized directly for the analysis of asymmetric coupled lines as they are no longer independent of each other. Instead, a more generalized approach that uses two independent modes of excitation, known as the normal modes, comprising π- and c-modes are used [1]. The π- and c-modes, respectively, are quite similar in principle to the odd- and even-modes as will be discussed next.

For this analysis, traveling waves on the TLs are expressed in terms of forward and backward voltages and currents as depicted in Figure 5.11. The characterization of π- or c-modes can be carried out through eight relevant parameters: the propagation constant of each mode, γ_c and γ_π, the characteristic impedance of each mode at each line in the presence of the other line, and two ratios, R_π and R_c, of the voltage or current on one line in reference to the other for both modes. The characteristic impedance of the c-modes are represented by $Z_{C,\text{line1}}$ and $Z_{C,\text{line2}}$ for lines 1 and 2, respectively. Similarly, the characteristic impedance of the π-modes are represented by $Z_{\pi,\text{line1}}$, and $Z_{\pi,\text{line2}}$ for lines 1 and 2, respectively. Hence, not all of these parameters are independent, and only six parameters are. This is because the mode characteristic impedance can be calculated for one line only and those of the other line can be expressed in terms of the R_c or R_π parameters. Table 5.1 summarizes the eight parameters in terms of the forward and backward voltages and currents of either of the lines in the presence of the other line. As shown, each mode is characterized by four parameters. The complex propagation constants γ_π and γ_c are defined in (5.70 and 5.71).

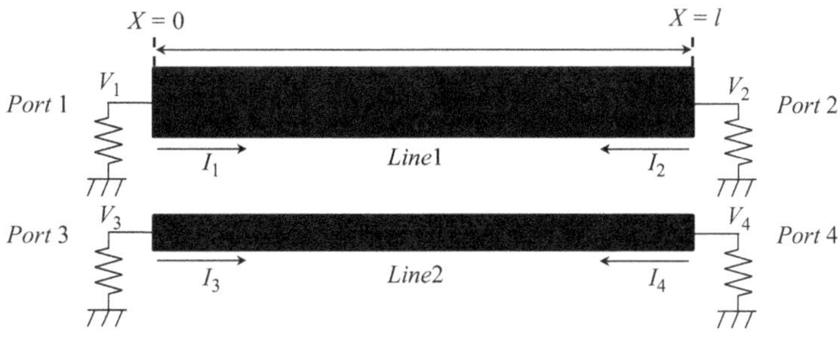

Figure 5.11 A four-port network showing asymmetric coupled lines

Table 5.1 Normal-mode parameters of asymmetrical coupled lines [1]

π-mode	**c-mode**
$Z_{\pi,\text{line1}} = \frac{V^+_{\pi,\text{line1}}}{I^+_{\pi,\text{line1}}} = \frac{V^-_{\pi,\text{line1}}}{I^-_{\pi,\text{line1}}}$	$Z_{c,\text{line1}} = \frac{V^+_{c,\text{line1}}}{I^+_{c,\text{line1}}} = \frac{V^-_{c,\text{line1}}}{I^-_{c,\text{line1}}}$
$Z_{\pi,\text{line2}} = \frac{V^+_{\pi,\text{line2}}}{I^+_{\pi,\text{line2}}} = \frac{V^-_{\pi,\text{line2}}}{I^-_{\pi,\text{line2}}}$	$Z_{c,\text{line2}} = \frac{V^+_{c,\text{line2}}}{I^+_{c,\text{line2}}} = \frac{V^-_{c,\text{line2}}}{I^-_{c,\text{line2}}}$
$R_\pi = \frac{V^+_{\pi,\text{line2}}}{V^+_{\pi,\text{line1}}} = \frac{V^-_{\pi,\text{line2}}}{V^-_{\pi,\text{line1}}}$	$R_c = \frac{V^+_{c,\text{line2}}}{V^+_{c,\text{line1}}} = \frac{V^-_{c,\text{line2}}}{V^-_{c,\text{line1}}}$
γ_π	γ_c

The two modes are related in terms of their currents and voltages as follows:

$$R_\pi = \frac{V^+_{\pi,\text{line2}}}{V^+_{\pi,\text{line1}}} = \frac{V^-_{\pi,\text{line2}}}{V^-_{\pi,\text{line1}}} = -\frac{I^+_{c,\text{line1}}}{I^+_{c,\text{line2}}} = -\frac{I^-_{c,\text{line1}}}{I^-_{c,\text{line2}}} \tag{5.38}$$

and

$$R_c = \frac{V^+_{c,\text{line2}}}{V^+_{c,\text{line1}}} = \frac{V^-_{c,\text{line2}}}{V^-_{c,\text{line1}}} = -\frac{I^+_{\pi,\text{line1}}}{I^+_{\pi,\text{line2}}} = -\frac{I^-_{\pi,\text{line1}}}{I^-_{\pi,\text{line2}}} \tag{5.39}$$

Equations (5.38) and (5.39) can be used to relate the ratio parameters R_c and R_π to the mode characteristic impedances as:

$$R_c R_\pi = -\frac{Z_{\pi,\text{line2}}}{Z_{\pi,\text{line1}}} = -\frac{Z_{c,\text{line2}}}{Z_{c,\text{line1}}} \tag{5.40}$$

In the case of identical transmission lines, $R_c = 1$, and $R_\pi = -1$. From (5.40) one can conclude that only six quantities out of the eight are required to characterize asymmetric coupled lines. In a similar manner to the even- and odd-mode analysis, voltages and currents on the two coupled lines can be expressed as an algebraic sum of forward and backward traveling waves of the c- and π-mode as follows:

$$V_{\text{line1}}(z) = A_1 e^{-\gamma_c z} + A_2 e^{\gamma_c z} + A_3 e^{-\gamma_\pi z} + A_4 e^{\gamma_\pi z} \tag{5.41}$$

$$V_{\text{line2}}(z) = A_1 R_c e^{-\gamma_c z} + A_2 R_c e^{\gamma_c z} + A_3 R_\pi e^{-\gamma_\pi z} + A_4 R_\pi e^{\gamma_\pi z} \tag{5.42}$$

$$I_{\text{line1}}(z) = \frac{A_1}{Z_{c,\text{line1}}} e^{-\gamma_c z} - \frac{A_2}{Z_{c,\text{line2}}} e^{\gamma_c z} + \frac{A_3}{Z_{\pi,\text{line1}}} e^{-\gamma_\pi z} - \frac{A_4}{Z_{\pi,\text{line2}}} e^{\gamma_\pi z} \tag{5.43}$$

$$I_{\text{line2}}(z) = \frac{A_1 R_c}{Z_{c,\text{line1}}} e^{-\gamma_c z} - \frac{A_2 R_c}{Z_{c,\text{line2}}} e^{\gamma_c z} + \frac{A_3 R_\pi}{Z_{\pi,\text{line1}}} e^{-\gamma_\pi z} - \frac{A_4 R_\pi}{Z_{\pi,\text{line2}}} e^{\gamma_\pi z} \tag{5.44}$$

where A_1 and A_3 are the phasor coefficients of the forward propagating c- and π-modes, respectively, and A_2 and A_4 are the phasor coefficients of the backward propagating c- and π-modes, respectively. The Z-parameters of the four-port network in Figure 5.11 can be found in terms of the c- and π-mode impedances as:

$$\begin{aligned} &Z_{11} = Z_{22},\ Z_{33} = Z_{44} \\ &Z_{12} = Z_{21},\ Z_{34} = Z_{43} \\ &Z_{13} = Z_{31},\ Z_{24} = Z_{42},\ Z_{14} = Z_{41},\ Z_{23} = Z_{32} \end{aligned} \tag{5.45}$$

where the elements of a 4×4 Z-matrix can be populated using the following set of equations [1]:

$$Z_{11} = \frac{Z_{c1} \coth \gamma_c l}{(1 - R_c/R_\pi)} + \frac{Z_{\pi 1} \coth \gamma_\pi l}{(1 - R_\pi/R_c)} \tag{5.46}$$

$$Z_{33} = \frac{Z_{c1} R_c^2 \coth \gamma_c l}{(1 - R_c/R_\pi)} + \frac{Z_{\pi 1} R_\pi^2 \coth \gamma_\pi l}{(1 - R_\pi/R_c)} \tag{5.47}$$

$$Z_{12} = \frac{Z_{c1}}{(1 - R_c/R_\pi)\sinh\gamma_c l} + \frac{Z_{\pi 1}}{(1 - R_\pi/R_c)\sinh\gamma_\pi l} \tag{5.48}$$

$$Z_{13} = \frac{Z_{c1}R_c \coth\gamma_c l}{(1 - R_c/R_\pi)} + \frac{Z_{\pi 1}R_\pi \coth\gamma_\pi l}{(1 - R_\pi/R_c)} \tag{5.49}$$

$$Z_{14} = \frac{Z_{c1}R_c}{(1 - R_c/R_\pi)\sinh\gamma_c l} + \frac{Z_{\pi 1}R_\pi}{(1 - R_\pi/R_c)\sinh\gamma_\pi l} \tag{5.50}$$

$$Z_{34} = \frac{Z_{c1}R_c^2}{(1 - R_c/R_\pi)\sinh\gamma_c l} + \frac{Z_{\pi 1}R_\pi^2}{(1 - R_\pi/R_c)\sinh\gamma_\pi l} \tag{5.51}$$

5.3.3.1 Approximate normal-mode parameters

The analysis of asymmetric coupled lines through the normal-mode parameters are somewhat complicated and requires lengthy field-theoretical analysis in order to obtain such parameters. Therefore, Ikalainen and Matthaei [4] have presented an approximation technique that works best for weakly coupled lines and uses the even- and odd-mode method on approximated symmetrical coupled lines derived from the asymmetric structure in order to find the distributed line parameters (i.e., both lines self and mutual inductance and capacitance) and consequently the normal-mode parameters. The steps of finding these parameters are summarized as follows: Step 1, under the assumption that the mutual inductance and capacitance between the asymmetrical line configuration shown in Figure 5.12(a) is the same as that of the symmetrical lines of width $(W_1 + W_2)/2$ separated by a distance S as depicted in Figure 5.12(b), the even- and odd-mode impedances and effective dielectric constants are calculated for this case of symmetrical coupled lines. Step 2, the self-inductance and self-capacitance of line 1 in the presence of line 2 can be calculated with the assumption that line 2 has the same width as line 1 as shown in Figure 5.12(c) and thus the even- and odd-mode parameters for the symmetrical

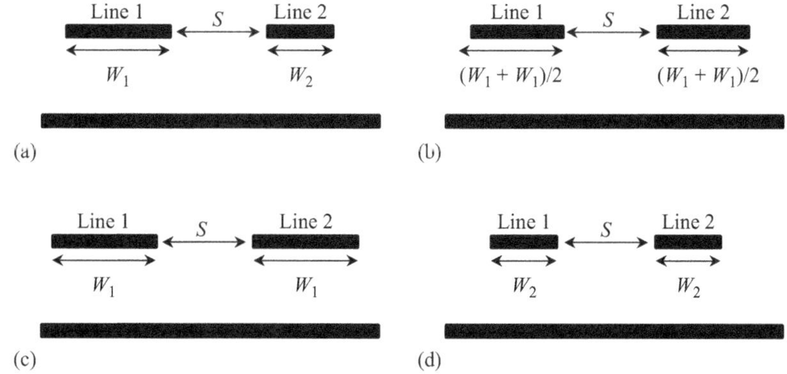

Figure 5.12 Four configurations used to approximate the analysis of the asymmetric coupled line in (a) using the symmetrical coupled lines in (b), (c), and (d)

coupled lines of width W_1 can be calculated. Step 3, in a similar fashion assuming the width of line 1 equals to that of line 2, the self-inductance and self-capacitance of line 2 can be calculated through the even- and odd-mode parameters of symmetrical coupled lines of width W_2 as depicted in Figure 5.12(d).

From [4] let $L_1 = L_2$ and $C_1 = C_2$ under the assumption of symmetric coupled lines, the self and mutual distributed parameters can be obtained as follows:

$$L_{\text{line},i} = \frac{1}{2\omega}(\beta_{\text{even}} Z_{\text{even}} + \beta_{\text{odd}} Z_{\text{odd}}) \tag{5.52}$$

$$C_{\text{line},i} = \frac{1}{2\omega}(\beta_{\text{odd}}/Z_{\text{odd}} + \beta_{\text{even}}/Z_{\text{even}}) \tag{5.53}$$

$$L_m = \frac{1}{2\omega}(\beta_{\text{even}} Z_{\text{even}} - \beta_{\text{odd}} Z_{\text{odd}}) \tag{5.54}$$

$$C_m = \frac{1}{2\omega}(\beta_{\text{odd}}/Z_{\text{odd}} - \beta_{\text{even}}/Z_{\text{even}}) \tag{5.55}$$

and $i = 1$ or 2 refers to the line number.

The inductive and capacitive coupling coefficients which are the ratio between the transferred energy between the lines to the total energy stored in the lines and can be defined as [5]:

$$K_L = \frac{L_m}{\sqrt{L_{\text{line1}} L_{\text{line2}}}} \tag{5.56}$$

$$K_C = \frac{C_m}{\sqrt{C_{\text{line1}} C_{\text{line2}}}} \tag{5.57}$$

Upon finding the distributed elements of the asymmetric coupled lines, one can calculate the normal mode parameters as described by [6] as follows:

$$Z_{c,line1} = \frac{1}{Y_{c,line1}} = \frac{1}{\gamma_c} \frac{z_1 z_2 - z_m^2}{z_2 - z_m R_c} \tag{5.58}$$

$$Z_{c,\text{line2}} = \frac{1}{Y_{c,\text{line2}}} = \frac{R_c}{\gamma_c} \frac{z_1 z_2 - z_m^2}{z_1 R_c - z_m} \tag{5.59}$$

$$Z_{\pi,\text{line1}} = \frac{1}{Y_{\pi,\text{line1}}} = \frac{1}{\gamma_\pi} \frac{z_1 z_2 - z_m^2}{z_2 - z_m R_\pi} \tag{5.60}$$

$$Z_{\pi,\text{line2}} = \frac{1}{Y_{\pi,\text{line2}}} = \frac{R_\pi}{\gamma_\pi} \frac{z_1 z_2 - z_m^2}{z_1 R_\pi - z_m} \tag{5.61}$$

The voltage and current ratio parameters for each of the normal modes are defined as:

$$R_c = \frac{(a_2 - a_1) + \sqrt{(a_2 - a_1)^2 + 4b_1 b_2}}{2b_1} \tag{5.62}$$

$$R_\pi = \frac{(a_2 - a_1) - \sqrt{(a_2 - a_1)^2 + 4b_1 b_2}}{2b_1} \tag{5.63}$$

where

$$a_1 = y_1 z_1 + y_m z_m \tag{5.64}$$

$$a_2 = y_2 z_2 + y_m z_m \tag{5.65}$$

$$b_1 = y_m z_1 + y_2 z_m \tag{5.66}$$

$$b_2 = y_m z_2 + y_1 z_m \tag{5.67}$$

and

$$\begin{aligned} z_1 &= j\omega L_{\text{line},1} \\ z_2 &= j\omega L_{\text{line},2} \\ z_m &= j\omega L_{\text{line},m} \end{aligned} \tag{5.68}$$

$$\begin{aligned} y_1 &= j\omega C_{\text{line},1} \\ y_2 &= j\omega C_{\text{line},2} \\ y_m &= -j\omega C_{\text{line,m}} \end{aligned} \tag{5.69}$$

Finally, the complex propagation constant of each mode can be calculated as:

$$\gamma_c^2 = \frac{a_1 + a_2}{2} + 0.5\sqrt{(a_1 - a_2)^2 + 4b_1 b_2} \tag{5.70}$$

$$\gamma_\pi^2 = \frac{a_1 + a_2}{2} - 0.5\sqrt{(a_1 - a_2)^2 + 4b_1 b_2} \tag{5.71}$$

5.3.4 Additivity of electric and magnetic coupling

In applications of crosstalk analysis, it has been demonstrated that under certain conditions the additivity property of electric and magnetic field coupling [7] can be applied to represent the EM coupling effects. In return, this aids in identifying the dominant coupling type and consequently the proper shielding method. The additivity property of coupled lines accommodates the simultaneous inductive and capacitive coupling due to multiple noise sources that can be caused by multiple coupled lines.

Considering the case of symmetric parallel coupled lines represented by the per unit length RLC model cell depicted in Figure 5.13, the even- and odd-mode characteristic impedances and propagation constants in terms of R_{11}, C_{11}, and L_{11} which are, respectively, self-resistance, self-capacitance, and self-inductance of one of the lines are expressed as:

$$Z_{\text{even}} = \sqrt{\frac{R_{11} + j(L_{11} + L_{12})\omega}{jC_{11}\omega}} \tag{5.72}$$

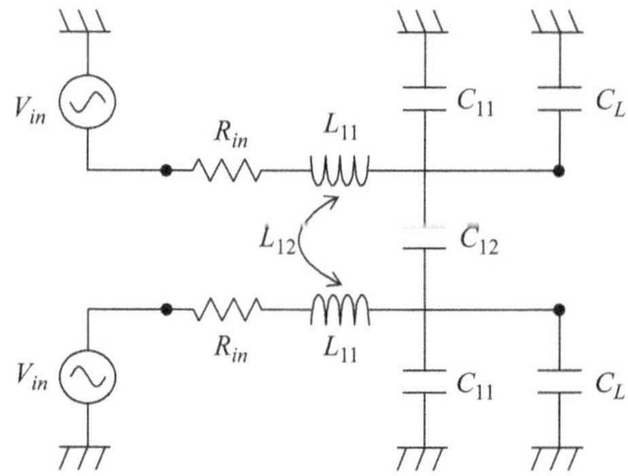

Figure 5.13 Equivalent RLC model of a coupled transmission line terminated with a capacitive load, C_L

$$Z_{odd} = \sqrt{\frac{R_{11} + j(L_{11} - L_{12})\omega}{j(C_{11} + 2C_{12})\omega}} \tag{5.73}$$

$$\gamma_{even} = \sqrt{(R_{11} + j(L_{11} + L_{12})\omega)jC_{11}\omega} \tag{5.74}$$

$$\gamma_{odd} = \sqrt{(R_{11} + j(L_{11} + L_{12})\omega)j(C_{11} + 2C_{12})\omega} \tag{5.75}$$

C_{12} and L_{12} are the mutual capacitance and inductance, respectively. Signals traveling in the coupled transmission line can be expressed as the linear combination of the even and odd mode voltages [8] as:

$$V_{coupled} = \frac{V_{even}(\text{RLC}) - V_{odd}(\text{RLC})}{2} \qquad 5.76$$

To calculate the mode voltages, assumptions and approximations on the V_{even} and V_{odd} are employed to allow simple expressions with tolerable approximation error. For inductive coupling only, C_{12} is assumed to be equal to zero, and similarly, for capacitive coupling only, L_{12} is equal to zero. In addition, microelectronic interconnections are electrically short lines and can be thought of as an *RC* distributed model where L_{11} and L_{12} both can be neglected. Therefore, the condition $R_{11} >> (L_{11} - L_{12}).\omega$ can be assumed in the odd-mode. With this assumption, (5.72)–(5.75) can be rewritten for only the capacitive model as:

$$Z_{even}(C) = \sqrt{\frac{R_{11}}{jC_{11}\omega}} \tag{5.77}$$

$$Z_{odd}(C) = \sqrt{\frac{R_{11}}{j(C_{11} + 2C_{12})\omega}} \tag{5.78}$$

$$\gamma_{even}(C) = \sqrt{jR_{11}C_{11}\omega} \tag{5.79}$$

$$\gamma_{odd}(C) = \sqrt{jR_{11}(C_{11} + 2C_{12})\omega} \tag{5.80}$$

Similarly, by eliminating the capacitive coupling effects, (5.72)–(5.75) for the inductive only model can be rewritten as:

$$Z_{\text{even}}(L) = \sqrt{\frac{R_{11} + j(L_{11} + L_{12})\omega}{jC_{11}\omega}} \tag{5.81}$$

$$Z_{\text{odd}}(L) = \sqrt{\frac{R_{11}}{jC_{11}\omega}} \tag{5.82}$$

$$\gamma_{\text{even}}(L) = \sqrt{(R_{11} + j(L_{11} + L_{12})\omega)jC_{11}\omega} \tag{5.83}$$

$$\gamma_{\text{odd}}(L) = \sqrt{(R_{11} + j(L_{11} + L_{12})\omega)jC_{11}\omega} \tag{5.84}$$

From the characteristic impedances of both the even- and odd-modes in each of the inductive and capacitive models, it can be observed that:

$$V_{\text{even}}(\text{RLC}) = V_{\text{even}}(L) \tag{5.85}$$

$$V_{\text{odd}}(\text{RLC}) = V_{\text{odd}}(C) \tag{5.86}$$

$$V_{\text{even}}(C) = V_{\text{odd}}(L) \tag{5.87}$$

By substituting (5.85)–(5.87) into (5.76), the coupled voltage can be represented by decoupling the capacitive and inductive coupling voltages as:

$$\begin{aligned} V_{\text{coupled}}(RLC) &= \frac{V_{\text{even}}(L) - V_{\text{odd}}(C)}{2} \\ &= \frac{V_{\text{even}}(C) - V_{\text{odd}}(C) + V_{\text{even}}(L) - V_{\text{odd}}(L)}{2} \\ &= V_{\text{coupled}}(C) + V_{\text{coupled}}(L) \end{aligned} \tag{5.88}$$

While the discussed analysis in this section focused on one coupled line structure, additivity of EM coupling in multicoupled lines has also been reported recently in [9] by following a similar approach, analytical RLC-coupled line models. However, due to the much aggressive approximation followed in developing the model derivation, the multicoupled line models may have limitations in their accuracy when compared to simple two coupled lines.

5.4 EM fields and shielding

5.4.1 Electromagnetic field boundary conditions

Understanding the governing rules of the electric and magnetic boundary conditions, especially, on conductive surfaces is key to developing the proper absorbing wave boundaries in computational electromagnetic codes and the physical realization of EM shields in RF modules. Electric and magnetic field vectors impinging upon an interface between two different media can be decomposed into their orthogonal (tangential and normal) components. The conditions that the fields must

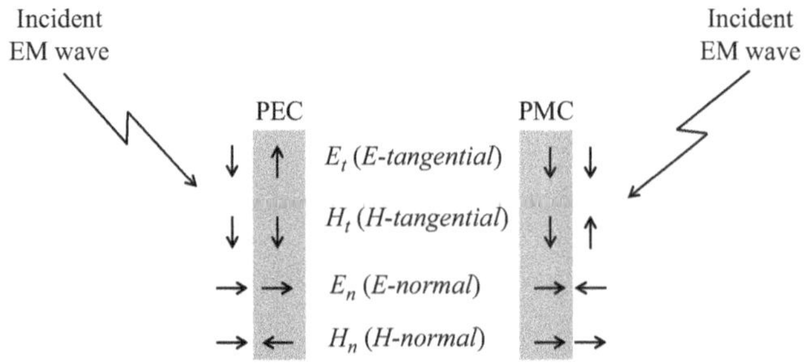

Figure 5.14 Illustration of the electric and magnetic field vectors on PMC and PEC interface

satisfy at the interface between the two media are called the boundary conditions. To illustrate the fundamental rules and behavior of electric and magnetic field vectors along a metallic interface, consider Figure 5.14, which depicts the directions of the field components on two types of conductive interfaces: a perfect electric conductor (PEC) and a perfect magnetic conductor (PMC). A PEC is an ideal conducting material with conductivity equals to infinity. Practically good conductors such as copper, gold, and aluminum can be assumed PECs. On the other hand, there is no physical material that has yet been discovered that behaves as a PMC; however, the concept is mathematically beneficial for applying symmetry conditions (e.g., defining a plane of symmetry of a magnetic wall as part of the even-, odd-mode analysis) and for developing equivalence and duality theorems.

The boundary conditions relating the field components on the two sides of the interface can be inferred from Maxwell's integral equations as derived in [10] resulting in the following general four boundary conditions assuming sources with electric and magnetic surface currents, J_s, J_{ms}, and electric and magnetic surface charges, ρ_s, ρ_{ms}, respectively:

$$\vec{E}_{t1} - \vec{E}_{t1} = \hat{n} \times \left(\vec{E}_1 - \vec{E}_2\right) = -J_{ms} \tag{5.89}$$

$$\vec{H}_{t1} - \vec{H}_{t2} = \hat{n} \times \left(\vec{H}_1 - \vec{H}_2\right) = J_s \tag{5.90}$$

$$\vec{D}_{n1} - \vec{D}_{n2} = \left(\vec{D}_1 - \vec{D}_2\right) \cdot \hat{n} = \rho_s \tag{5.91}$$

$$\vec{B}_{n1} - \vec{B}_{n2} = \left(\vec{B}_1 - \vec{B}_2\right) \cdot \hat{n} = \rho_{ms} \tag{5.92}$$

where $\hat{n}$ is the unit vector normal to the boundary surface pointing from medium 1 to medium 2.

The electric boundary conditions are determined by Gauss' law and Maxwell—Faraday's law for the electric field, while the magnetic boundary conditions are determined by Gauss' law for the magnetic field and Ampere's law. Equation (5.91) implies that the boundary condition appropriate for a surface of field discontinuity containing an electric surface charge. Similarly, the condition in (5.92) is appropriate for a surface of field discontinuity containing a magnetic surface charge. Both integral and differential forms of Maxwell's equations are related through Stokes' Theorem and Gauss' Theorem. Therefore, one can then use either form according to the problem being solved. Static problems simplify Maxwell's equations into decoupled electrostatic and magnetostatic equations.

In the case of nonconductive media with two regions comprising distinct permittivities and permeabilities, the tangential electric and magnetic field components on one side of the material boundary are continuous and equal to the tangential component on the other side of the boundary. The normal components of the electric and the magnetic field vectors are discontinuous at the boundary surface as they rely upon the media relative permittivity and permeability, respectively. The boundary conditions when assuming no surface charge density at the boundary can be summarized as follows:

$$\frac{\vec{H}_{1n}}{\vec{H}_{2n}} = \frac{\mu_{r2}}{\mu_{r1}} \quad \text{and} \quad \vec{B}_{1n} = \vec{B}_{2n} \tag{5.93}$$

$$\vec{H}_{1t} = \vec{H}_{2t} \quad \text{and} \quad \frac{\vec{B}_{1n}}{\vec{B}_{2n}} = \frac{\mu_{r1}}{\mu_{r2}} \tag{5.94}$$

$$\frac{\vec{E}_{1n}}{\vec{E}_{2n}} = \frac{\varepsilon_{r2}}{\varepsilon_{r1}} \quad \text{and} \quad \vec{D}_{1n} = \vec{D}_{2n} \tag{5.95}$$

$$\vec{E}_{1t} = \vec{E}_{2t} \quad \text{and} \quad \frac{\vec{D}_{1t}}{\vec{D}_{2t}} = \frac{\varepsilon_{r1}}{\varepsilon_{r2}} \tag{5.96}$$

For a PEC, the field within the material is zero by definition, and the electric charge can reside only on the surface of the conductor. Thus, in the case of a PEC, J_{ms} and ρ_{ms} are zeros so that the boundary conditions are reduced to:

$$\hat{n} \times \vec{E} = 0 \tag{5.97}$$

$$\hat{n} \times \vec{H} = J_s \tag{5.98}$$

$$\vec{D} \times \hat{n} = \rho_s \tag{5.99}$$

$$\vec{B} \times \hat{n} = 0 \tag{5.100}$$

In the case of a PMC, J_s and ρ_s are zeros, and the boundary conditions can be simplified as:

$$\hat{n} \times \vec{E} = -J_{ms} \tag{5.101}$$

$$\hat{n} \times \vec{H} = 0 \tag{5.102}$$

$$\vec{D} \times \hat{n} = 0 \tag{5.103}$$

$$\vec{B} \cdot \hat{n} = \rho_{ms} \tag{5.104}$$

In both the PEC and PMC cases, the normal vector, $\hat{n}$, points out of the conductor and into the adjacent region of nonzero fields. From the constraints provided by the boundary conditions for a medium on the EM field components and according to Faraday's law, the total tangential components of the E-field intensity vector on a PEC surface and the total magnetic field intensity vector across a PMC surface are zeros, respectively.

5.4.2 Near-, transition-, and far-field regions

The region around a radiating element (e.g., antenna) can be divided into two primary zones: near- and far-zone. Some of the engineers and researchers have interest in understanding the behavior of EM-fields at the near-zone, and as a result, in some of the literature the near-zone is segmented into two additional regions: reactive (nonradiative) and transitioning-zone (radiative). Those field boundaries are not fixed in space and can move closer or farther from the radiating source depending on the frequency of the propagating wave and the amount of error that an application can tolerate. Depending on how far the EM fields are from the radiating source, their intensities decrease inversely with distance, in addition, to the change in their characteristics. With a sufficient distance from the radiator, E- and H-field components become perpendicular and appear as a planar-wave to an observation point located in the far-field region.

Figure 5.15 shows the two main regions associated with a radiating source divided in terms of the source wavelength and assuming a surface type antenna. For a wire type antenna, the far-field can be assumed at a distance of three times the

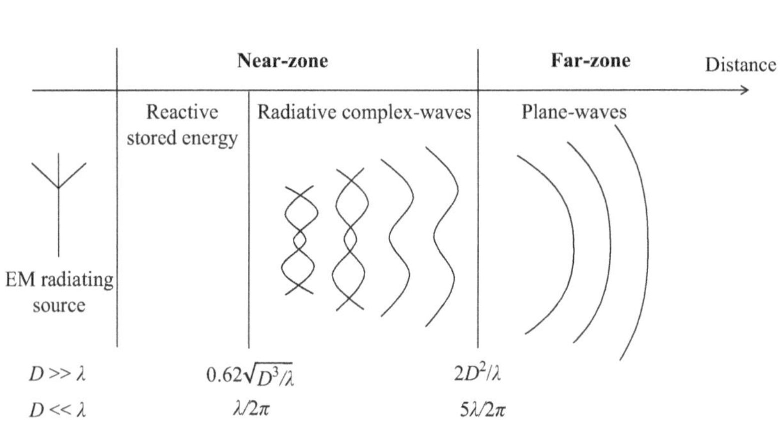

Figure 5.15 Field regions of an EM radiating source illustrating two principal zones as a function of wavelength. The maximum linear dimension (D) of the radiating source aperture is for large antennas where $D >> \lambda$ and for electromagnetically small antennas $D << \lambda$ to resemble a spherical wave from a point source. The length, D, of a small antenna is not important and field zones can be approximated directly based on the operating wavelength

operating wavelength. As was pointed out, there is no distinct boundary at which the field characteristics start to change as it depends on the EM environment surrounding the radiating element and its effective aperture. Among the many definitions that describe these regions, the one presented in Figure 5.15 is the most common model used to distinguish between the field zones.

In the near-zone reactive region, *E*- and *H*-fields are the strongest among other regions and typically one of the fields dominate depending on the type of the radiating element. While fields are confined in the reactive region, it tends to start radiating in the radiative region, also called the Fresnel-zone or transition-zone. In the far-field region, the radiation pattern (i.e., angular distribution of the energy) doesn't change significantly with distance; hence, it is also called plane-waves. Plane-waves are easy to analyze, unlike near radiative fields that exhibit complex radiation patterns and change rapidly with distance.

According to Schelkunoff's derivations of Maxwell equations for an infinitesimal electric dipole and magnetic loop antennas [11], each of the three field regions is dominated by a distance term as follows: fields at near-zone are significantly influenced by the $1/r^3$ term, the transitioning-zone is impacted mostly by the $1/r^2$ term, and the far-zone is dominated by $1/r$, where r is the distance from the radiating source to an observation point. This is logical since the terms $1/r^3$ and $1/r^2$ attenuate rapidly with increasing distance and, therefore, the $1/r$ term dominates in the far-field attenuation.

Although analysis of far-fields is of interest to many engineers who are working on the design of antennas for long-distance wireless communication systems, analysis of near-fields yet is an interesting topic for computational EM engineers, for example, those looking for methods to efficiently extrapolate the far-fields from near-field data through Green's function in cases where modeling fields in the far-zone leads to prohibitive computational overhead. To some extent, near-fields are also studied by electromagnetic compatibility (EMC) engineers to understand the coupling mechanism in PCB interconnects and more importantly between the integrated transceiver modules in a system.

5.4.3 Electromagnetic shielding

Generally, electronic systems such as wireless devices, radar systems, and television stations act as potential sources of EMI. To mitigate external intentional and unintentional EMI sources from penetrating a device or to prevent internal RF emissions from escaping a device, metallic enclosures of electronic circuits and modules are key to address EMI issues [11]. Regardless the size of the electronic system, shielding against EMI is of interest to both military and commercial applications. The main differences are the shielding frequency range and the higher shielding levels required for military use in comparison to the requirements for commercial applications. Ideally, for high levels of attenuation, the shield must have no penetrations (e.g., seams, slots, holes, and penetrating cables); however, this may not be feasible for many applications.

Microscale electronics (e.g., RFIC modules) are eventually integrated into a larger system to enable, for example, wireless communication functionality or other intended tasks. Due to the congestion of modern communication systems with multifunctional RF modules, it is important to ensure their reliability when sharing a confined space in a system. EM shielding is thus a fundamental and effective way to mitigate undesired EM radiated emissions between the integrated modules. An RF module can be fully enclosed or partially enclosed (perforated) by the EM shield. To achieve the desired hardness against EMI several aspects need to be considered such as:

- Operating frequency;
- Enclosure Q-factor;
- Enclosure size and shape;
- Number of apertures or seams;
- The dielectric fill density of the enclosure;
- Characteristics of the EM fields (near- or far-field); and
- The thickness of the shield.

The frequency of operation is a primary feature of the EM environment that determines the number of possible modes supported within the shield cavity. The cavity Q-factor would impact the insertion loss and multipath effects where high Q tends to enhance multipath. The number and shape of apertures will impact the coupling type. The cavity fill will determine the nature of loss and amount of scattering in the volume under consideration. Shielding cavity, if sufficiently large, could cause cavity resonances indicating a possible EMI problem. The choice of electric versus magnetic shields, the shapes/sizes of ground and signal planes, driver current amplitudes, and so forth, can be determined during the design and optimization phases of the layout.

Shielding effectiveness (SE) is a useful expression that is used to describe the shielding material immunity. SE is a positive number and the higher the SE value, the better the shield. As it will be discussed next, SE is a function of the shielding material parameters, the frequency of operation, and the angles of the incident fields. The exact expression of SE at an arbitrary point located at the far-field can be calculated in terms of the ratio of the incident and transmitted electric and magnetic field strengths as follows:

$$\mathrm{SE}(E)_{\mathrm{dB}} = 20\log_{10}\left(\frac{|E_{\mathrm{inc}}|}{|E_{\mathrm{trans}}|}\right) \tag{5.105}$$

$$\mathrm{SE}(H)_{\mathrm{dB}} = 20\log_{10}\left(\frac{|H_{\mathrm{inc}}|}{|H_{\mathrm{trans}}|}\right) \tag{5.106}$$

where E_{inc} and H_{inc} are the amplitudes of the incident electric and magnetic field intensities measured at a point in the absence of the shield. E_{trans} and H_{trans} are the field quantities at the same point, however, with the shield interposed. It is worth pointing out that SE(E) and SE(H) are equal for far-field or plane-waves and unequal for the near-field. The SE expression in (5.105) and (5.106) under the far-field condition assumes a good shielding conductor possessing very low intrinsic impedance (η_{shield}) than the free space (η_0) with a thickness (t) much greater than the skin depth (δ) of the shield metal. The SE in terms of these parameters can be approximated as:

$$\mathrm{SE}(E)_{\mathrm{dB}} = \mathrm{SE}(H)_{\mathrm{dB}} \cong 20\log_{10}\left(\left|\frac{\eta_0}{|4\eta_{\mathrm{shield}}|}\right|e^{t/\delta}\right) \tag{5.107}$$

Plane-wave SE can also be described as the sum of three loss terms; each term is calculated in decibel. The terms are: penetration or absorption loss (A), the first reflection loss at the shield surface (R), and the internal multiple reflection loss within the shield (B). This decomposition is known as Schelkunoff's approach to the EM shield that is stemmed from the transmission line model and expressed as:

$$\mathrm{SE}_{\mathrm{dB}} = A + R + B \tag{5.108}$$

Indeed, there are plenty of scopes to develop other methods for characterizing the SE of composite materials by describing the reflection and absorption contributions to a shield by means of reflected and absorbed power densities which are easily related to quantities that can be evaluated through standard measurements. SE was analyzed by Schelkunoff using wave impedances in a similar fashion to that of the TL theory where he applied concepts such as reflection and transmission coefficients to the shielding analysis. The TL model provides an exact solution for the SE due to a normally incident plane-wave impinging on a planar material. However, this model can only provide approximate solutions in the case of non-plane waves or near-field sources. In such cases, careful attention must be paid to the evaluation of the wave impedance to ensure sufficient accuracy of the approximated SE result [12]. A planar wave which exists in the far-field has wave impedance similar to that of the free space (i.e., 377 ohms).

The below discussion will demonstrate in details the SE three terms identified by Schelkunoff.

Both magnetic and electric lossy materials can be used to absorb both radiated H- and E-fields. The absorption loss measured in decibel can be expressed by:

$$A(\mathrm{dB}) = 8.686(\mu\pi\sigma f)^{1/2}d \tag{5.109}$$

As can be noted from (5.109), the absorption loss is a function of the physical properties of the shielding material including permeability (μ), conductivity (σ), thickness (d) measured in meter, and the frequency of the incident wave (f) in Hertz. While an electric field shield relies mainly on the conductivity of the shield, a magnetic shield depends on the material permeability.

Next, the R parameter represents the loss due to the first reflection of the E- and H-fields (E_{ref} and H_{ref}) at the outer shield surface as depicted in Figure 5.16. The term R depends on the shape of the shield. Schelkunoff analyzed the *SE* for three shapes: planar, cylindrical, and spherical using the transmission line model. For planar and cylindrical shield shapes, R in decibel can be calculated as:

$$R(\mathrm{dB}) = 20\log_{10}\left(\frac{\sqrt{|k+1|}}{4|k|}\right) \tag{5.110}$$

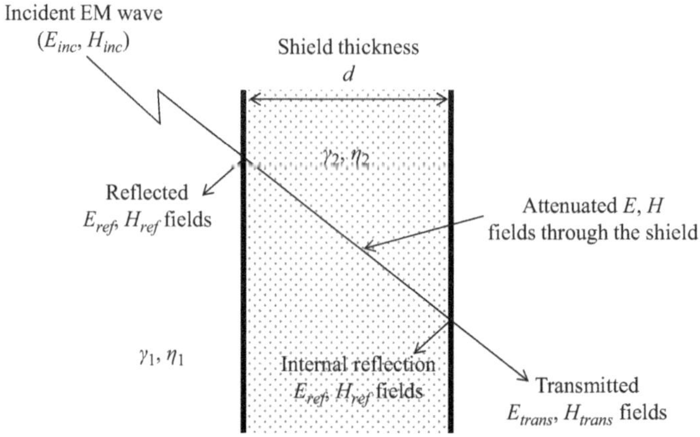

Figure 5.16 EM field reflection and absorption upon impinging on a shielding slab defined by its wave impedance (η_2) and propagation constant (γ_2). The media surrounding the shield is defined by η_1 and γ_1

and for a spherical shape, R is calculated as:

$$R(\text{dB}) = 20\log_{10}\left(\frac{(k+1)\left(\dfrac{k}{2}+1\right)}{3|k|}\right) \tag{5.111}$$

where k represents the intrinsic impedance ratio ($\eta_{\text{shield}}/\eta_0$) of the shield to the incident wave impedance. k values much greater or less than one indicates large impedance mismatch and therefore large reflections. k for the E-field can be calculated through the following expression:

$$k = \frac{1}{2\pi df}\sqrt{\mu_0\varepsilon_0} \tag{5.112}$$

while k for the H-field has the following relationship:

$$k = 2\pi df\sqrt{\mu_0\varepsilon_0} \tag{5.113}$$

The B parameter is a correction term that represents the multiple internal reflections within the shield boundary. For shields with thick and good conductors, B is a negligible term, and for shield thicknesses that are thin compared to their skin depth, the B term is a negative value (in dB) which can significantly reduce the shielding effectiveness. By ignoring the B term, the shielding equation in (5.108) reduces to $\text{SE} = A + R$.

By calculating the ratio of the E- and H-fields of infinitesimal electric and magnetic sources, respectively, one can calculate the impedance of the wave. Figure 5.17 depicts the opposing wave impedance behavior between a short dipole antenna (electric source) and a small loop antenna (magnetic source) in space and at a single

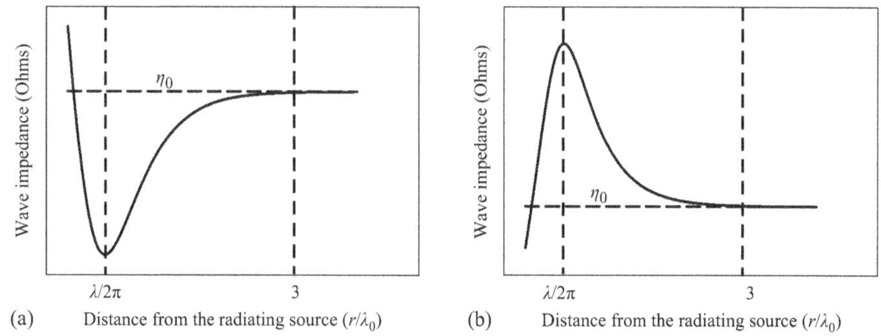

Figure 5.17 Wave impedance behavior as a function of distance generated by (a) an electric Hertzian dipole and (b) an infinitesimal magnetic dipole

frequency as a function of the distance indicated by the operating wavelength. The results in this graph help in understanding the reason why SE(E) $\neq$ SE(H) at the near-field region, while approximately equal for waves observed at the far-field. At a distance of three times the wavelength, both wave impedances merge to that of the free space. The free space wave impedance is calculated using the free space permeability (μ_0) and permittivity (ε_0) as:

$$\eta_0 = \sqrt{\frac{\mu_0}{\varepsilon_0}} \tag{5.114}$$

At the far-field region, the waves generated from these radiating sources are almost planar, and both electric and magnetic fields possess orthogonal components in the direction of the wave propagation.

In the near-field region, the wave impedance of the electric and the magnetic sources are different. The magnetic source generates a low impedance field and increases to the free space impedance as the distance increases to the far-field region. In contrast, the electric source generates a high impedance field and decreases to the free space impedance as the distance increases [11]. Both electric and magnetic wave impedances exhibit minima and maxima, respectively, at a distance equal to approximately $\lambda/2\pi$.

The reflection loss is the main contributor to the shielding mechanism for low frequency plane-waves, whereas at high frequency the absorption loss is more predominant. Concerning RF module shielding to mitigate mostly near-field coupling, Figure 5.17 provides a useful insight into identifying the prober shield for the proper wave or source type. At near-field, the EM wave is a complex combination of both E- and H-fields, and usually, it is easy to shield against the E-field in a similar fashion to that used in far-field, unlike the H-field. Near-field magnetic coupling is difficult to mitigate at low frequencies since the reflection and absorption losses are quite small. Alternative means of mitigating the magnetic coupling can be done through diverting the magnetic flux using high permeability materials such as nickel (Ni), or by implementing a conductor loop such that the incident magnetic field on the loop induces a magnetic flux opposing to the original field and thereby results in reduced coupling.

5.4.4 Interconnects shielding by via fence

Crosstalk is a well known EMC and SI problem, especially, in RFICs comprising digital circuits, DC lines, and interconnects that are placed in proximity to each other within a compact module package. Crosstalk is unwanted EM coupling that could catastrophically affect the reliable operation and SI of both digital and analog circuits. SI refers to ensuring that a signal, when sent through a pair of traces, will arrive at the receive side with the desired level, timing, and wave-shape. For instance, crosstalk, when occurs, may impact the efficiency, phase, stability, and noise figure of active devices in a module. It can also cause frequency response degradation including insertion loss and bandwidth. Usually, it is very difficult to determine the actual location of the coupling sources on the module that causes crosstalk. The amount of coupling is dependent on the physical dimensions of the coupled structures, their separation, the operating frequency, and the mode of operation. Normally, the first line of defense is to mitigate the coupling source by reducing the current strength, slewing the rise/fall time of the digital signal (e.g., by adding series resistors or ferrites), increasing the spacing between the traces, or shortening/straightening the high-frequency trace. Such countermeasures may provide a limited improvement, and may not even be an option for given design constraints. Another way of reducing crosstalk is to exploit the inherent parasitic effects (i.e., mutual capacitance and mutual inductance) among the coupled interconnects. Usually increasing the capacitive coupling along the coupled surface of the microstrip traces is used to reduce far-end crosstalk [13,14]. Also, guard trace configurations can be employed as an electric shield between the coupled traces and can be used in single and multilayer substrate technologies. Figure 5.18(a) illustrates the structure of a two adjacent and uniform coupled traces, both of the same length and width. Figure 5.18(b) is the same configuration as in Figure 5.18(a), however, with a guard trace that is terminated at both ends by matched loads to prevent any standing waves. For additional shielding, one may also ground the guard trace with via fence (metal-filled via holes) as depicted in Figure 5.18(c). The diameter of each via and the spacing between them on the guard trace must follow a simple rule defined as [15]:

$$D < \frac{\lambda}{5}, \quad P \le 2D \tag{5.115}$$

where P and D are the via pitch and via diameter, respectively. λ is the operating wavelength. This rule is sufficient to ensure low coupling between the adjacent traces that are guarded with the via fence.

With the aid of full-wave EM simulation, results of the far-end crosstalk (i.e., S_{41}) over a sweep of frequency, from 1 to 6 GHz, demonstrates the difference between the uniform adjacent microstrip traces with and without a guard trace and with a guard trace grounded by a via fence. The guard trace configuration with via fence shows an improvement in reducing the crosstalk by about 4 dB, in average.

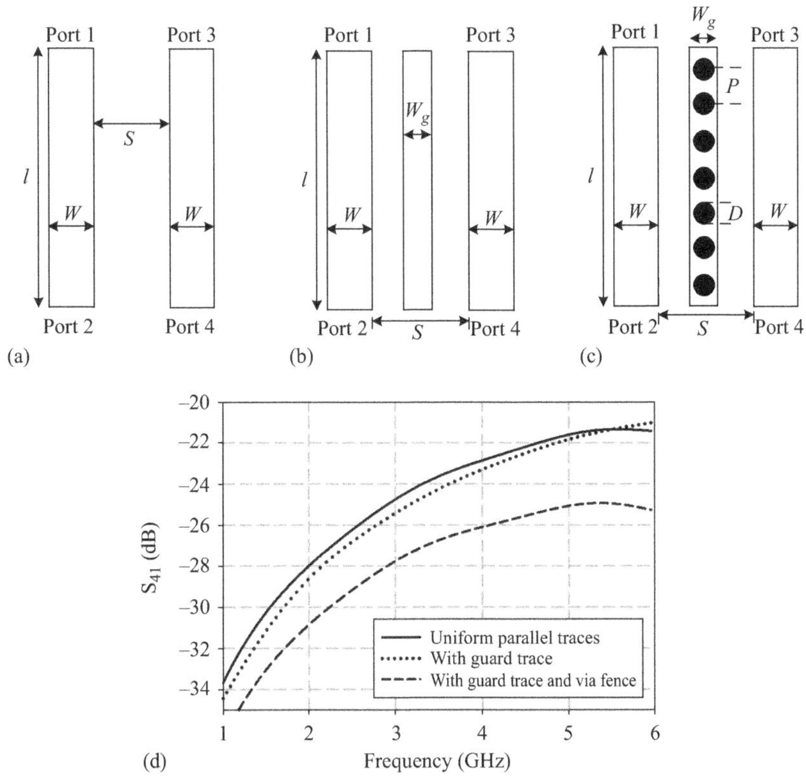

Figure 5.18 A pair of uniform coupled microstrip traces (a) without a guard trace; (b) with a guard trace; (c) with a guard trace grounded with via fence; and (d) the far-end crosstalk response (between port 1 and port 4) as a function of frequency for the two microstrip coupled lines with/without guard trace and via fence

The Far-end (i.e., forward coupling) crosstalk is of particular interest due to the stronger coupling as compared to that at the near-end (i.e., backward coupling) in cases when the traces are well-spaced and carrying signals with different phase velocities.

Additional grounded vias can be added to the guard trace to help further reduce the separation between the vias and thereby reduce crosstalk. However, in some cases, large number of vias degrades the SI of the signals on the adjacent traces and makes routing additional traces not possible, especially, in a crowded layout.

5.5 Anisotropic laminate materials

A medium in which an EM field propagates can be characterized by three media parameters (permittivity, permeability, and conductivity) which can be functions of position and time. In a linear isotropic medium, the characteristic of a propagating

EM wave is independent of their propagation direction where the medium has its characteristic parameters as scalar quantities. In an anisotropic medium, however, the permittivity and permeability are tensors.

In the design of RF circuits and modules, the knowledge of the precise substrate permittivity and permeability are critical for attaining accurate design and EM field analysis as it influences the relationships between the magnetic induction (B) and magnetic field intensity (H); and between the electric displacement field (D) and the electric field intensity (E) in an EM environment. Wave propagation in an isotropic media is where D and E are parallel, and the same holds for B and H. Hence, for a homogeneous isotropic dielectric and magnetic materials, the permittivity and permeability of a media are scalar quantities as expressed by the following constitutive relations:

$$D = \varepsilon_0 \varepsilon_r E \tag{5.116}$$

$$B = \mu_0 \mu_r H \tag{5.117}$$

where ε_0 and μ_0 are the permittivity and permeability in vacuum. ε_r and μ_r are the dimensionless relative permittivity and permeability of the medium.

The inhomogeneity of the substrate material used in RF modules occurs mostly due to the use of composition materials (e.g., composite Teflon and fiberglass or alumina). This causes two distinct possible polarization directions in anisotropic planar media where waves across these polarized directions propagate with different velocities. Assuming that the permeability and conductivity are scalar quantities, the permittivity tensor is expressed in a symmetric matrix form as:

$$\begin{bmatrix} D_x \\ D_y \\ D_z \end{bmatrix} = \varepsilon_0 \begin{bmatrix} \varepsilon_x & 0 & 0 \\ 0 & \varepsilon_y & 0 \\ 0 & 0 & \varepsilon_z \end{bmatrix} \begin{bmatrix} E_x \\ E_y \\ E_z \end{bmatrix} \tag{5.118}$$

where ε_x, ε_y, and ε_z are called the principal dielectric constants, and the orthogonal axes of the coordinate system are the principal axes of the medium. In a lossless material, all elements of the ε matrix are real, and the tensor is symmetric—meaning that $\varepsilon_{ij} = \varepsilon_{ji}$. One of the simplest types of anisotropy where two of the three diagonal permittivity tensor elements are equal and the third is different. In this case, the material is called a uniaxial dielectric.

Similarly, the permeability tensor is expressed in a symmetric matrix form as:

$$\begin{bmatrix} B_x \\ B_y \\ B_z \end{bmatrix} = \mu_0 \begin{bmatrix} \mu_x & 0 & 0 \\ 0 & \mu_y & 0 \\ 0 & 0 & \mu_z \end{bmatrix} \begin{bmatrix} H_x \\ H_y \\ H_z \end{bmatrix} \tag{5.119}$$

where μ_x, μ_y, and μ_z are the principal permeability constants.

Accurate characterization of a substrate dielectric constant is important for the design of high-frequency precision microwave circuits. However, most of the laminate materials are not isotropic, and laminate suppliers do not typically provide the anisotropy data of their materials. Furthermore, many microwave design tools do not allow anisotropy. Therefore, due to the absence of anisotropic material data, one way to determine whether the dielectric material in use is uniaxial-anisotropic or isotropic is by carrying out a two-step measurement procedure using a resonant structure [16]. One measurement is for calculating the horizontal electric field and another for the orthogonal electric field, thus two substrate dielectric constants can be calculated based on the direction of propagation. The same measurement concept of tensor material holds for magnetic permeability and material conductivity.

5.6 PCB and laminate design considerations

Crosstalk between closely spaced signal interconnects occurs due to the mutual capacitance and mutual inductance couplings. Identifying the coupling type at the circuit traces is important to choose the suitable shielding technique. Coupling mitigation methods for inductive and capacitive coupling are different and most of the times are in conflict.

The media of an RF module and its related characteristics have a significant impact on shaping the EM fields within the module environment and thereby the overall system performance. In particular, multilayer laminates are essential components in the design of compact RF modules that usually encompass routing traces and matching elements as embedded passives. New laminate fabrication processes are capable of producing more than ten layers with high accuracies in terms of layer-to-layer registration and material stack thicknesses. The increasing demand on new laminate technologies exhibiting new thicknesses of metal layers and new dielectric material stacks to allow shrinkage of the total package sizes has led into packing the routing traces closer to each other and scaling down the width of the metal traces along with the layers thicknesses to some extent. However, this exacerbates the influence of EM parasitic coupling between the routing traces, especially, the capacitive coupling.

High impedance sources contribute significantly to the near-zone electric field crosstalk, and low impedance sources (e.g., long trace loops) contribute to the near-zone magnetic-field coupling. With the knowledge of such facts, one can potentially estimate the distribution/pattern of the near-zone electric and magnetic emissions so that countermeasures can be taken to protect a given design. To identify crosstalk due to electric/capacitive coupling, one can look at the high voltage or the low current location on the module interconnects to identify the source of such coupling. Electric coupling will appear as a current source in parallel with the victim trace, while a magnetic coupling will appear as a voltage source at the victim trace.

Capacitive coupling is usually encountered in closely spaced traces and can be mitigated by shielding the aggressor or victim trace using grounded conductive walls (e.g., adding via fence between the traces and connecting the shielding fence to the ground of the module). The shield must be grounded to provide a low impedance path to the electric field. Additionally, capacitive coupling increases with increasing frequency and decreases by increasing the separation between the traces.

Inductive coupling, however, can be identified by finding the high current location on the traces since magnetic-field lines are generated around current carrying conductors. The cross-sectional exposure of the coupled traces is an important factor in determining the amount of inductive coupling which increases by increasing the number of the crossing field lines. Similar to capacitive coupling, inductive coupling increases with frequency and decreases by increasing the separation between the traces. Moreover, inductive coupling can be alleviated by twisting the traces to eliminate the net flux; however, this may not be a practical option for planar PCB traces. Another shielding technique could be surrounding the aggressor trace with a magnetic material to divert the magnetic flux. Knowing that the circulation of a current in a conductor creates a magnetic field at which it can couple into a nearby circuit, it is recommended that interconnecting traces take the straightest path between the circuits, when possible, to reduce any current carrying loops.

To observe the impact of electric and magnetic coupling on the performance of the overall RF module, one can vary the relative permittivity and relative permeability of the media surrounding the module components using, for example, a full-wave EM simulator. Doing so can provide a good insight into identifying the type of coupling rendering the module components more sensitive and consequently taking the proper countermeasures. In general, all metals can provide electric-field shielding. However, metals possessing high permeabilities can only be significant for magnetic-field shielding.

RF transmitting modules derived by power amplifiers use DC power supplies to bias the operation of the amplifiers. A DC current is then carried from the supply to the amplifier through the module interconnects. Knowing that a steady current produces a steady magnetic field as discovered by Oersted who linked electricity with magnetism, such DC lines may also create coupling problems in dense modules.

Other crosstalk mitigation techniques could be accomplished by limiting the trace height in relative to the ground plane to help reduce unwanted radiated EM coupling since the larger the distance, the more tendency of a signal trace to radiate and couple into a circuit in proximity. Using thin dielectric materials that make the traces close to the ground plane or having high dielectric constant material reduces EM radiated emissions. Additionally on this subject, using coplanar waveguide (CPW) for the signal interconnects in a module provides several advantages when compared to microstrip traces. The basic structure of CPW consists of a signal trace surrounded by in-plane side conductors that are usually grounded and a ground plane underneath. CPW traces possess lower coupling to the adjacent lines since the in-plane surrounding grounds are closer than any other signal lines. Other benefits of CPW also include low radiation losses, reduced dispersion, more convenient grounding for the surface mount devices, and low sensitivity of the characteristic impedance of the CPW traces to changes that could occur on the substrate thickness. The impedance of a CPW trace depends on the ratio of the center trace width to the gap width (i.e., the gap between the signal line and the in-plane ground), unlike in microstrip traces where the characteristic impedance relies on the ratio of the trace width to the substrate height.

References

[1] Mongia R., Bahl I., Bhartia P., and Hong J. *RF and Microwave Coupled-Line Circuits*. 2nd edn. Norwood, MA: Artech House; 2007

[2] Gruodis A.J., "Transient analysis of uniform resistive transmission lines in a homogeneous medium." *IBM Journal of Research and Development*. 1979;**23**(6):675–81

[3] Degerstrom M.J., Gilbert B.K., and Daniel E.S. "Accurate resistance, inductance, capacitance, and conductance (RLCG) from uniform transmission line measurements." *IEEE-EPEP 18th Conference on Electrical Performance of Electronic Packaging*; San Jose, CA, 27–29 Oct 2008. pp. 77–80

[4] Ikalainen P.K., and Matthaei G.L. "Wide-band, forward-coupling microstrip hybrids with high directivity." *IEEE Transactions on Microwave Theory and Techniques*. 1987;**35**(8):719–25

[5] Oliver B.M. "Directional electromagnetic couplers." *Proceedings of the IRE*. 1954;**42**(11):1686–92

[6] Tripatbi V.K. "Asymmetric coupled transmission lines in an inhomogeneous medium." *IEEE Transactions on Microwave Theory and Techniques*. 1975; **23**(9):734–39

[7] Lorival J.E., Deschacht D., Quere Y., Le Gouguec T., and Huret F. "Additivity of capacitive and inductive coupling in submicronic interconnects." *International Conference on Design and Test of Integrated Systems in Nanoscale Technology*; Tunis, 2006. pp. 300–304

[8] Agarwal K., Sylvester D., and Blaauw D. "A simplified transmission-line based crosstalk noise model for on-chip RLC wiring." *IEEE Asia and South Pacific Design Automation Conference*; Yohohama, Japan, 2004. pp. 859–65

[9] Vishnyakov V., Friedman E.G., and Kolodny A. "Multi-aggressor capacitive and inductive coupling noise modeling and mitigation." *Microelectronics Journal*. 2012;**43**(4):235–43

[10] Kong J.A. *Electromagnetic Wave Theory*. Cambridge, MA: EMW Publishing; 2005

[11] Paul C. *Introduction to Electromagnetic Compatibility*. 2nd edn. Hoboken, NJ: Wiley-Interscience; 2006

[12] Celozzi S., Araneo R., and Lovat G. *Electromagnetic Shielding*. Hoboken, NJ: John Wiley & Sons, Inc.; 2008

[13] Lee K., Lee H., Jung H., Sim J., and Park H. "A serpentine guard trace to reduce the far-end crosstalk voltage and the crosstalk induced timing jitter of parallel microstrip lines." *IEEE Transactions on Advanced Packaging*. 2008;**31**(4):809–17

[14] Eged B., Mernyei F., Novak I., and Bajor P. "Reduction of far-end crosstalk on coupled microstrip PCB interconnects." *IEEE Instrumentation and Measurement Technology Conference (Cat. No. 94CH3424-9), IMTC/94 Conference Proceedings*; Hamamatsu, Japan, 1994. pp. 287–90

[15] Deslandes D., and Wu K. "Design consideration and performance analysis of substrate integrated waveguide components." *32nd European Microwave Conference*; Milan, Italy, 2002. pp. 1–4

[16] Rautio J.C. "A proposed uniaxial anisotropic dielectric measurement technique." *IEEE MTTS International Microwave Symposium*; Guadalajara, Mexico, 19–20 Feb 2009. pp. 59–62

Chapter 6

CAD of RF and microwave circuits and modules

6.1 Introduction

Computer-aided design (CAD) and simulators are essential tools that help in expediting the process of implementing theoretical concepts into practical designs and offer an insight into the behavior of a complex problem. CAD facilitates testing design ideas before putting them into practice and more importantly before building a physical prototype. In fact, it is becoming imperative for design engineers to carry out simulations on standalone circuits as well as on large modules before deciding on manufacturing them.

The development of CAD tools gained momentum shortly after the beginning of the computer era in the 1950s to aid in integrating linear models of microwave components to solve complex circuit problems [1]. In recent years, however, the demand for multifunctional and reconfigurable radio frequency (RF) front-end modules incorporating as many circuits as possible in a small form factor has become more stringent. Moreover, the advent of long-term evolution (LTE) as a rapidly growing global standard for both paired and unpaired spectrum has enabled unprecedented high data rates and capacity. Accordingly, LTE was further developed to introduce LTE-advanced with several improvements including new concepts such as carrier aggregation (CA) and enhanced utilization of multi-input, multioutput (MIMO) antenna techniques for high spectral efficiency per coverage area and high capacity broadband transmission in both frequency division duplex (FDD) and time division duplex (TDD). Such advances and many others have justified the need for developing advanced analysis methodologies and optimization approaches to tackle the emerging complexities.

The complexity of modern RF modules comprising multidomain devices (i.e., digital and analog) and the limitations of standalone simulation methods imposes high demands on analysis techniques that support efficient analysis of mixed-signal electronics. Such methodologies must have the capability to carry-out concurrent system-level analysis with the ability to bring down the computational overhead since RF engineers desire to investigate all possible scenarios for any given design change in a system. In this context, having the suitable CAD tools is becoming inevitable to capture the possible multidomain interaction effects and the proximity coupling between the shared circuits in a system. Standalone CAD tools are incapable of meeting these requirements. For instance, full-wave computational electromagnetic (CEM) solvers lack accurate models for digital or nonlinear devices; transmission line methods fail to account for radiated fields; and circuit solvers (e.g., PSPICE or its derivative HSPICE) are limited to quasi-static frequencies. Therefore, the new direction in research and industry on RF and microwave system-level analysis is to develop hybrid schemes combining traditional solvers and models in an integrated fashion to solve problems of higher complexity.

To carry out concurrent analysis between the different domains, domain decomposition schemes are available where linear and nonlinear; electric and thermal; or electric and acoustic domains of a given problem are analyzed separately. Next, results from each of the domains are integrated through a circuit solver such as Keysight's advanced design system (ADS), and finally, system-level tuning can then be carried out by optimizing the module parameters within the circuit solver framework.

Using an optimization approach that relies on standalone CAD tools in a brute-force manner to manually adjust the element values of a circuit network until achieving the desired system-level performance can be somewhat impractical, especially, when dealing with several design variables. In this regard, circuit modeling techniques play a critical role in improving the speed of simulation, maintaining the accuracy of the entire system-level performance, enabling the automation of optimization algorithms, and allowing degrees of freedom represented by the number of the input parameters that can be modeled. Traditional modeling methods rely primarily on measurement data which are time-consuming and costly. Alternatively, simulation-based models are more favorable efficient approach, especially, those based on lumped-element models created using high fidelity simulation data as they are responsive and representative of the component's electrical and physical behavior. Modeling of electronic components is a subject that will continue to evolve as the need for efficient and accurate models is always in an increase to cope with the complexity of the emerging RF modules.

Other approaches that link the multidomains concurrently and allow full two-way coupling mechanism between the domains without the use of low order models are also available. One of the CAD tools that provide this capability is COMSOL Multiphysics, a commercial simulation platform, where physical effects from different domains can be coupled simultaneously. Domain-coupling includes interaction between the structural mechanics, material piezoelectricity, thermal effects, and EM parasitics. While this approach seems more accurate and resembles the real world problem in capturing all effects and interaction from all domains, it may not be perceived as a viable option in allowing fast redesign cycles due to the excessive simulation time needed for a converged solution. However, with the emerging evolution of CAD software packages and the developments on the interfacing capabilities between the different computational methods within the same platforms, a true concurrent multidomain analysis can become more user-friendly, efficient, and accurate in the foreseeable future.

In this chapter, the discussion will be on the generic aspects of CAD and modeling of RF and microwave circuits and the commonly used simulation methods in terms of their inherent assumptions and limitations.

6.2 Computational electromagnetic-based simulators

Several computational EM techniques have evolved over the past few decades to improve simulation efficiency, accuracy, and the capability of handling different problems with different electrical sizes from microscale elements such as bond wires to macroscale objects such as radar cross section of a ship, as an example. Having a general background of the capabilities and the limitations of available CEM tools helps design engineers in making the suitable tool selection for the problem under modeling. From this perspective, this section briefly summarizes commonly used CEM tools and highlights their different methods, each with its peculiar advantages and drawbacks. This section also highlights the types of application best suited to the EM simulation methods based on the model geometry (either planer or three dimensional), broadband or narrowband, and electrical size.

6.2.1 Commercially available CEM tools

CEM techniques can be divided in terms of their mesh structure into two main groups: volumetric methods and boundary or surface methods. In the volumetric-based methods, the whole modeling space enclosed by the EM boundaries is discretized and differential equations are then calculated to obtain the EM field quantities. In the surface-based methods, the boundaries themselves are discretized and surface integral equations are computed to obtain the near-field EM quantities. Green's function can then be solved to transform the near-field into far-field quantities. Other techniques employ methods from both categories to yield efficient analysis, for instance, for simulating electrically large structures.

It is becoming customary that vendors of CEM tools offer different solvers under the same software package which allow users to choose between them depending on the EM problem being solved. The selection of a proper EM tool relies on many factors such as model accuracy, simulation time, computational overhead, and easiness of the graphical user interface (GUI). Most of today's CEM algorithms are highly parallelizable. Thus, computer systems with multiple processors can highly benefit from this feature, not only to support the core computational algorithms but also to efficiently support post-processing algorithms when calculating, for example, *S*-parameters data, radiation patterns, and current distributions.

Commercial field solvers employ well-established numerical methods for solving Maxwell's equations. Among the several numerical methods, the finite element method (FEM) is commonly used for analyzing volumetric three-dimensional (3D) structures and is best suited for solving highly inhomogeneous materials. As well, the method of moment (MoM) is used for planar circuits, especially, for radiating and coupling structures. While FEM and MoM are frequency domain methods, other techniques such as the finite integration technique (FIT) and the finite difference time-domain (FDTD) are widely used time domain methods. The main advantage of time-domain simulators is the ability to achieve a broadband solution using a single simulation run. Additionally, in recent years, there has been an increasing interest in simulation techniques that are capable of solving electrically

Table 6.1 Some of the commonly used linear CEM tools used in the design of RF and microwave components and modules

Tool	Computation method	Solver type	Mesh type	Commonly used mesh shape	Computational overhead
ANSYS HFSS	FEM	Frequency	Volume	Tetrahedral	High
CST MWS Studio	FIT	Transient	Volume	Hexahedral, tetrahedral	Medium
	FEM	Frequency	Volume	Tetrahedral	High
IE3D	MoM	Frequency	Surface	Rectangles, triangles	Low
KEYSIGHT Momentum	MoM	Frequency	Surface	Rectangles, triangles	Low
Sonnet EM	MoM	Frequency	Surface	Rectangles, triangles	Low
COMSOL	FEM	Frequency	Volume	Tetrahedral	High
	Boundary element method (BEM)	Frequency	Surface	Rectangles, triangles	Low
FEKO	MoM	Frequency	Surface	Rectangles, triangles	Low
	FEM	Frequency	Volume	Tetrahedral	High
	FDTD	Time	Volume	Yee cell	Low
	MLFMM	Frequency	Surface	Rectangles, triangles	Low
GEMS	FDTD	Time	Volume	Yee cell	Low

large EM structures. This resulted in several specialized techniques such as the multilevel fast multipole method (MLFMM) which is a reduced order of the MoM.

As shown in Table 6.1, a plethora of numerical analysis methods and tools have been developed to solve complex and diverse EM problems. Table 6.1 summarizes common commercially available CEM tools and some of their main features. As can be observed, the decision on which simulation method is appropriate for a given EM problem is highly application-dependent and very subjective.

6.2.2 Limitations of different CEM methods

This section highlights some of the well-known limitations associated with the CEM methods in handling complex EM problems. Particularly, those based on time- and frequency-domain methods will be discussed.

Time-domain techniques offer accurate EM analysis but may suffer from computational inefficiencies and convergence problems for modules incorporating high Q-factor resonant elements. High Q elements are characterized by stiff differential equations that require having small mesh size, and thereby a large number of mesh elements to be solved. In return, this exacerbates the computational overhead of a time-domain simulator. To give a glimpse into an EM model complexity, a 3D problem can have six field components (i.e., electric and magnetic fields in the three Cartesian coordinates) and three material coefficients (i.e., ε, μ, and σ) in combination with a time-stepping procedure per unit cell in the spatial grid. However, time-domain methods (e.g., FDTD) are fairly stable and do not rely primarily on matrices to invert, unlike MoM and finite-elements techniques that can be prone to singularity errors (i.e., division by zero errors). Frequency domain results can be obtained from a time domain solvers by applying Fourier transform.

Frequency-domain methods possess excellent convergence which allows for accurate and reliable results in relatively short computation times. Moreover, frequency-domain methods can handle high Q and strong coupling structures more suitably than time-domain-based methods. Among the several frequency-domain methods, the MoM does not require absorbing boundary conditions (ABCs) for carrying out a simulation that makes it better suited to solve open space problems such as antennas and EM scatterers from electrically large radiating structures. In essence, Green's function in MoM allows near-to-far-field transformation without the need to extend or discretize the modeling space to the far-field region [2], unlike FEM and FDTD methods which are more suitable for solving 3D structures such as MICs. However, frequency-domain methods incorporating Green's functions and coupling integral equations are relatively more difficult to implement when compared to the time-domain algorithms formulated using differential equations that are relatively easy to solve. In both domains, the computational cost is proportional to the number of excitation ports assigned to the modeling problem.

6.2.3 Boundary conditions

The use of volumetric CEM tools requires a methodology to resemble the open space surrounding an EM problem where waves propagate indefinitely in the outwards direction to the free space. Given the finite computational resources in accommodating large EM problems, waves can be mathematically truncated or dissipated using what is called analytical absorbing boundary conditions (ABCs). ABCs are artificial boundaries used to limit the computation volume. An alternative to ABCs when simulating electrically large structures is the finite element-boundary integral (FE-BI) which is typically used in a hybrid simulation setup combining the FEM and integral-equation (IE) solvers. In certain applications that handle strong radiation at the boundaries such as beam propagation, other radiation boundary conditions exist where waves continue to propagate past the boundary are called transparent boundary condition (TBC) or radiation boundary conditions (RBC). In contrast to the commonly used ABC methods, TBC requires no adjustable parameters that makes it problem independent and robust.

An ideal ABC will enable a wave to only travel outwards through it without reflections (i.e., $|\Gamma| = 0$); however, practical analytical ABCs can provide attenuation of the reflected wave in the range of -100 dB [3] depending on the angle of incidence. The better the ABC performance, the less influence of the reflected waves on the accuracy of the modeled results. One of the basic ABC implementations that can be used to mitigate or truncate radiated waves from reflecting into the modeling space is an enclosure realized as a perfect electric conductor (PEC). Another form of an ABC is the perfectly matched layer (PML), which is a means of absorbing outwards waves using a numerical medium that is not necessarily realizable. Berenger in 1994 presented the PML concept using split-field formulation into two unphysical fields within the PML region. The main idea falls under using an analytic continuation of wave-propagation equations to a complex contour that makes the oscillating waves exponentially decay at the PML region. Uniaxial perfectly matched layer (UPML) is an anisotropic absorbing boundary which can be thought as a simplified version of the PML. Unlike PML, UPML boundary conditions are realizable because their implementation is established by a Maxwellian formulation rather than a mathematical model.

6.2.4 Volumetric and surface mesh

Surface meshing supports a broad range of planar microwave circuits. The surface mesh cell can be available in different shapes such as triangular and quadrilateral elements. In 2.5D solvers, volumetric meshing can be generated from a surface mesh to analyze planar or semi-3D objects. Complex 3D structures such as multi-layer PCBs and multichip modules, however, are best handled by specialized volumetric meshing techniques to capture key structural features. The 3D mesh can take several basic structural forms such as tetrahedral, cubic, pyramidal, and hexahedron elements.

Depending on the computational methods used, the determination of the smallest mesh size must follow a stability constraint which has to be fulfilled in every single mesh cell in the modeling space. A good example here could be how the mesh size is defined in time-domain solvers that satisfy the so-called Courant–Friedrichs–Levy (CFL) criterion and is expressed as:

$$v_m \Delta t \le \left[\frac{1}{\Delta x^2} + \frac{1}{\Delta y^2} + \frac{1}{\Delta z^2}\right]^{-1/2} \tag{6.1}$$

where v_m is the maximum wave phase velocity. In time-domain solvers, the field components advance in time by one time-step (Δt) at every iteration as well as in the spatial grid by Δx, Δy, and Δz. From (6.1), the smaller the mesh size, the smaller the time step. The smallest features size in the model determines the smallest mesh size that needs to be simulated and accordingly Δt. Having a large time step would seem advantageous and imply improved simulation speed at which the signal can propagate faster in the computational volume; however, the time step cannot be increased arbitrarily since the Courant stability condition controls it. Additionally, the mesh size cannot be arbitrarily increased as it is constrained by the influence of numerical dispersion on the modeled results, meaning that each frequency component propagates at a slightly different velocity in the mesh volume. In other words, a fine mesh size will result in a nondispersive solution.

Some structures vary greatly in their aspect ratios such as narrow slots, thin wires, and sheets which require special treatment, for instance, using variable mesh cell either in size or shape to fit such features of interest. Scaling the spatial cells to mesh objects with high aspect ratios is computationally expensive. Some solvers support automatic and adaptive mesh generation methods as in ANSYS HFSS. The desired simulation accuracy is usually set by the user and can be achieved by refining the mesh size which implies that the size of the meshing elements is reduced to capture more details of the problem under modeling. Mesh construction and mesh seeding can also be set up manually by the user.

6.3 Circuit theory-based simulators

Several software packages are available for the analysis of electrical circuits and support linear and nonlinear analysis. Linear simulators exist with different implementations of the basic nodal analysis. Simulation program with integrated circuit emphasis (SPICE) analysis platforms developed in the 1970s at the University of California at Berkley are among one of the most widely used tools. Nodal analysis is a well-known method in circuits' theory that is used to determine the voltage at each node relative to a reference node or ground. In nonlinear circuit simulators, different techniques are employed such as harmonic balance (HB), transient time-domain, and envelope methods. Discussion of the linear and non-linear analysis methods are discussed in Sections 6.3.1 and 6.3.2.

6.3.1 Linear circuit simulators

This section discusses how a circuit CAD tool (e.g., SPICE or HSPICE simulator) assembles linear equations from the circuit description or netlist. In other words, how a graphical circuit schematic is interpreted by a circuit simulator to formulate its equations in a matrix form and subsequently to solve the matrix in time or frequency domain using the so-called modified nodal analysis (MNA).

6.3.1.1 Modified nodal analysis

MNA exists to overcome limitations of the classical nodal analysis method for linear electrical circuits and to provide a more compact formulation. The basic nodal analysis is incapable of efficiently processing voltage and current-dependent sources [4].

Lumped element circuit simulators utilize topological and element laws to analyze a circuit performance. Topological laws such as Kirchhoff's voltage law (KVL) and Kirchhoff's current law (KCL) are independent of the circuit elements (i.e., RLC elements) that are expressed by the element equations. From the KCL equation in (6.2), the algebraic sum of all branch currents flowing into a node is equal to the sum of currents leaving that node:

$$\sum_{\text{node}} i_{\text{out}} = \sum_{\text{node}} i_{\text{in}} \tag{6.2}$$

And from the KVL, the algebraic sum of all branch voltages around a loop in a circuit is equal to zero:

$$\sum_{\text{loop}} V_{i,j} = 0 \tag{6.3}$$

The circuit in Figure 6.1 is a simple example of all resistive elements load used to describe the general formulation of circuit equations in a matrix form using nodal analysis.

Applying the KCL equation in (6.2) results in:
at node 1

$$(v_1 - 0)g_1 + (v_1 - v_2)g_2 = J \tag{6.4a}$$

and at node 2

$$(v_2 - 0)g_3 + (v_2 - v_1)g_2 = 0 \tag{6.4b}$$

Rearranging (6.4a) and (6.4b) yields:

$$(g_1 - g_2)v_1 + (-g_2)v_2 = J \tag{6.4c}$$

and

$$(-g_2)v_1 + (g_2 - g_3)v_2 = 0 \tag{6.4d}$$

Equations (6.4c) and (6.4d) can be rewritten in a matrix form as:

$$\underbrace{\begin{bmatrix} g_1 + g_2 & -g_2 \\ -g_2 & g_2 + g_3 \end{bmatrix}}_{\text{Conductance matrix}} \cdot \underbrace{\begin{bmatrix} v_1 \\ v_2 \end{bmatrix}}_{\text{Voltage vector}} = \underbrace{\begin{bmatrix} J \\ 0 \end{bmatrix}}_{\text{Source vector}} \tag{6.5}$$

For a purely resistive circuit, the matrix solution can be considered as a DC solution since it does not depend on frequency or time. The general matrix form can be written as $G \times V = B$, where G is the real conductance matrix, V is the unknown voltage vector to be found, and B is the source vector. Circuits with reactive elements can be in the following matrix form $A \times V = B$, where A is the admittance matrix. The A matrix is also referred to as the MNA matrix.

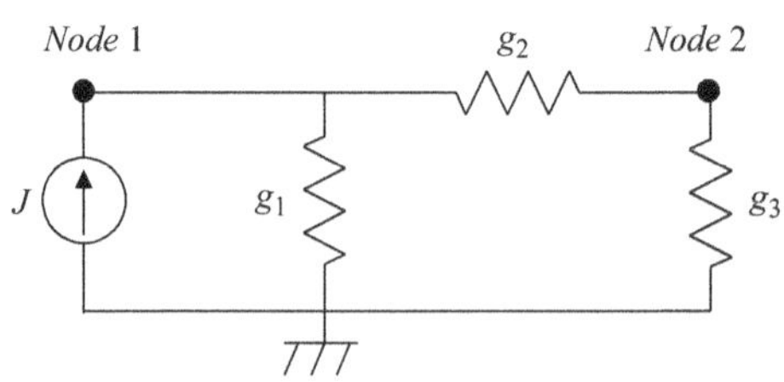

Figure 6.1 A sample circuit schematic of all resistive elements

We can observe from Figure 6.1 that elements connected to ground appear once in the MNA matrix and elements connected to a node n appear in the nth row and nth column of the matrix. Moreover, a source connected to the nth node will appear in the nth row in the source vector.

An MNA matrix is symmetric for all passive circuits where $a_{i,j} = a_{j,i}\ \forall\, i \neq j$. These observations regarding the pattern of the matrix equations can be utilized in scripting a computer code to populate the matrix elements from the software circuit schematic window and this process is referred to as the approach of *stamps*.

Table 6.2 summarizes the stamps of common circuit elements connected between two nodes or between a node and a ground. The term $S = j\omega$ is associated with the reactive elements and represents the Laplace complex frequency variable assuming a sinusoidal source.

From Table 6.2, one can observe that a matrix solution at DC or zero frequency using matrix inversion methods will lead to division by zero in the case of a circuit with an inductor, for example. Thus, a modified stamp for inductors to construct the MNA matrix is needed. Similarly, dealing with a voltage source introduces a

Table 6.2 Stamps of linear RLC elements

Passive element	Connection type	Stamp of the element
Resistor	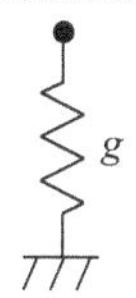	$A(i, i) = A(i, i) + g$
		$A(i, i) = A(i, i) + g$ $A(i, j) = A(i, j) - g$ $A(j, i) = A(j, i) - g$ $A(j, j) = A(j, j) + g$
Capacitor	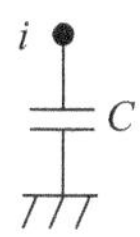	$A(i, i) = A(i, i) + SC$
	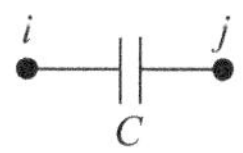	$A(i, i) = A(i, i) + SC$ $A(i, j) = A(i, j) - SC$ $A(j, i) = A(j, i) - SC$ $A(j, j) = A(j, j) + SC$
Inductor	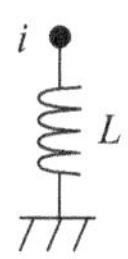	$A(i, i) = A(i, i) + 1/SL$
	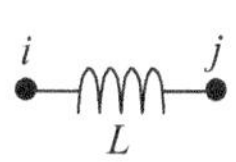	$A(i, i) = A(i, i) + 1/SL$ $A(i, j) = A(i, j) - 1/SL$ $A(j, i) = A(j, i) - 1/SL$ $A(j, j) = A(j, j) + 1/SL$

problem because the MNA matrix equations are constructed to follow KCL. Therefore, a modified stamp will be needed for voltage sources as well.

Consider the circuit in Figure 6.2 with a current source, the equivalent MNA matrix of this circuit following the stamps in Table 6.2 is demonstrated in (6.6):

$$\begin{bmatrix} g_1 + sc & -sc & 0 \\ -sc & sc + g_2 + g_3 & -g_3 \\ 0 & -g_3 & g_3 + 1/sL \end{bmatrix} \cdot \begin{bmatrix} v_1 \\ v_2 \\ v_3 \end{bmatrix} = \begin{bmatrix} J \\ 0 \\ 0 \end{bmatrix} \tag{6.6}$$

Table 6.3 lists the stamps of independent and dependent current and voltage sources. In this table, the total number of columns is referred to as TNC, while the total number of rows is denoted as TNR.

In the case of a voltage source, we add an extra node and introduce a new variable called I representing the current following through the voltage source E. Considering the circuit in Figure 6.3 with a voltage source, the equivalent MNA matrix of this circuit is demonstrated in (6.7):

$$\begin{bmatrix} g_1 & 0 & 0 & 1 \\ 0 & g_2 + g_3 & -g_3 & -1 \\ 0 & -g_3 & g_3 + sc & 0 \\ 1 & -1 & 0 & 0 \end{bmatrix} \begin{bmatrix} 0 \\ 0 \\ 0 \\ I \end{bmatrix} = \begin{bmatrix} 0 \\ 0 \\ 0 \\ E \end{bmatrix} \tag{6.7}$$

As can be noted, the larger the number of nodes in the circuit schematic, the larger the MNA matrix and the slower the solution will be. Efficient matrix inversion methods are generally employed to provide faster frequency-domain solutions such as the LU decomposition method. Matrix sparsity is an important indicator of how fast the solution can be obtained. Having a sparse matrix enables a faster and efficient solution when compared to a dense matrix. The sparsity of a matrix is the ratio between the number of zero elements and the total number of elements.

As a good practice, toggling-off non-used elements in a circuit schematic window usually helps in reducing the MNA matrix size and thereby leads to a faster simulation.

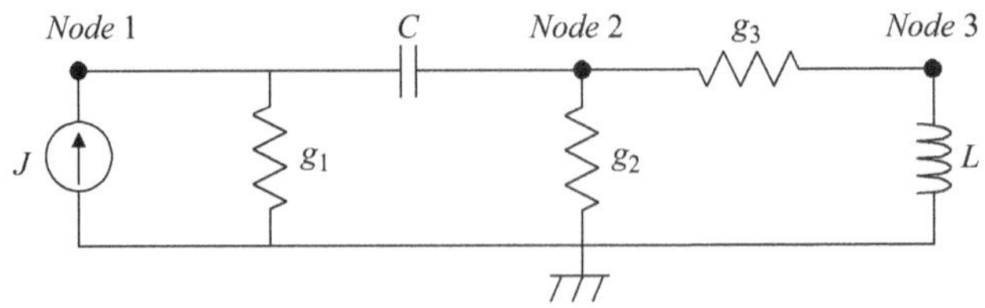

Figure 6.2 An example of a circuit schematic with resistive, inductive, and capacitive elements

Table 6.3 Stamps of current and voltage sources

Source	Connection type	Stamp
Independent current	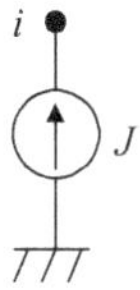	$B(i) = J$
	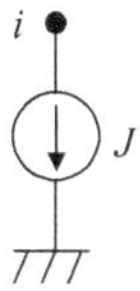	$B(i) = -J$
	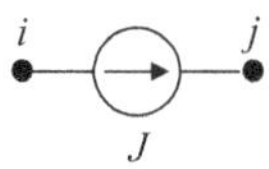	$B(i) = -J$ $B(j) = J$
Independent voltage	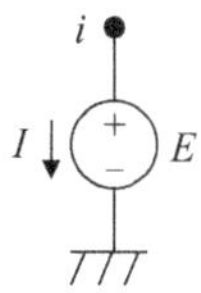	$B(TNR+1) = E$ $V(TNR+1) = I$ $A(i, TNC+1) = +1$ $A(\mathrm{TNR}+1, i) = +1$
	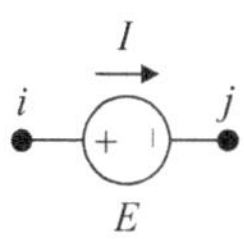	$B(TNR+1) = E$ $V(TNR+1) = I$ $A(i, TNC+1) = +1$ $A(TNR+1, i) = +1$ $A(j, TNC+1) = -1$ $A(TNR+1, j) = -1$
Dependent current (voltage-controlled current source)	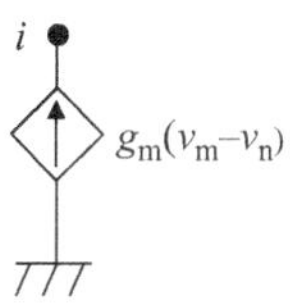	$A(i, m) = g_m$ $A(i, n) = -g_m$
	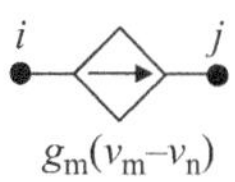	$A(i, m) = g_m$ $A(i, n) = -g_m$ $A(j, m) = -g_m$ $A(i, n) = g_m$
Dependent voltage (voltage-controlled voltage source)	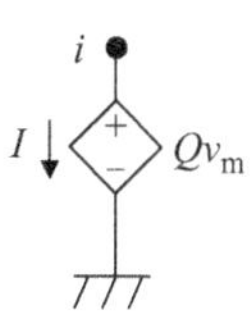	$A(i, TNC+1) = +1$ $A(TNR+1, i) = +1$ $A(TNR+1, m) = -Q$ $V(TNR+1) = I$
	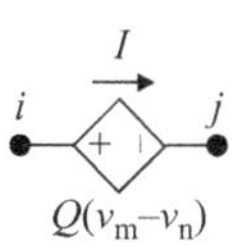	$A(i, TNC+1) = +1$ $A(TNR+1, i) = +1$ $A(j, TNC+1) = -1$ $A(TNR+1, j) = -1$ $A(TNR+1, m) = -Q$ $A(TNR+1, n) = Q$ $V(TNR+1) = I$

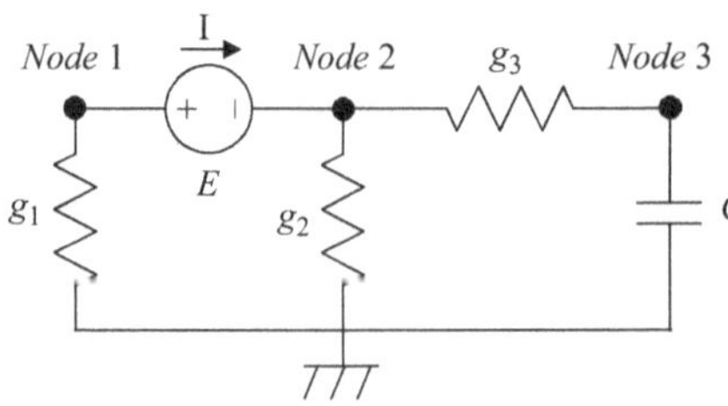

Figure 6.3 An example of a circuit schematic with resistive, independent voltage source, and capacitive elements

6.3.2 Nonlinear circuit simulators

Nonlinear simulation methods of active microwave circuits have witnessed a considerable development during the past decade. The increasing power-level requirements of transceiver modules pose significant challenges in analyzing the strong nonlinear behavior of devices and components under large signal excitation. The small-signal *S*-parameter analysis does not provide any linearity information of the circuit since the input signals assumed in such analysis are essentially small in amplitude and thereby low-level harmonics are produced. To predict the nonlinear behavior of a circuit, advanced analysis methods have been developed and considered in today's CAD tools comprising transient and steady-state analysis techniques. Among the several methods, harmonic balance, transient time-domain, and envelope simulation analysis are the commonly used ones.

6.3.2.1 Harmonic balance

Harmonic balance (HB) is a frequency-domain iterative analysis method that calculates the steady-state response due to periodic large signal excitations of a nonlinear electrical circuit such as RF mixers, amplifiers, and oscillators [5]. Under large signal conditions and a finite number of sinusoid excitations, HB is beneficial in analyzing several nonlinear device parameters including but not limited to third-order intercept (TOI) points, intermodulation distortion (IMD) components, load-pull contour analyses of power amplifiers, and nonlinear noise analysis. HB allows a sweep of both signal power and frequency. The core principle of frequency-domain nonlinear simulators relies upon the fact that the frequency components of the output of a nonlinear circuit are based on the knowledge of the frequency contents of the input signal(s). Thus, the calculations of harmonics and different IMD cases can be carried out through trigonometric expansion and power series techniques [1].

HB becomes computationally expensive when the signal spectral contents increases and when these contents are nonharmonic frequencies [6]. Nevertheless, requiring the analysis to reach a steady-state response entails long simulation time which makes HB an unattractive option for an optimization problem. Typically, multiple numbers of simple single-tone HB simulations are faster and efficient than one large HB simulation [7].

6.3.2.2 Transient time-domain

Unlike the HB method, the transient time-domain method does not require a prior knowledge of the spectral contents of the input signal(s). It is also beneficial in capturing real phenomena such as oscillations, instability, and subharmonic generation. Transient time-domain-based simulators, however, do not perform at their best when analyzing the linear nature of RF and microwave circuits due to the excessive amount of time needed to reach the steady-state solution. A steady-state solution requires that the circuit is simulated until the transient solution decays to a user-defined convergence level. To explain, time-domain solvers employ a time-domain signal for port excitation, typically a Gaussian pulse, which propagates throughout the computational volume of the model structure. The width of the excitation signal is determined based on the desired frequency contents of interest. As the signal continues to propagate, its energy content decays due to the various loss mechanisms in the model (such as dissipation due to materials resistivity or radiation loss). In a sense, the speed of the decaying energy will depend on the amount of loss inherent to the model, and this can result in excessive simulation time when dealing with high Q or narrow-band circuits [8]. Such limitations have led to the development of another time-domain steady-state solution method, known as the shooting method [5,9], which is capable of handling highly nonlinear autonomous circuits and circuits that rely mostly on time-dependent elements such as switches and mixers. The underlying principal relies upon finding the steady-state solution for the sinusoidal excitations by assuming and searching for all variables that are periodic and repeats themselves every one period. However, it does not support multitone excitation and it is not advantageous when dealing with microwave high-Q and distributed elements that comprise a large number of segments [9,10].

Nonlinear time-domain analysis is usually carried-out only around the RF carrier and its harmonic frequencies, otherwise, transient time-domain analysis usually suffers from convergence problems in cases of analyzing closely spaced frequencies (e.g., intermodulation frequencies). This is because the minimum simulation time step must be compatible with the highest frequency present in the analysis and at the same time to solve for the lowest frequency the simulation must run for a considerable period of time [8].

Solving a nonlinear equation can be performed using one of several iterative numerical techniques or a combination between them such as, but not limited to, Bisection, Secant, False Position, Fixed-Point, and Newton's method. These techniques as many others differ between them in the computational cost and speed. The general analysis approach is that the representation of a differential equation of a nonlinear circuit can be expressed by an algebraic representation which can then be used to form matrix equations suitable for implementation in a CAD tool.

6.3.2.3 Envelope method

Envelope simulation techniques are hybrid (i.e., frequency- and time-domains) simulation methods that have aroused due to the continual demand for efficient system-level simulations especially for high-order digitally modulated RF signals and complex nonlinear responses. The underlying principle relies on the fact that a modulated signal can be expressed in time-domain as the product of an envelope and a carrier [11]. In essence, the envelope of a modulated signal is sampled and expressed in time-domain as amplitude and phase signals in a form that can be used as inputs for HB analysis [12]. Envelope simulation allows the amplitude and phase at each spectral frequency to vary in time, thus, harmonics are no longer represented by a constant as in HB.

Envelope methods are more efficient than the transient methods when there is no overlap between the adjacent carrier frequencies of the envelope spectra. Indeed, convergence issues can arise when spectra overlap occurs as the envelope noise at the overlap regions are double counted [11] leading to inaccurate results.

6.4 Domain decomposition for system level co-simulations

As electronic modules continue to take the central role in the operations of modern communication systems, their reliability related failures and accidents may cause significant losses to various industries. Understanding these challenges and trying to emulate the complexity of such modules with all degrees of freedom using brute-force simulation methods requires an extremely large computational overhead which hinders our ability to carry out accurate and efficient system-level multi-domain analysis. In this context, co-simulation techniques based on domain decomposition methods are considered the key to reducing the product design cycles and providing an accurate estimate of the system-level performance of new product development. A domain decomposition based on cascading circuit models described by network port parameters has been the widely adopted approach in most of today's CAD tools. The intrinsic idea exploits domain decomposition schemes as will be discussed in Sections 6.4.1–6.4.4 that allow concurrent and efficient analysis of electronic systems.

6.4.1 Linear and nonlinear simulations

Concurrent analysis of linear and nonlinear domains can be carried out by linking the linear domain models of a given problem into a circuit simulator (e.g., PSPICE) where diverse library models of nonlinear components are available and can be utilized in conjunction with the linear models. This scheme offers an efficient, yet accurate and concurrent mixed-signal analysis for analyzing complex RF modules. Here, the frequency dependent nature of the linear structure requires the network-parameters (e.g., *S*-parameters) to be computed at each frequency of interest; hence, dispersion characteristics are well-preserved. Such an approach provides flexibility in the choice of individual solvers, selecting independent optimization of different component structures, and an overall tractable hybrid analysis for design engineers.

Figure 6.4 illustrates a commonly followed co-design approach by RF module design engineers where the linear domain components (cavities, die packages, PCB

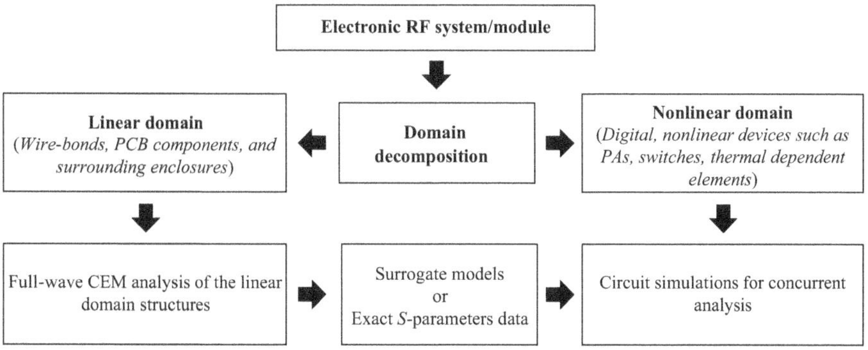

Figure 6.4 Domain decomposition approach for concurrent simulation of linear and nonlinear circuits

traces, etc.) are characterized by a CEM tool to extract their network parameters (e.g., *S*-matrix) [13,14]. A compact frequency-domain model or an *S*-parameters matrix can then be imported into a circuit simulator (e.g., ADS) to link nonlinear and digital loads to the linear model enabling a concurrent mixed signal analysis.

6.4.2 Electro-thermal simulations

Electro-thermal simulations are essential when the electrical properties of a device material are highly dependent on temperature variations. For example, operating at non-design temperatures may lead to different biasing conditions as in active devices or drift in the operating frequency as in piezoelectric-based devices. Moreover, the temperature rise of a device can be an indicator of electromigration failures for SoC metallization structures and microinductor interconnects. The fact that in transmitting modules comprising PAs with power efficiency (η) of 33% will have twice as much of its power dissipated as heat alludes to the importance of electro-thermal analysis for both on-chip and module-level in order to avoid potential reliability issues [15].

The underlying principle used in the analysis of the electro-thermal behavior of a device starts by calculating the power dissipation between the device nodes and then solving thermal equations to calculate the temperature rise that can be fed back into the device electrical equations (i.e., temperature dependent equations/models) to note the effect on the electrical performance. Figure 6.5 illustrates an electro-thermal simulation setup where power dissipation data are used to solve a set of thermal equations and provide temperature data to the device electrical model. The importance of being able to simulate an electro-thermal behavior is to identify temperature hotspot locations in the device or the module in terms of heat density (W/m^3) or heat flux (W/m^2) so that countermeasures can be taken.

Device heating is attributed to three main mechanisms: self-, mutual-, and transient-heating. In basic circuit simulator packages, the temperature is a fixed constant parameter usually at room temperature which may lead to inaccurate simulation results. Some of the advanced circuit simulators include defined electro-thermal models

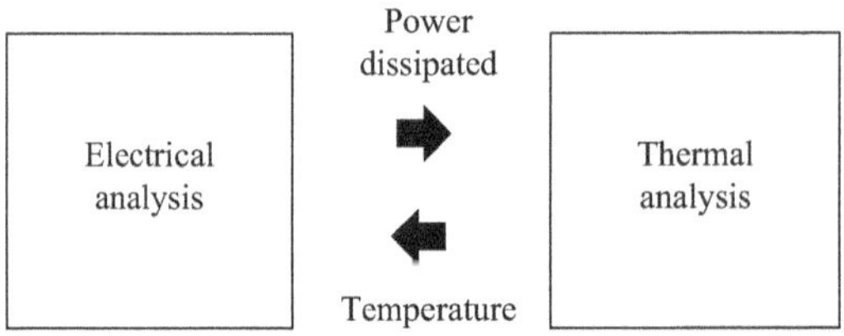

Figure 6.5 Scheme of a concurrent electro-thermal simulation

for common devices. However, such models are limited to only self-heating effects and do not account for mutual heating between the devices or the transient evolution of the heat transfer. COMSOL Multiphysics and Keysight's ADS are useful tools that enable simultaneous analysis between the two domains, electrical and thermal.

To this end, two approaches are commonly employed in the analysis of an electro-thermal problem. The first approach relies on a concurrent iterative analysis between the two domains resulting in a very accurate, however, computationally expensive solution. The second approach incorporates a reduced order thermal model represented by a thermal resistance or an RC network [16] which can be obtained via measurement or FEM thermal simulation of the device layout. The thermal models are then used to generate thermal parameters that can be imported into an electrical circuit simulator to solve the electro-thermal problem. The second approach is relatively less accurate, especially in the design of RF PAs, as it does not account for mutual thermal coupling from neighboring nodes.

The linkage between the thermal and electrical domains is described through the following heat transfer relation:

$$\nabla(k\nabla T) = -p_{\text{loss}} \tag{6.8}$$

The negative sign is to indicate that the heat transfer occurs from a high-temperature point toward a lower temperature point. T is the temperature and k is the thermal conductivity that represents the rate of heat transfer per unit area for a temperature gradient of 1 °C/m. Hence, k varies with temperature and spatial location which makes the thermal analysis a nonlinear problem. p_{loss} is the dissipated power representing the heat flow.

The reduced order thermal models approach uses dissipated power data at each node to extract the thermal resistance matrix (R_{th}). The following relationship can be used to identify the node temperature that is fed back to the electrical model of the device:

$$T_{\text{device}} = R_{\text{th}} \cdot P_{\text{dissipated}} + T_{\text{ambient}} \tag{6.9}$$

$$R_{th} = \begin{bmatrix} R_{11} & R_{12} & \cdots & R_{1N} \\ R_{21} & R_{22} & & \vdots \\ \vdots & & \ddots & \vdots \\ R_{N1} & \cdots & \cdots & R_{NN} \end{bmatrix} \tag{6.10}$$

where T_{device} is the device temperature and $T_{ambient}$ is the ambient temperature surrounding the device. The diagonal resistance (R_{nn}) elements of the matrix in (6.10) correspond to the individual device self-heating, while the off-diagonal elements represent the thermal coupling from neighboring nodes.

6.4.2.1 Thermal management considerations

The increasing requirements for boosting the output power of RF transmitting modules and the demand for small form factor is becoming mainstream. These trends place increased challenges on the thermal management of complex RF modules. On the positive side, such challenges aid in accelerating research progresses on high-conductivity materials, particularly, composite materials leveraging high thermal conductivities. During the past few decades, advances in thermal management methods enabled effective heat-sinking techniques to cope with the high thermal densities in electronic products [17]. Having a poor thermal design limits the performance of an integrated circuit. It can also cause detrimental reliability effects embodied in metal electromigration, contact degradation, and increased stresses between stacked materials due to their different coefficient of thermal expansion (CTE). Aiming for a reduced size device worsens its thermal performance as thermal resistance tends to increase. Heat-spreading materials are typically selected to provide high thermal conductivity (measured in W/m-k) and low CTE (measured in ppm/°C) at a lower cost. Materials with low CTE and high thermal conductivity are made possible through composites between different alloys. Table 6.4 lists some of the common materials used in microelectronic packaging.

Since minimizing the thermal resistance path between the coolant and the heat source is the intended goal and aside from the fact that the type of metallization layers in an RF chip/module plays a major role in the heat transfer process, thermal resistance is also influenced by the dielectric material surrounding the device metal structures. The type of dielectric stack-up controls how fast the device temperature drops and how efficiently the heat travels out of the device. For instance, low

Table 6.4 Thermal conductivity and expansion coefficients of common materials and alloys used in microelectronics packaging

Alloy	Thermal conductivity (W/mk)	CTE (ppm/ °C)
Copper (Cu)	390	17
Aluminum (Al)	200	24
Molybdenum (Mo)	139	5
Tungsten (W)	174	4.6
CuW	190	8
CuMo	170	8
Nickel (Ni)	91	13.3
Silicon (Si)	150	5
Silver (Ag)	430	19.7

dielectric constant materials tend to have relatively lower thermal conductivities when compared to those with high dielectric constants. Therefore, a designer can make tradeoffs between the electrical and thermal performance by properly selecting the dielectric material substrate of the design when feasible. Adding thermal vias and heat spreading bars on a PCB board comprising TX devices such as PAs are essential parts of the design. In addition, solder joints on the chip level can serve as efficient heat transfer routes to the heat-sinks as their materials have relatively high thermal conductivities [18]. In general, the thermal conductivity of the PCB traces and vias, the number of thermal vias, and the type of the dielectric material housing them are all important factors that need to be considered in a PCB design. According to Wiedemann–Franz law, both electric and thermal conductivities are related quantities. In essence, an increase of the conductor electric conductivity implies an increase in its thermal conductivity at finite ambient temperature.

The impact of the thermal effects on the electrical performance can be quantified using the temperature coefficient of velocity (TCV) which can be a positive or a negative number. TCV can also be used to determine another quantity, which is the temperature coefficient of frequency (TCF) for the material. Both TCV and TCF are associated mostly with electro-acoustic piezoelectric-based devices that are sensitive to variations in temperature. As a measure to the amount of shift in frequency for resonant devices due to change in temperature, the TCF is simply given by:

$$\mathrm{TCF} = \frac{1}{f_0}\frac{\Delta f_0}{\Delta T} = \frac{1}{f_0}\frac{f - f_0}{T - T_0} \tag{6.11}$$

where f_0 is the resonant frequency, Δf_0 is the change in frequency position due to different operating temperature (ΔT), T_0 is the ambient room temperature, and T and f are the new temperature and frequency values other than T_0 and f_0, respectively. The TCF result is expressed in parts per million per degree centigrade (ppm/°C) which represents a 1 Hz change for each MHz of operating frequency.

As an example, consider an acoustic wave device that has a resonance frequency of 2400 MHz at a room temperature of 25 °C. If an increase in the device temperature to 80 °C results in frequency drift of 2 MHz down in frequency, the TCF is then calculated as:

$$\mathrm{TCF} = \frac{1}{f_0}\frac{\Delta f_0}{\Delta T} = \frac{1}{2400\mathrm{MHz}}\frac{(2398 - 2400) \times 1\mathrm{e}6}{80 - 25} = -15.1515 \tag{6.12}$$

6.4.3 Electro-acoustic simulations

Electro-acoustic wave devices find several applications in the RF front-end modules especially for portable wireless systems where size, weight, and high performance are of great interest. The most common products of the electro-acoustic-based technologies are frequency selective filters, microphones, oscillators, and sensors. Electro-acoustic devices rely primarily on the piezoelectric effect or the so-called electro-mechanical coupling that converts an RF input to an acoustic wave and, conversely, converts an acoustic wave to an RF output. Because the speed of an acoustic wave is approximately five orders of magnitude slower than the speed of light, the relatively short-wavelength of an acoustic wave at RF and microwave frequencies makes it possible to realize extremely small resonant structures and microelectronic RF devices.

As the demand on developing electro-acoustic wave devices has become very rapid in recent years, the need for accurate and efficient electro-acoustic analysis imposed by the need for a very fast-paced product development cycles has escalated to an unprecedented manner. One commonly used approach followed in the analysis of an electro-acoustic performance of a device is by incorporating models from the two separate domains, electric and acoustic domains, which are then co-simulated concurrently via a circuit solver tool. Specifically, the electrical model handles the EM parasitics due to the package housing the acoustic components and can be obtained with the aid of a CEM tool, while the acoustic model is to describe the acoustic behavior of the piezoelectric components in the device and can be extracted from FEM simulation data or measurement data.

6.4.4 Module integration

CEM tools support port-based network analysis that facilitates the integration of multidomain RF modules. Models of active and passive microwave components can be ported into a circuit solver for complex module analysis. Port-based analysis allows exploiting the computational advantages of CEM tools and circuit solvers without compromising accuracy and also allows choosing the best solver for each domain.

A realistic EM environment of an RF module involves linear components such as surface mount devices, interconnect traces on a multilayered PCB, and bondwires, in addition to nonlinear devices comprising PAs, LNAs, and switches. Furthermore, EM interaction from a circuit can couple to proximity circuit elements within the module via direct or indirect coupling paths. It is worth mentioning that decomposing an RF module into simpler EM problems that are characterized electromagnetically in isolation and not taking into account the EM interaction between them may render overall system-level results far from the expected performance seen at the time of system evaluation, especially when operating in the gigahertz frequency range. Although the segmentation approach appears convenient when a large module is handled by different design teams and requires less computational resources than performing a true full module simulation, such approach does not represent the realistic EM environment in which an RF module operates. A true module-level simulation must be conducted for better estimation of the key system performance parameters such as self-isolation and cross-isolation levels, impedance matching between the successive transceiver stages, receiver sensitivity, noise figure, and power-handling capability. For instance, inductance and Q-factor values of surface mount inductors or embedded inductors can be considerably de-tuned once incorporated in a module environment due to the introduced EM parasitics. Thereby, many EM issues remain undetected during the design and production phases of a module, if not analyzed properly, leading to unpredictable failures in actual operation.

Key challenges that encounter simulating the overall system performance of a module include:

- Lack of accurate models of some of the devices and components in the system hinders the ability to simulate the actual system behavior, especially, in complex hybrid multichip modules. A small inaccuracy in each of the incorporated models can exacerbate linearity and noise effects.
- Nonoptimal module integration between the successive module components can lead to impedance mismatches between the cascaded devices yielding unnecessarily poor system insertion loss, analog/mixed-signal isolation issues, and significant parasitic and distributed effects.
- Substrate effects can have a significant impact on the impedance of the module interconnects and on the on-chip inductors [19].
- Unlike capacitors, surface-mount and embedded inductors are aggressive module components that present a major ohmic loss because of the flow of current across the inductor's coil and their skin effect. Also, the magnetic field generated by the inductor coils can easily couple into a nearby circuit and influence their performance.
- Several other important aspects such as die-to-die EM coupling, layout optimization of the die component to adjust EM parasitics within the die, SMD-to-SMD coupling, and layout optimization of a module laminate are all necessary to ensure optimal electrical performance.

In summary, the more inclusion of model details to capture their parasitic effects and loss mechanisms, the more accurate the analyzed performance will be. Disagreement between simulation and measurement data usually occurs due to ignoring small geometric details in the model, not considering the surrounding EM environment correctly, using approximate material parameters, especially for frequency dependent materials, and inaccuracies due to overlooking critical setting of the simulation tool being used. These effects combined add yet another dimension of complexity to the parasitic extraction of large module interconnects. A commonly followed approach in the design of microelectronic devices and modules is illustrated in Figure 6.6.

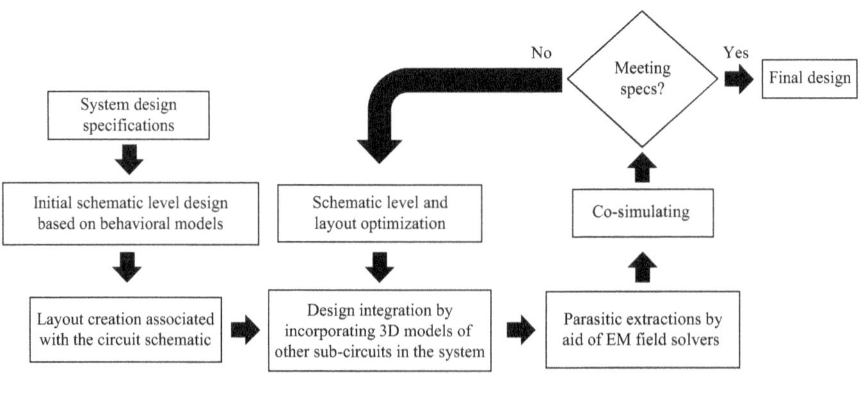

Figure 6.6 Design flow of a multidomain RF module

6.5 RF and microwave modeling

Accurate linear models of RF components are either based on measurement data or high fidelity simulation data. In most cases, simulation-based models are low cost and time efficient preferred option that can be obtained by solving Maxwell's equations or other constitutive differential equations using one of the computer-oriented numerical methods such as the FEM or MoM methods. Such fine models usually consume a considerable amount of time and also computationally expensive. Generating reduced order behavioral models for linear and nonlinear RF circuits and components is the key to reduce the simulation time and enable investigating several design iterations. A large number of design iterations are usually needed due to the insufficient information regarding the physical layout in the early stages of designing a complex module. The increasing trend toward following a general simulation workflow management approach similar in concept to what is known as system assembly modeling (SAM), gives great flexibility during the product development and rapid response to market changes. One way to achieve the desired module-level performance is through an optimization process to determine the optimal parameters. However, direct use of CEM tools which are computationally expensive is not a viable option to optimize a complex module. Alternatively, employing approximate models such as empirical models, polynomial curve fitting, equivalent lumped circuits, machine learning, or statistical-based models of an accurate electromagnetic simulation or measurement data provide an efficient solution to optimize a large module. Having accurate models of the various module components helps system engineers to make proper system performance and cost tradeoff decisions. Model accuracy is greatly enhanced when they are based on extensive real measured data or simulation data of the RF components and their EM surroundings to capture the parasitic effects.

Most early models were based on equivalent circuit models or black box models which do not provide a clear representation of the device physical parameters. Although equivalent lumped circuit models are attractive for well-established technology process, they do not provide flexibility for an immature process that is subject to frequent changes. Alternatively, physical models which are descriptive in terms of the device physical or geometrical structure facilitate the development of technologically based processes and provide a deep understanding of the intrinsic physical operation of the device in addition to the key parameters contributing to the model characteristic. Statistically based models are also flexible to optimize and to run different modeling experiments based on. The accuracy of the final model obtained using any of the different types of modeling methods is usually gauged using one of the cross-validation methods where an independent test set of data, which are not part of the model training process, are used for validation.

Typically, a uniformly smooth device response can be modeled using any model approximation methods. For a complex and highly nonlinear device response, careful attention should be paid to the selection of the appropriate modeling method. One common procedure followed in modeling complex nonlinear data is by converting the data into a different form that could result in a smoother response data (e.g., real and imaginary, phase and amplitude, Z-parameters or S-parameter, etc.). With all the developments that were made in boosting the model accuracy of active and passive RF and microwave components, the pursuit of accurate and simple models requiring minimal number of data points to represent a given device will continue to be a field of interest.

6.5.1 Polynomial curve-fit-based modeling

For many years, RF models used in simulation tools have been developed using closed-form mathematical descriptions of the modeled devices. These models are also known as compact models. One form of such models is the polynomial curve fitting which is a set of interpolation techniques used to construct a polynomial function of a given data points obtained by measurement or simulation. The use of high order polynomial function does not necessarily result in a more accurate modeled data. An over fitted curve can result in a poor representation of the actual data. Therefore, the selection of the order of the polynomial function is usually conducted as part of the modeling process in an iterative fashion to select the proper model order that results in the smallest error possible between the actual and modeled data. In curve fitting methods, all data points are available at the time of fitting, while in regression or machine learning methods, only a small set of data is available for training the model as will be discussed in Section 6.5.2. In addition, regression methods allow for statistical inference.

6.5.2 Machine learning-based modeling

Machine learning algorithms were originally designed to generate highly accurate regression models for computationally expensive simulation code requiring as little data points and as little user interaction as possible [20]. The outcomes are the generation of accurate surrogate models for use in circuit solver that can concurrently analyze multidomain systems in a fast manner. There are several machine learning algorithms such as artificial neural networks (ANNs), support vector machine (SVM), radial basis function (RBF), etc. Despite the fact that having different algorithms to select from are beneficial in finding the suitable method based on the behavior/pattern of the data being modeled, ANN among others is the most prominent method. ANN models have recently gained attention as unconventional alternatives to modeling, where models are first trained to learn a given system's behaviors from measured/simulated data. There are different training methods used to teach the ANN network using the same set of data. These methods are known as training algorithms and selecting the proper training method plays a significant role in determining the efficiency and accuracy of the generated final models. ANN trained models provide fast, yet accurate answers to the task learned. Machine learning models in general and ANN, in particular, can be more versatile than analytical models and more accurate than empirical models at instances where there is no sufficient information about the problem input–output relationship. These models learn the input/output relationships of a problem through the optimization process of training, during which, their weights are iteratively adjusted following a general multilayer perceptron (MLP) configuration as illustrated in Figure 6.7.

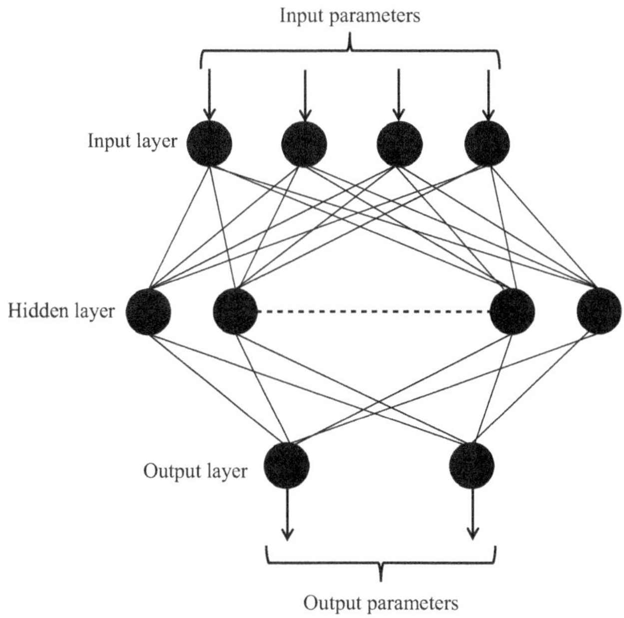

Figure 6.7 Basic configuration of a multilayer perceptron (MLP) artificial neural network

One of the most popular training algorithms is based on the back propagation (BP) method. Other algorithms include quasi-Newton, conjugate-gradient, simplex method, genetic algorithm, Levenberg–Marquardt, and simulated annealing [20]. Moreover, these training algorithms can be classified based on their ability to avoid getting trapped at local minima into local or global training. Also, they can be classified as gradient-based or nongradient-based depending on whether or not gradient information is utilized in the training process. The example in Figure 6.7 shows a simple MLP3 network with four inputs and two outputs. The activation function of a neuron is a relay function and its main purpose is to convert an input signal to an output signal at a given layer so that it can be fed as an input to the next layer. The output of a neuron is the weighted sum of its inputs. The activation function can be realized in several forms such as a bipolar sigmoid, an arc-tangent, or a hyperbolic tangent function. The sigmoid function is expressed as:

$$S(\gamma) = \frac{1 - e^{-\gamma}}{1 + e^{-\gamma}} \tag{6.13}$$

where γ and S are the input and the output of the neuron. The arc-tangent function is given by:

$$S(\gamma) = \frac{2}{\pi}\arctan(\gamma) \tag{6.14}$$

while, the hyperbolic tangent is given by:

$$S(\gamma) = \frac{e^{\gamma} - e^{-\gamma}}{e^{\gamma} + e^{-\gamma}} \tag{6.15}$$

All of these functions are continuous, monotonic, bounded, and differentiable. Also, all functions exhibit a smooth switching function with S equals negative one and positive one for negative and positive values of γ, respectively. The transition between zero and one occurs around $\gamma = 0$. The commonly used activation function in ANN is sigmoid because of its computational simplicity and it suites well applications that produce bounded output values in the range of $[-1, 1]$. Training data are usually scaled prior to entering into the input layer to suit the neural network operation. Indeed, data scaling is an important step because the input parameter values may differ from one modeling problem to another by orders of magnitude. This in return could affect the learning quality of the ANN model as the output values may become sensitive to the input values with higher magnitude. Scaling schemes could be linear or log-arithmetic scaling depending on the applied data. In a similar fashion, the output results are obtained from the output layer of the ANN after scaling them back to obtain the actual outputs.

ANN models feature the following advantages:

- ANNs exhibit exceptional generalization and good extrapolation capabilities.
- The universal approximation theorem states that a simple ANN structure with a single hidden layer can approximate any arbitrary nonlinear, continuous, and multidimensional input–output function to any level of the desired accuracy.
- ANN enables fast, flexible methods of producing training data using local, global, and adaptive sampling techniques (e.g., error-based, density-based, gradient-based, or hybrid).
- ANNs are generic, flexible, and scalable, and this broadens the scope of their application.

The size of training data is an important factor that determines the efficiency of the modeling process. Although only samples from the design space are used as training data, they are computationally or experimentally expensive. To cut-down the training data requirement, one may apply the SVM concept, which is gaining popularity as a competitive alternative to ANNs. SVM-based modeling is another method that establishes its efficiency by means of mapping two models of the problem with different fidelities. In essence, coarse model data are augmented with fine model data to describe the problem. As an example, empirical circuit theory data can represent the coarse model data of a given problem while full-wave EM simulation data of the same problem represent the fine model data. Here, the fine model data are used to calibrate the coarse data when needed to improve the modeling accuracy and speed.

In general, an ANN or SVM modeling without the incorporation of the non-linear power of the activation function would be a linear regression model, which has limited capability when trying to learn complex nonlinear data.

6.5.3 Automatic model generation

With today's advances in CAD tools, it is becoming possible to link different tools to automate a design or a modeling task. Using the power of scripting and easiness of interfacing between the well-developed software packages, it becomes feasible to automatically generate models for RF components which are frequently used in the design of large modules (e.g., PCB embedded inductors, MIM capacitors, resistors, matching networks, etc.). Such models can be implemented in product development kits (PDKs) for use by the circuit designers rather than incorporating ideal elements that do not resemble the real behavior of the actual component. These library models can help in guaranteeing manufacturability-driven designs and reliable systems while ensuring lower cost and shorter product time-to-market.

Training data required for the development of these models can be obtained through measurements or from CEM solvers in an automated fashion to yield models with high-level of accuracy. One of the comprehensive modeling tools is the MATLAB®-based SUrrogate MOdeling (SUMO) toolbox developed by a research team at Ghent University [21]. SUMO provides a framework to implement, test, and benchmark new modeling and sampling algorithms [22,23]. Well-distributed, sufficient, and accurately measured/simulated training data is crucial for generating highly accurate models. Therefore, to automatically produce an optimal yet minimum-sized data set, an adaptive sampling algorithm is required. The final model once generated can be integrated into a standard circuit tool such as Keysight's ADS [24] allowing seamless multidomain analysis.

Figure 6.8 illustrates an automatic model generation framework that involves several tools operating in harmony to provide final models based on full-wave EM simulation data with little user interaction once the flow is developed and tested.

It is worth pointing out that different PCB material stack and thicknesses require generating new models of the passive RF components to capture the correct parasitics influencing the modeled element values and its broadband behavior over frequency.

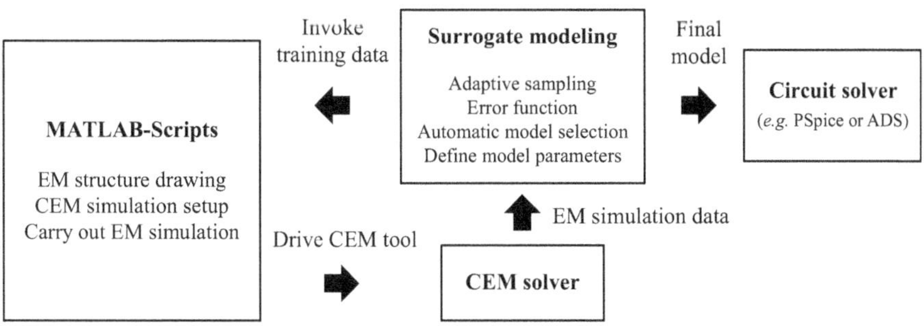

Figure 6.8 Block diagram of the automatic model generation process

6.6 Power handling analysis

Wireless communication service providers are always enthusiastic about the idea of increasing the product transmit power up to the permitted levels as a simple way to boost communication range between TX and RX devices and thus reducing the number of cell towers needed for coverage. Recently, a new class of TDD-LTE cellular networks that enables high power user equipment (HPUE) is permitted under the third generation partnership project (3GPP) standards [25]. Although HPUE permits wide coverage, the continual demand on compact size transceiver modules brings more challenges regarding their power handling capabilities that may impact the product lifespan, breakdown, and frequency de-tuning.

In general, applying an RF power to an electronic device causes an overall rise in the device temperature due to power dissipation across the lossy elements of the device. Also, this may be accompanied with several hotspot locations where the device exhibit high current densities. To address thermal related high-power issues, heat sinks or passive cooling mechanisms must be implemented through the packaging technology while taking into account maintaining a small form factor device.

Two types of power analysis are commonly used. One with peak RF input power that usually occurs over a small period of time and is used to test the ability of standalone components to withstand pulsed and sudden changes in power levels that may lead to an RF breakdown. And another with continuous wave (CW) RF input power that occurs over an extended period of time to test the device reliability and long-term effects such as aging and functionality degradation. The pass/fail criteria in each of the two tests are usually determined after performing actual laboratory measurements to note the maximum power a device can handle at which it breaks down. Accordingly, power thresholds are recorded for use in subsequent power simulations with the appropriate scaling factor depending on the device under evaluation.

When performing a power handling capability test on a standalone device, it is important to run the test under matching conditions similar to that in its intended module. Otherwise, a large portion of the input power will be reflected at the device ports, and thereby less power is propagating through the device leading to an inaccurate test. Other factors that may impact the power handling results are the ambient temperature and moisture surrounding the device under test.

As an example of calculating the volume energy density and area power density of a parallel plate capacitor, consider a capacitor with stored energy measured in Joule is defined as:

$$w_c(t) = \frac{1}{2} C v_c^2(t) \tag{6.16}$$

Hence, the stored energy is always a positive number. The capacitance formula in terms of the plate physical dimensions is defined as:

$$C = \varepsilon \frac{A}{d} \tag{6.17}$$

where A is the parallel plate area and d is the separation distance between the plates. The voltage across the capacitor can be expressed in terms of the associated electric field (E) as:

$$V = Ed \tag{6.18}$$

Thus, the energy density (per unit volume) can be calculated as:

$$w = \frac{1}{2}\varepsilon E^2 \tag{6.19}$$

From (6.19) it is demonstrated that the energy density in a dielectric medium is proportional to the permittivity of the medium. Thus, high dielectric constant materials are more susceptible to high-power related breakdowns when losing its insulating ability.

The energy stored per unit time is the instantaneous power. Thus the capacitor's instantaneous power, $p_c(t)$, in Watts can be given by:

$$p_c(t) = \frac{dw_c(t)}{dt} = \underbrace{C\frac{dv_c(t)}{dt}}_{i_c(t)} v_c(t) \tag{6.20}$$

where $i_c(t)$ is the current through the capacitor. The area power density (in W/m^2) is simply the average/real power dissipated through the resistive part of the capacitor (e.g., contact resistance) in the form of thermal energy and is given by:

$$P_{\text{density}} = \frac{\text{Re}\{p_c(t)\}}{A} \tag{6.21}$$

6.7 Electromagnetic compatibility analysis

Ensuring reliable operation of a system within a given EM environment without introducing unwanted EM disturbance or being disturbed by other coupling sources has led to the development of specialized analysis methods such as electromagnetic compatibility (EMC) analysis. A given design is usually analyzed for electromagnetic interference (EMI) vulnerability by relying on experimental methods to ensure compliance. These experimental methods are, however, time-consuming, costly, and difficult or impossible to reproduce in computer simulations. Such limitations hinder our ability to carry out accurate and efficient EMI/EMC analysis of electronic modules in realistic operating conditions and thus lead to higher costs, longer product time-to-market, and sometimes cancellation of a product.

Most of the important developments and practices in EMI/EMC engineering were made possible by the advent of the efficient CEM algorithms and tools along with the beginning of the modern computing power. This evolution immensely aided in providing the capabilities needed for emulating the complexity of modern electronic systems and carrying out efficient and accurate EMC analysis with affordable computational resources.

The recent development on the EMI/EMC analysis approaches employs hybrid techniques that exploit the domain decomposition idea illustrated in Figure 6.9 and allows concurrent analysis. Under this scheme, traditionally, a time-domain technique based on model-reduction method [26–32] is employed, where the linear domain (cavities, cables, PCB traces, etc.) is solved using finite-difference-time-domain (FDTD) methods (e.g., Sigrity's product, Speed2000 [33]) to extract the PCB port characteristics. Consequently, a compact time-domain macromodel of the linear domain is obtained. By linking the macromodel to a circuit simulator which includes nonlinear and digital loads, concurrent EMI/EMC analysis can be carried out. The time-domain techniques offer accurate analysis, however, suffer from computational inefficiencies and have convergence problems for circuit elements characterized by stiff differential equations. In this context, the partial element equivalent circuit (PEEC) method is used to transform an entire EM structure into a circuit compatible form (i.e., RLC network) that can be combined with nonlinear elements under the same MNA matrix [34,35]. Although PEEC has been incorporated in many commercial products (e.g., Simlab PCB-Mod product from FlowCAD [36], SIwave-TPA from ANSYS [37], GeckoEMC [38], EMIStreem from NEC [39]), main challenges of the PEEC method as a linear modeling approach are the computational inefficiencies due to large MNA matrix for high-frequency analysis and the difficulty in modeling nonlinear and anisotropy properties of magnetic materials [40]. Accordingly, a new domain decomposition method, known as the hybrid Scattering (or *S*-) matrix approach [17,41,42], has been introduced based on field extraction and port analysis methods [43] for accurately analyzing both near- and far-zone EMI sources and their impact on a given EM structure. This scheme offers an efficient yet accurate and concurrent EMI/EMC analysis by combining the linear and nonlinear domains of a given EMI/EMC

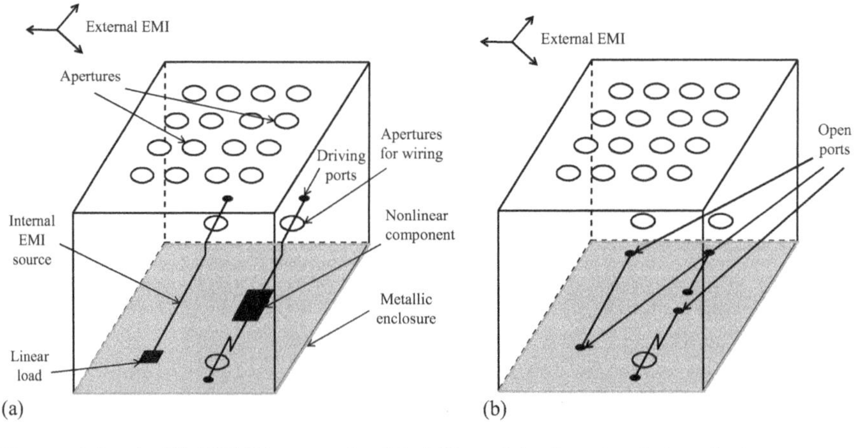

Figure 6.9 An EMI/EMC example: (a) full model of a mixed signal electronic system and (b) model-problem in (a) with open ports (linear domain only)

problem via a hybrid port, representing an external EMI source. Here, the frequency dependent nature of the modal waves of a structure requires S-parameters to be computed at each frequency of interest; hence, dispersion characteristics of TLs are preserved [41]. However, this scheme is limited to a single EMI source and valid only when the circuit structure supports the dominant propagating TL modes on its interconnects. Furthermore, the approach was first introduced to support plane wave excitation to a TL, where a distinct forced mode associated with the incident plane wave and natural TL modes due to reflections from the line ends are induced. A near-zone source, however, induces a continuous spectrum of forced modes and it is difficult to employ curve fitting approaches such as the generalized pencil of function (GPOF) or Prony's methods for extracting the TL modes [42–44] due to the large computational overhead. Both GPOF and Prony's techniques are linear methods that can be used to estimate the parameters of a waveform shaped as a sum of complex exponentials. As a solution to these challenges, FFT methods can be employed rather than the GPOF to identify the TL modes in the presence of the continuous spectrum of forced modes introduced by the near zone sources [42,45]. Moreover, it is well-known that a continuous spectrum of plane waves can represent a near-zone source and to capture the ever-changing effects, analysis based on Weyl's identity [46] are essential. Weyl's identity states that a continuous spectrum of plane waves can represent a spherical wave.

6.8 Monte Carlo yield analysis and optimization

Several modern, highly integrated RF and microwave modules suffer from large variations due to tolerances of their manufacturing processes, particularly at micro/nanoscale fabrication. Due to a large number of possible variations, it is difficult or even impossible to verify the impact of all manufacturing variations through high-fidelity simulations. As a result, achieving a high production yield is becoming a difficult task especially during an aggressive project timeline. In most cases, design engineers rely on lessons learned from previous mistakes when designing for a given electronic product. Following this approach, design issues are detected only after a prototype is built and tested. This process leads to multiple and slow design iterations, higher manufacturing costs, and increased product time-to-market. Therefore, it is highly desirable to be able to predict the possible design failures at early stages of the product development in order to identify the optimum design variant that can withstand manufacturing variations and pass the specifications.

In this context, Monte Carlo analysis [47] is a viable design centering simulation method that helps efficiently avoid the design specifications and consequently yield loss. Monte Carlo analysis is a set of statistical algorithms and techniques based on the random sampling of the design input parameters describing the manufacturing variability. Among several tolerance analysis techniques, Monte Carlo is a powerful and comprehensive approach that provides a statistical model capable of predicting failures, sensitivity, or reliability of the model parameters by conducting a reduced number of experiments to cover the design space. It is, therefore, possible to conclude its analogy with machine learning-based algorithms and response surface models in acquiring a reduced number of training data points to generate a parametric model that can then be utilized to predict outcomes. The reliability of the analysis is tied to the estimates the designer makes. In a sense, after a nominal design has been established and has passed the design specifications, a designer needs to identify which design parameters have the most influence on the design specifications and what are the critical elements that impact these specifications. Omitting any of the central design parameters could result in total module performance failure.

The worst-case analysis achieved by varying the design parameter values to their extremes in a way that produces the worst possible result is referred to as corner simulations. Using Monte Carlo analysis, interpolation estimates between these corners and the nominal design is possible by adaptively and randomly sampling the design space to provide device/module performance statistics due to its components variations either as a probability density function (PDF) or cumulative distribution function (CDF). In essence, the more random points that can be used to sample a problem, the better the model approximation is. Figure 6.10 illustrates a common scheme applicable for tolerance analysis and estimation of production yield.

Normal distribution (also called the Gaussian distribution) is generally used to characterize the uncertainty associated with the estimated design samples. According to the central limit theorem (CLT) in probability [47], the properties of a sufficiently large number of samples of an independent random variable tend toward a normal distribution regardless of the distribution from which they are sampled.

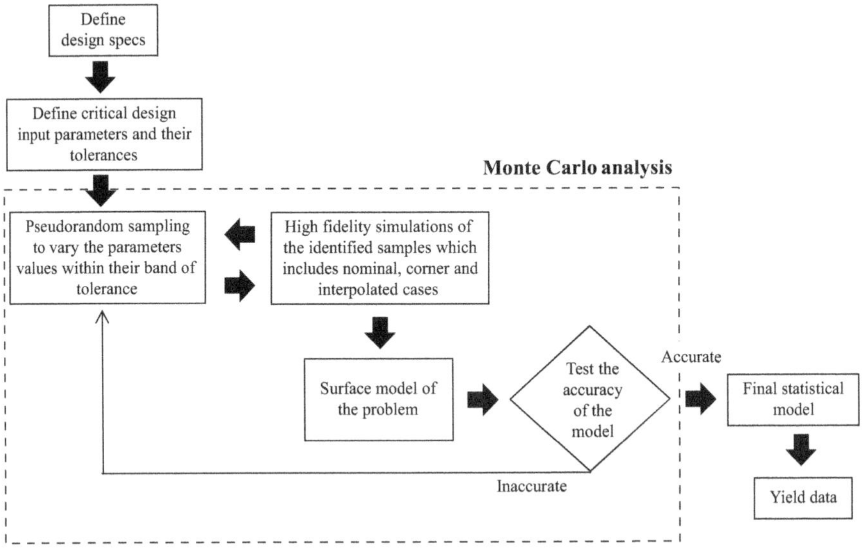

Figure 6.10 A possible scheme of Monte Carlo tolerance analysis for an electronic device

Usually, the behavior of a small sample size is unpredictable, and with a large sample size, the sample statistics (i.e., mean and variance) obtained from the Monte Carlo experiments can vary from one experiment to another. Therefore, the results can then vary about the actual population mean within an error [48]. The margin of error (E) under a given confidence level (CL) or a confidence interval ($Z_{a/2}$) can be calculated in two ways depending on the knowledge of the population standard deviation (σ) as follows:

$$E \geq \begin{cases} Z_{a/2}\dfrac{\sqrt{P(1-P)}}{\sqrt{N}}, \text{ unknown } \sigma \\ Z_{a/2}\dfrac{\sigma}{\sqrt{N}}, \text{ known } \sigma \end{cases} \tag{6.22}$$

where P is population proportion and $0 \leq P \leq 1$. In most cases, the actual population statistic parameters are unknown, and it is difficult to specify σ for the population. Alternatively, the sample statistic can be calculated when the sample data are in hand and, therefore, a population proportion can be used to estimate the sample size. If P or σ are not known, $P = 0.5$ can be assumed that results in the largest number of samples. The sample size (N) under a certain confidence level and margin of error can be rewritten by rearranging (6.22) as:

$$N \geq \begin{cases} \left(\dfrac{Z_{a/2}}{E}\right)^2 [P(1-P)] \\ \left(\dfrac{\sigma Z_{a/2}}{E}\right)^2 \end{cases} \tag{6.23}$$

The standard deviation measures the spread of data from the mean value and the farther the deviation from the mean, the greater the standard deviation number is. In the analysis above, the higher the standard deviation number, the higher the confidence level, which means a wide spread of data points are used to conclude the yield result.

Finally, after performing MC analysis to determine the distribution functions, the input parameters of a design can then be centered so that optimal parameter values are obtained to yield the least sensitive results using yield optimization algorithms.

6.9 Circuit elements sensitivity

The sensitivity of a circuit is a measure of the rate of change of the circuit outputs due to variation in its input parameters. A simple means of finding the sensitivity of an electric circuit can be thought as the partial derivative of the circuit output response with respect to its design input variables as expressed mathematically by:

$$S = \frac{\partial f(x, y, \ldots)}{\partial x} = \lim_{\Delta x \to 0} \frac{f(x + \Delta x, y, \ldots) - f(x, y, \ldots)}{\Delta x} \tag{6.24}$$

where S is the sensitivity, $f(x)$ is the characteristic of the circuit, and x is one of the design parameters.

Evaluating the sensitivity of an electronic circuit provides a means to make suitable performance tradeoffs and to identify the critical components in the design. One of the powerful methods used in circuit sensitivity analysis and design optimization is the Lagrangian method based on a backward adjoint approach. The adjoint or auxiliary network sensitivity method [49] can be employed to work in conjunction with the Tellegen's theorem [1,49] to obtain an efficient analysis. The underlying principle relies on defining an auxiliary network that establishes a simple linkage to the original network, for example, performing a transpose operation on the nodal admittance matrix of a circuit or reciprocity relations between the networks. This auxiliary step lends itself to reduced expressions for the sensitivity equations. Furthermore, it enables sensitivity calculation at cost function independent of the number of design variables. Adjoint variables are gradients of the cost function with respect to the design input parameters. The adjoint analysis is efficient when calculating sensitivities of a small number of outputs with respect to a large number of inputs. In this context, Tellegen's theorem is a useful network analysis method and very general in a sense that it is valid for any linear or nonlinear, active or passive networks. The theory indicates that the summation of instantaneous powers for an N number of branches in an electrical circuit is zero. Considering an electrical circuit consisting of N number of branches with i_N instantaneous currents flowing through them and satisfies KCL and V_N instantaneous voltages across them and satisfies KVL, then the Tellegen's equation is expressed as:

$$\sum_{k=1}^{N} V_k I_k = 0 \tag{6.25}$$

6.10 Six Sigma

Six Sigma is a quality measure that has become more pervasive in the industry and is used to statistically measure the quality of manufactured products or to improve, analyze, and control fabrication or assembly processes. It can also be used as a benchmark of a specific product or a process capability in comparison to others. The product can be a complete system, a module, a discrete component, or a component within a module. Similarly, a process can be a fabrication process, an assembly process, or all their involved steps. This data-driven approach for eliminating defects and ensuring quality improvement was pioneered at Motorola in 1987 [50]. The Greek letter sigma refers to the standard deviation from the mean in a Gaussian distribution of a given process and indicates how much variation exists in a process (i.e., how often defects are likely to occur). The standard deviation can be calculated as:

$$\sigma = \sqrt{\frac{\sum (x_i - \bar{x})^2}{n-1}} \tag{6.26}$$

where $\bar{x}$ is the average (statistical mean) of the samples, and n is the sample size. Using statistical analysis, we usually draw inference and estimate characteristics of the whole population (mean and variance) from the samples. The variability of a component can be described by a range between the extreme values of a variable, a standard deviation, or a variance.

The Six Sigma is a common measure that implies six standard deviations between the mean and the nearest tolerable specification limit in the process. Six Sigma means that a process must not produce more than 3.4 defects per million parts, almost free of defects. In other words, this metric demonstrates quality levels at 99.9997% performance for products or processes. Three sigma results in 66,807 defects per million and four sigma gives 6,210 defects per million. Hence, increasing sigma results in a nonlinear decrease in the number of defected parts per million (ppm). Also, the increase in product quality as sigma increases translates into a cost decrease. As a matter of fact, complying with the six sigma production standard in the case of a system on a module that contains several parts does not imply that only 3.4 defective systems per million have been produced. The 3.4, in this case, represents the number of possible errors. If a system consists of hundreds of different components that are fabricated using different processes and is assembled using several procedures, then a large number of possible errors is considered per each system. Therefore, another terminology can be used to describe such possible errors is the defects per million opportunities (DPMO). Despite the fact that different companies can tailor the use of Six Sigma for their own product quality purpose that is also driven by what is important to the customer, companies that have embraced the Six Sigma programs have achieved the highest quality, lowest cost products, and shorter product cycle time. In this context, when dealing with mass-produced products, their performance must be evaluated in an efficient manner. Therefore, automating the measurement process and the easiness of extracting the measured performance data is essential to observe the quality and variability of the process efficiently.

Adopting the Six Sigma approach involves design, product, or process engineers to have the basic knowledge of using statistical tools in order to obtain reduced defect targets. Whether the quality improvement is targeting a product or a process, figures of merit (FOM) are defined following their statistical distribution. The distribution statistics for a process is usually plotted so that the x-axis carries the range of variations on the measured performance of the design or process parameter of interest and the y-axis carries the number of parts corresponding to each of the x-axis variables. The larger the number of samples, the closer the distribution to the normal distribution. The normal distribution has a bell shape density curve with a peak at the mean value that drops off symmetrically at both sides away from the mean. The width of the drop off around the mean is governed by the distribution standard deviation parameter. The normal distribution function with a variable (x), a mean (μ), and a variance (σ^2) is a statistic distribution with probability density function described by:

$$f(x) = \frac{1}{\sigma\sqrt{2\pi}} e^{-\frac{1}{2}\left(\frac{x-\mu}{\sigma}\right)^2} \tag{6.27}$$

The cumulative distribution function can be written as the integral of its probability density function and is given by:

$$F(x) = \int_{-\infty}^{x} f(t)dt = \int_{-\infty}^{x} \frac{1}{\sigma\sqrt{2\pi}} e^{-(t-\mu)^2/(2\sigma^2)} dt \tag{6.28}$$

The integral of the normal cumulative distribution function in (6.28) does not exist in a simple closed form. However, it can be computed numerically. Figure 6.11 illustrates the PDF curves of a normally distributed function for different standard deviations and Table 6.5 lists the different C_{pk} values for different deviation values from the mean assuming 1.5σ shift in the mean from the center of the specification lines. As can be observed, when sigma increases, DPMO decreases and C_{pk} increases.

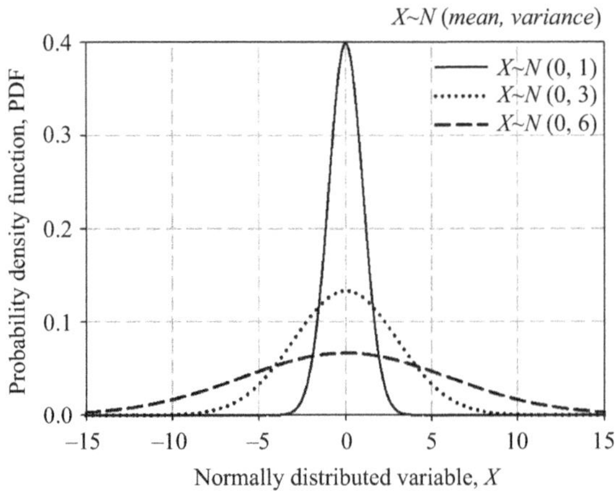

Figure 6.11 Normal PDF curves for different standard deviations at the same mean value

Table 6.5 Sigma, DPMO, and C_{pk} equivalency conversion table assuming normally distributed samples and 1.5 sigma shift

Deviation (σ)	**DPMO**	C_{pk}
$\pm 1\sigma$	691,462	0.33
$\pm 2\sigma$	308,538	0.67
$\pm 3\sigma$	66,807	1.00
$\pm 4\sigma$	6,209	1.33
$\pm 5\sigma$	232.6	1.67
$\pm 6\sigma$	3.4	2.0

The DPMO is calculated as:

$$\text{DPMO} = \frac{\text{Total number of defects}}{\text{Number of samples} \times \text{ number of defect opportunities per sample}} \times 1{,}000{,}000 \tag{6.29}$$

Ideally, a device manufacturer would like to have a probability distribution with σ that approaches zero (meaning that no process variation or part defects) and to relate that to the Six Sigma performance, there are two important metrics, namely, the process capability index (C_p) and the process bias (C_{pk}) that are commonly used. The two metrics are quite similar. The smaller the standard deviation, the greater the C_p and C_{pk} statistics are. Moreover, under ideal conditions, C_p and C_{pk} have exactly the same value.

The C_p metric is the ratio between the allowed tolerance which is the width between the specification limits and the width that includes approximately all the samples and in this context is defined as six-times the standard deviation. C_p is then calculated as:

$$\begin{aligned} C_P &= \frac{\text{USL} - \text{LSL}}{6\sigma} = \frac{\text{CPU} + \text{CPL}}{2} \\ \text{CPU} &= \frac{\text{USL} - \mu}{3\sigma} = \frac{Z_U}{3} \\ \text{CPL} &= \frac{\mu - \text{LSL}}{3\sigma} = \frac{Z_L}{3} \end{aligned} \tag{6.30}$$

where LSL and USL are the lower specification limit, and the upper specification limit, respectively. Data appearing below LSL or above USL are usually the most interesting values concerning the probability of occurrence outside of customer specifications. Z_U and Z_L are derived from normalizing the x values of the observed measurement space to the process standard deviation using the following transformation:

$$Z = \frac{x - \mu}{\sigma} \tag{6.31}$$

When CPU and CPL are equal, this implies that the process is centered between the specification limits. Having a centered process is desirable; however, in practice having the greatest number of units inside the specification limits may not be achievable for a given actual process. This in return can result in obtaining a sample mean different than the actual mean (i.e., the middle of the specification limits) that is usually found to be 1.5σ or greater.

The C_{pk} metric is a number which indicates how close a process to its specification limits. The larger the C_{pk} number, the low the possibility that any part will be outside the specification limit. The C_{pk} is given by:

$$C_{pk} = \min(\text{CPU, CPL}) \tag{6.32}$$

The farther the sample mean value is from the center of the specification limits, the lower the C_{pk} is. Also, the greater the difference between the decided sample mean and the actual mean, the more different the C_p and C_{pk} become. For a process to be of Six Sigma quality, the following conditions must hold:

$$\begin{aligned} C_p &\geq 2 \\ C_{pk} &\geq 1.5 \end{aligned} \tag{6.33}$$

References

[1] Steer M.B., Bandler J.W., and Snowden C.M. "Computer-aided design of RF and microwave circuits and systems." *IEEE Transactions on Microwave Theory and Techniques*. 2002;**50**(3):996–1005

[2] Petre P., and Sarkar T.K. "Planar near-field to far-field transformation using an array of dipole probes." *IEEE Transactions on Antennas and Propagation*. 1994;**42**(4):534–37

[3] Bérenger J.-P. "A historical review of the absorbing boundary conditions for electromagnetics." *Forum for Electromagnetic Research Methods and Application Technologies (FERMAT)*. 2015;**9**(4):1–28

[4] Ho C.-W., Ruehli A., and Brennan P. "The modified nodal approach to network analysis." *IEEE Transactions on Circuits and Systems*. 1975;**22**(6):504–09

[5] Nakhla M.S., and Vlach J. "A piecewise harmonic balance technique for determination of the periodic response of nonlinear systems." *IEEE Transactions on Circuits and Systems*. 1976;**23**(2):85–91

[6] Flores J.L., and Suárez A. "On the use of the auxiliary generator technique to improve non-linear microwave circuit analysis and design." *ARMMS RF & Microwave Society Conference*; Steventon, Oxfordshire, UK, April 2013, pp. 1–10

[7] Keysight Advanced Design System (ADS). "Harmonic balance simulation." Feb 2011. Available at: http://edadownload.software.keysight.com/eedl/ads/2011_01/pdf/cktsimhb.pdf [Accessed April 15, 2019]

[8] Skelboe S. "Computation of the periodic steady-state response of nonlinear networks by extrapolation methods." *IEEE Transactions on Circuits and Systems*. 1980;**27**(3):161–75

[9] Kundert K. "Simulation methods for RF integrated circuits." *1997 Proceedings of IEEE International Conference on Computer Aided Design (ICCAD)*; San Jose, CA, 1997. pp. 752–65

[10] Kundert K.S., and Sangiovanni-Vincentelli A. "Simulation of nonlinear circuits in the frequency domain." *IEEE Transactions on Computer-Aided Design of Integrated Circuits and Systems*. 1986;**5**(4):521–35

[11] Keysight Advanced Design System (ADS). *Circuit Envelope Simulation*. 2011. Available at: http://edadownload.software.keysight.com/eedl/ads/2011/pdf/cktsimenv.pdf [Accessed April 15, 2019]

[12] Condon M., and Dautbegovic E. "A novel envelope simulation technique for high-frequency nonlinear circuits." *33rd European Microwave Conference Proceedings*, Munich, Germany; vol. 2, Oct 2003. pp. 619–622

[13] He Q., and Jiao D. "Co-simulation of linear electromagnetic structures and non-linear devices in the time-domain finite-element reduction-recovery method." *IEEE Antennas and Propagation Society International Symposium*; Charleston, SC, 2009. pp. 1–4

[14] Bayram Y., and Volakis J.L. "Hybrid S-parameters for transmission line networks with linear/nonlinear load terminations subject to arbitrary excitations." *IEEE Transactions on Microwave Theory and Techniques*. 2007; **55**(5):941–50

[15] Shakouri A., Kang S.M., Bar-Cohen A., and Courtois B. "Scanning the special issue on on-chip thermal engineering." *IEEE Proceedings*. 2006;**94**(8): 1473–75

[16] Bagnoli P.E., Casarosa C., Dallago E., and Nardoni M. "Thermal resistance analysis by induced transient (TRAIT) method for power electronic devices thermal characterization – Part I: Fundamentals and Theory." *IEEE Transactions on Power Electronics*. 1998;**13**(6):1208–19

[17] Yovanovich M. "Four decades of research on thermal contact, gap, and joint resistance in microelectronics." *IEEE Transactions on Components and Packaging Technologies*. 2005;**28**(2):182–206

[18] Im S., Srivastava N., Banerjee K., and Goodson K. "Thermal scaling analysis of multilevel Cu/low-*k* interconnect structures in deep nanometer scale technologies." *22nd International VLSI Multilevel Interconnect Conference (VMIC) Proceedings*; Fremont, CA, 2008, pp. 525–30

[19] Roest D., Ymeri H., Vandenberghe S., *et al.* "Some measurement results for frequency-dependent inductance of IC interconnects on a lossy silicon substrate." *IEEE Electron Device Letters*. 2002;**23**(2):103–04

[20] Zhang Q.J., and Gupta K.C. *Neural Networks for RF and Microwave Design*. Norwood, MA: Artech House; 2000

[21] *SUMO Toolbox* [online]. Available from http://www.sumo.intec.ugent.be/SUMO [Accessed Feb 02, 2018]

[22] Couckuyt I., Crombecq K., Gorissen D., and Dhaene T. "Automated response surface model generation with sequential design." *First International Conference on Soft Computing Technology in Civil, Structural and Environmental Engineering*; Madeira, Portugal, Sept. 2009, pp. 1–16

[23] Couckuyt I., Gorissen D., Crombecq K., Deschrijver D., and Dhaene T. "The SUMO toolbox: a tool for automatic regression modeling and active learning." *2013 IEEE Africon*; Pointe-Aux-Piments, Mauritius, Sept. 2013, pp. 1–4

[24] Keysight Advanced Design System (ADS). *Ver. 2016.10*. Fountaingrove Parkway, CA: Agilent Technologies

[25] 3GPP. "TR 36.837 – V11.0.0 – public safety broadband high power user equipment (UE) for band 14." 2012. Available at: https://www.3gpp.org/ftp/Specs/archive/36_series/36.837/ [Accessed April 15, 2019]

[26] Kerns K. and Yang A. "Preservation of passivity during RLC network reduction via split congruence transformations." *IEEE Transaction on Computer-Aided Design*. 1998;**17**(7):582–91

[27] Gunupudi P., Nakhla M., and Achar R. "Simulation of high-speed distributed interconnects using Krylov subspace techniques." *IEEE Transaction on Computer-Aided Design of Integrated Circuits and Systems*. 2000;**19**(7):799–808

[28] Achar R., Gunupudi P., and Nakhla M. "Passive interconnect reduction algorithm for distributed/measured networks." *IEEE Transaction Circuits and Systems*. 2000;**47**(4):287–301

[29] Zheng G., Zhang Q.J., Nakhla M., and Achar R. "An efficient approach for simulation of measured subnetworks with complex frequency hopping." *Proceedings of IEEE/ACM International Conference on Computer Aided Design (ICCAD)*; San Jose, CA, Nov. 1996. pp. 23–26

[30] Yu Q., Wang J.M.L., and Kuh E.S. "Multipoint multiport algorithm for passive reduced-order model of interconnect networks." IEEE *International Symposium on Circuits and Systems*; Monterey, CA, 1998, pp. 74–77

[31] Odabasioglu A., Celik M., and Pillage L.T. "PRIMA: Passive reduced-order interconnect macromodeling algorithm." *IEEE Transaction Computer-Aided Design*. 1998;**17**(8):645–54

[32] Li E.-P., Liu E.-X., Li L.-W., and Leong M.-S. "A coupled efficient and systematic full-wave time-domain macromodeling and circuit Simulation method for signal integrity analysis of high-speed interconnects." *IEEE Transaction on Advanced Packaging*. 2004;**27**(1):213–23

[33] Miersch E., Goekcen M., and Steinecke T. *Time and frequency analysis of signal and power noise of a microcontroller (μC) plus its packaging (LQFP+PCB)* [online]. http://www.sigrity.com/papers/2009/Time%20and%20Frequency%20Analysis%20Paper%20Feb%202009.pdf [Accessed Feb 03, 2018]

[34] Kovacevic I., Muesing A., Friedli T., and Kolar J. "Electromagnetic modeling of EMI input filters." *7th International Conference on Integrated Power Electronics Systems*; Nuremberg, Germany, March 2012, pp. 1–9

[35] Gope D., Ruehli A., and Jandhyala V. "Surface-based PEEC formulation for modeling conductors and dielectrics in time and frequency domain combined circuit electromagnetic simulations." *IEEE 13th Topical Meeting on Electrical Performance of Electronic Packaging*; Portland, OR, Oct. 2004, pp. 329–32

[36] *FolwCAD* [online]. Available from http://www.flowcad.eu/ [Accessed Feb 05, 2018]

[37] ANSYS SIwave – Turbo Package Analyzer (TPA) [online]. Available: http://www.ansys.com

[38] *GeckoEMC* [online]. Available from http://www.gecko-simulations.com/geckoemc.html [Accessed Feb 05, 2018]

[39] *NEC-EMIStream* [online]. Available from http://www.nec.com/en/global/prod/emistream [Accessed Feb 05, 2018]

[40] Kovačević I. "PEEC-based virtual design of EMI input filters." PhD dissertation, ETH Zurich, 2012
[41] Bayram Y. "Hybrid methods for mixed signal circuits subject to on & off-board electromagnetic interference." *PhD dissertation*. The Ohio State University, 2006
[42] Khan Z. A. "EMI/EMC analysis of electronic systems subject to near zone illuminations." *PhD dissertation*. The Ohio State University, 2007
[43] Lee D.C.K., Sowerby K.W., and Neve M.J. "Extracting fine multipath detail from measured data at 5.8 ghz." *IEEE 59th Vehicular Technology Conference*; Milan, Italy, May 2004, pp. 74–78
[44] Izmirli O. "Non-harmonic sinusoidal modeling synthesis using short-time high resolution parameter analysis." *Proceedings COST G-6 Conference Digital Audio Effects*; Verona, Italy, Dec. 2000, pp. 7–9
[45] Khan Z.A., Bayram Y., and Volakis J.L. "EMI/EMC analysis of printed circuit boards subject to near-zone illuminations." *IEEE Transactions on Electromagnetic Compatibility*. 2009;**52**(2):406–08
[46] Gao J., and Shanker B. "Time domain Weyl's identity and the causality trick based formulation of the time domain periodic Green's function." *IEEE Transactions on Antennas and Propagation*. 2007;**55**(6):1656–66
[47] Sadiku M. *Monte Carlo Methods for Electromagnetics*. 1st edn. Boca Raton, FL: CRC Press; 2009
[48] Sternstein M. *Statistics*. Barron's Series. *New York, NY*: Hauppauge; 1996
[49] Ilievski Z. Model Order Reduction and Sensitivity Analysis. *Eindhoven: Eindhoven University of Technology*; 2010
[50] Golio M. *The RF and Microwave Handbook*. Boca Raton, FL: CRC Press; 2001

Chapter 7

Components of RF front-end modules

7.1 Introduction

Driven by the original equipment manufacturers (OEMs) who make smartphones, the fourth generation (4G) and the fifth generation (5G) mobile technologies are changing the radio frequency (RF) front-end (RFFE) industry landscape. In particular, 5G wireless technologies which will enable much faster network speeds are expected to further shape the RFFE modules market by integrating more complex functionalities and contents into the RFFE modules. Moreover, 5G technology will witness an introduction of new business opportunities for the deployment of the commercial 5G network infrastructures and consequently the demand for software in communication networks to establish the required communication protocols. It is anticipated that 5G technology will become widely available for both mobile and cellular users by the year 2020 and reach full deployment by 2025. The 4G development activities over the past decade have led to impressive innovations at the devices level which in return paved the way for the development of the 5G device and system level. This highly complex industry with diverse products and technology challenges is dominated by a few major RFFE module suppliers (e.g., Broadcom, Skyworks Solutions, Qorvo, TDK, and Murata) who account for approximately 90% of the RFFE market.

The wireless RFFEs and components market is highly dynamic and has witnessed unprecedented developments over the past decade for mobile devices and especially for smartphones. The total available market (TAM) in this rapidly evolving business is expected to reach $22B by 2022 [1]. Such high growth would attract players in other semiconductor markets to consider and eventually participate in this market opportunity. The growth, however, in the RFFE market is not evenly distributed and some components manufacturers gain higher market shares than others. This is dependent upon on several factors including the type of the component or product, cost, and performance when compared to the existing competitors. Due to the number of incorporated filters in a typical RF module, filters represent the biggest business in the RFFE industry with the expectation that the value of this business to be more than triple from 2016 to 2022 [1]. The growth is driven by the demand on filtering solutions in response to the increasing number of RFFE module antennas as well as due to the demand for multiple carrier aggregation (CA) schemes to enable multiple long-term evolution (LTE) carriers that can be used simultaneously to provide high data rates and broadband capabilities. The second biggest portion of this industry is for power amplifiers (PAs) and low noise amplifiers (LNAs), and their market growth in the RFFE business is expected to be flat over the next few years. This is because the PA market growth is driven by the number of transmitting bands in an RFFE module, while the LNA market is mainly driven by the addition of new antennas. Given that one PA can support multiple transmitting bands and similarly an LNA can support multiple receiving bands; their numbers in any RFFE module is relatively less than the number of incorporated filters. The third biggest business is the RF switches, especially, antenna switches and their market growth is expected to double over the next few years [1]. Next in the business growth are the antenna tuners, with growth expected to reach 7.5-fold by the year 2022 [1] and this market is driven by the employment of diversity antennas and multiple-input-multiple-output (MIMO) antennas to allow higher data rates and higher effective range.

The basic RFFE module architecture generally includes any circuitries between the antenna and the digital baseband system. This comprises several active and passive elements and integrated circuits based on different semiconductor processes in each of the transmit/receive RF signal chain. All are designed in a way to attaining successful signal performance in their intended modules. In the RF receive (Rx) path, the RFFE components usually include filters, RF switches, LNAs, and mixers for down-conversion in order to process the signals received at the antenna into signals suitable for input into the baseband. The RFFE portion of the receiver can be thought of as an analog-to-digital conversion process. On the transmit (Tx) path, the RFFE chain can share several components with the receive path such as antennas, time-division duplexing (TDD) filters, and RF switches. In contrast with the Rx path, the Tx chain may include local oscillators and PAs.

This chapter will discuss the different components utilized in the design of modern RFFE transceiver modules, whether used to construct a simple or a highly integrated complex module.

7.2 RF power amplifiers

An RF PA is an indispensable component in almost any RF wireless communication system and is used to boost the RF signal power to sufficient levels for transmission through the interface media (e.g., air) from an RF transmitter to an RF receiver. RF PAs are mainly used in RF power applications that require an output power ranging from milli-watts to tens, or hundreds of watts. RF power applications can include wireless communication transmitters, radars, test instruments, and satellite communication systems. The performance of any wireless system greatly relies on the linearity and efficiency of its power amplifiers. This is because linearity and power efficiency are two opposing performance requirements that necessitate design trade-offs. An ideal PA is expected to boost the strength of the input signal level without changing the signal shape, format, or mode. However, PAs are naturally nonlinear devices, and signal amplification usually occurs with some distortion. In active devices, in general, careful selection of the device biasing point is crucial to find the best compromise between gain, NF, and linearity for any given application.

To give a glimpse of the history of PAs evolution, in 1906, Lee De Forest utilized the diode vacuum tube technology that was one of the electronic devices of interest during that era to develop the first amplifier, called the triode or the audion vacuum tube, by introducing a third electrode to control the current. The primary purpose of using the two electrode diode vacuum tubes was to rectify signals from alternating current (AC) to direct current (DC), however not to amplify them. In 1947, at Bell Telephone Laboratories, a breakthrough in the electronics industry had begun with the advent of the solid-state transistors replacing the bulky audion tubes. The solid-state transistors are smaller, lightweight, and more power efficient when compared to the vacuum tube solution. The name transistor is a simplified terminology of a "transfer resistor" which refers to an electrically controlled variable resistor. Therefore, a transistor amplifier can be thought of as a current or voltage-controlled device designed to convert a low power input signal into a higher power output signal. A hybrid between the two is also possible allowing current to voltage conversion as in a transimpedance amplifier. For instance, a field effect transistor (FET) can be described as a voltage controlled device because the voltage applied to the FET gate determines its output characteristics, while a bipolar junction transistor (BJT) is considered as a current controlled device because its base current controls the current flow from the emitter to the collector of the transistor. Moreover, amplifiers are used to increase the voltage of the signal, the current of the signal, or the power of the signal and can be classified based on their intended application into voltage amplifiers, current amplifier, and power amplifiers. The ratio between the output signal and the input signal whether the signal is a voltage, a current or a power signal is referred to as gain.

In voltage amplifiers the amplitude of the input alternating voltage signal is small, and the collector current is low (meaning that the collector load has high resistance) resulting in a low AC output power. Hence, voltage amplifiers don't dissipate much power and thereby don't produce dissipated heat. The physical sizes of voltage amplifier transistors are relatively smaller than those used in power amplifiers. Its main application is to strengthen the weak voltage signals and make them less affected by the imposed noise level.

Similarly, a current amplifier amplifies the input current by a fixed amount and delivers it to the succeeding circuit. The difference between a voltage amplifier and a current amplifier is in their input and output impedances. A current amplifier has a high input impedance (infinity in an ideal case) and a low output impedance (collector load is ideally zero), while a voltage amplifier has an input impedance that is ideally zero and an output impedance of infinity.

A transimpedance amplifier which is a hybrid between the current and voltage amplifiers has a current signal flowing into a low input impedance transistor and a voltage signal taken from low output impedance.

In power amplifiers, the amplitude of the AC input signal is large, and the collector current is high (collector load has low resistance) resulting in a high AC output power while maintaining minimal change on the output voltage. Most power amplifiers have a significant current gain, which results in power gain, while in voltage amplifiers, current gain is very low. Power amplifiers dissipate relatively higher power in the form of heat when compared to voltage amplifiers, and therefore, a cooling mechanism may be required. The main application is to deliver sufficient power to a given load. In a multistage amplifier, power amplification is implemented toward the final stages.

Transistors can be also classified with respect to their doped materials. For instance, BJTs can be configured into pnp or npn type. A BJT has three terminals: the emitter (E), the base (B), and the collector (C) connected to three separately doped regions forming two junctions. The pnp transistor type implies that an n-doped region is sandwiched between two p-doped regions with two junctions (i.e., pn and np junctions similar to that in diodes). The current travels from the p region to the n region, thus npn and pnp result in two current direction transistors. Similarly, a FET is a three terminal device (i.e., source, gate, and drain) that is categorized into p-type and n-type. In a FET there is a depletion region between the np junctions that exist without any bias.

In active devices such as PAs or LNAs, the optimal load that satisfies maximum output power or best noise performance, respectively, doesn't usually imply a conjugate impedance match at the output of the PA or at the input of the LNA. Most often, a conjugate impedance match and an optimal load presented to the input of an LNA or the output of a PA differ significantly. Therefore, design trade-offs are made to balance the impedance matching conditions between the successive RF stages and the desired output signal characteristics from such active devices. Incorrect design of active circuits can lead to catastrophic failures in the form of damaging the power source and other power-sensitive elements in the RF chain. Therefore, physically debugging the performance of active circuits requires careful attention.

RF power device technology used in producing integrated high-frequency PAs are mostly based on Silicon (Si), gallium arsenide (GaAs), or gallium nitride (GaN) semiconductor materials. Among other technologies, GaAs and GaN provide high-performance PA characteristics and both have their own pros and cons. For example, GaAs-based devices are constructed on a semi-insulating material and thereby low power loss can be obtained. Also, GaAs materials have high dielectric constant and can operate reliably at high temperatures. If compared to Si-based devices, GaAs is more expensive to fabricate, more fragile, and has lower thermal conductivity. On the other hand, GaN-based devices can withstand higher temperatures than GaAs, more power efficient, and has higher power density. However, GaN PAs haven't been out for a long time and therefore hasn't been adopted on a large scale. Table 7.1 lists the different properties of commonly used semiconductor materials in the implementation of PAs.

In terms of the fabrication topology, several options are available depending on the intended application. The list is as follows:

- bipolar junction transistors (BJTs),
- heterojunction bipolar transistors (HBTs),
- metal-semiconductor FETs (MESFETs),
- high electron mobility transistors (HEMTs),
- laterally diffused metal oxide semiconductors (LDMOS), and
- enhancement-mode MOSFETs.

Table 7.1 Different semiconductor material properties for PA device technology

Semiconductor technology	Si	GaAs	GaN	Diamond
Thermal conductivity (W/cm-K)	1.5	0.45	1.3	22
Electric breakdown field strength (MV/cm)	0.3	0.4	2	10
Dielectric constant	11.9	13.1	9	5.5
Electron mobility (cm^2/V.s)	1,500	8,500	1,250	2,200
Hole mobility (cm^2/V.s)	600	400	850	850
Bandgap (eV)	1.12	1.43	3.45	5.45

The choice of the topological and manufacturing process of a PA device technology is dependable upon several factors which include the cut-off frequency at which the power flowing through the PA begins to reduce, the required output power level over frequency, the voltage breakdown, thermal conductivity, power density, electron mobility that indicates if the PA can amplify signals into the gigahertz ranges, and the device linearity performance. PAs can also be classified based on the operating frequencies into audio PAs used to amplify the signals in the audio frequency range (i.e., 20 Hz to 20 kHz), also known as small signal PAs, and radio PAs that are used to amplify signals within the RF range (i.e., 3 KHz to 300 GHz) and also called large signal PAs. Moreover, PAs can also be classified based on the mode of operation (i.e., the location of the operating point on the IV characteristics of the transistor) to include several classes. Sections 7.2.1.1–7.2.1.7 discuss the different PA classes in terms of their advantages and disadvantages.

7.2.1 Classes of power amplifiers

The different electrical characteristics of different types of transistor amplifiers are referred to as PA classes. Different PA classes (classes A, B, AB, and C) vary by the length of their conduction angle (α) over a specified load line as illustrated in Figure 7.1. The conduction angle of a PA is defined as the portion of the input signal cycle (out of 360°) at which the PA conducts. Usually, the higher the conduction angle of the PA, the lower the power efficiency. A transistor load line is a line drawn between the saturation and cut-off points. Other amplifier classes such as D, E, F, G, S, T, etc. are known as switching amplifiers and are primarily utilized in digital circuits and pulse width modulation (PWM). In these classes, the conduction angle is zero and as can be concluded, the output in this set of amplifiers is derived into the saturation and cut-off regions of the transistors to switch the signal between the ON and OFF states.

The selection of a PA class over another is determined by several performance factors. These factors include linearity, bandwidth, gain, power efficiency, bias voltage, and cost.

7.2.1.1 Class A

A single output class A amplifier conducts the full 360° cycle of the input waveform that is the entire period of the input signal. Class A amplifier is very simple in terms

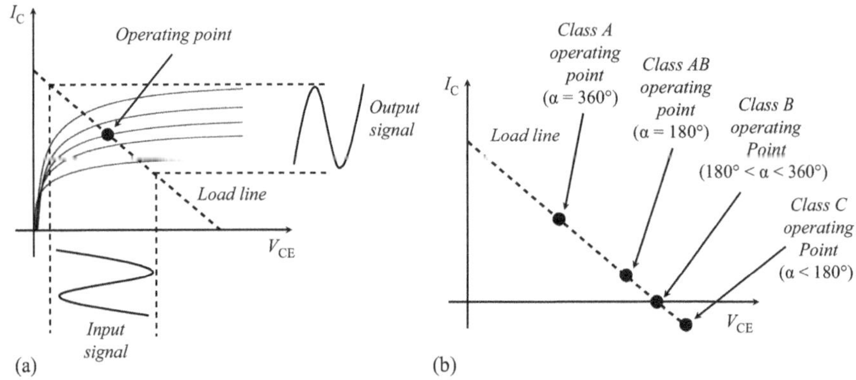

Figure 7.1 (a) An illustrative example of an IV characteristic of a BJT device showing a class A amplifier operating point on the device load line, and (b) amplifier classification into classes according to their mode of operation on the load line

of design and highly linear; however, has the lowest power efficiency among other PA classes with a maximum theoretical efficiency of 50%. The low efficiency in class A amplifiers is attributed to having a biasing point in the active region around a central *Q*-point of its characteristic curves as depicted in Figure 7.1(a) and (b). This implies that class A amplifiers will remain far from saturation or cutoff regions given that the transistors DC biasing is properly selected to ensure correct operation and low distortion. The low harmonic and intermodulation distortion in this class makes it suitable for precise amplification of amplitude-modulated (AM) signals.

7.2.1.2 Class B

Class B has relatively higher efficiency than class A with an ideal maximum efficiency of 78.54%. The PA in this class is biased at a lower operating point compare to that of the class A. The conduction angle in class B is 180°, and thereby, it is not as linear as class A. Unlike class A, class B amplifier can be used in a push–pull configuration as illustrated in Figure 7.2(a) where two class B transistors are necessary in order to reproduce the input signal, one to provide an output for the positive half cycle and another to provide an operation to the negative half cycle of the input signal. Using class B in a push–pull configuration may lead to a type of distortion known as crossover distortion as depicted in Figure 7.2(b). Crossover distortion occurs when the two amplifiers at the output of the push–pull configuration do not fully combine both halves of the waveform and are observed as a flattening of the sinusoidal waveform. This is because of the nonlinearity of the base-emitter junction of each transistor when the base-emitter voltage is less than 0.7 volt that causes misalignment between the zero voltage points of the two amplifiers when the signal crossover from one amplifier to the other. The crossover can be reduced by slightly forward biasing the base-emitter junctions, which leads to another class known as class AB amplifier as will be discussed next.

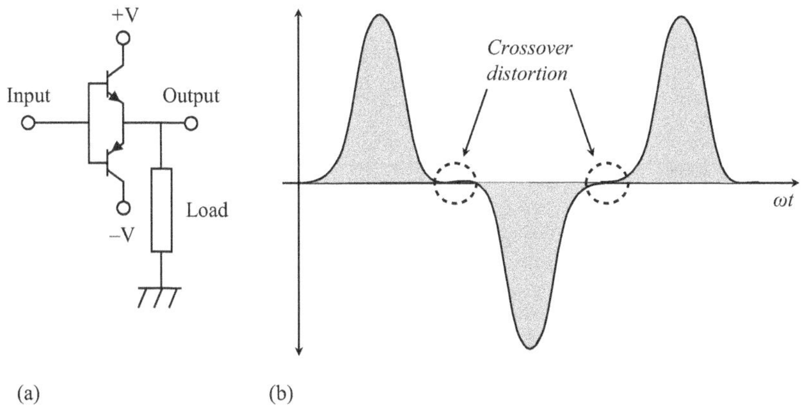

Figure 7.2 (*a*) *A circuit diagram of a push–pull circuit and* (*b*) *PA output waveform illustrating a crossover distortion that can be produced from a push–pull configuration*

The operation of class B amplifiers can also exist in other circuit arrangements such as transformer coupled push–pull amplifier and complementary symmetry amplifier circuit. The transformer coupled amplifier uses a center-tapped transformer at the input to produce opposite polarity to the two transistors and an output center-tapped transformer to drive the load in a push–pull mode of operation. In the complementary-symmetry amplifier circuit, complementary transistors (i.e., pnp and npn) are used to obtain a full cycle of operation without the need for center-tapped transformers.

7.2.1.3 Class AB

The class AB Amplifier, as the name indicates, is a combination of the class A and class B type amplifiers. The class AB amplifiers share the same circuit structure as class B. Its primary purpose is to overcome the crossover distortion problems seen in class B amplifiers while still using the push–pull operation and also to achieve an improved efficiency than class A. This is achievable by designing each of the push–pull transistors to conduct for slightly more than the half cycle of conduction in class B, however, much less than the full cycle of conduction of class A. The class AB amplifier is one of the most commonly used types of audio power amplifier designs. As can be concluded, a class AB amplifier has a theoretical maximum efficiency between 50 and 78.5%.

7.2.1.4 Class C

The operating point of a class C amplifier is located in the cutoff region as demonstrated in Figure 7.1(b) with a conduction angle less than 180° (typically around 90°). It exhibits high efficiency that can reach up to 80%, however, with poor linearity. This class of amplifiers is used with special tuned resonant circuits in order to construct the complete sine wave from the partial input wave. The complete output waveform is accomplished by using reactive (i.e., LC) circuits of a particular resonant frequency at the PA collector. Class C amplifiers are relatively nonlinear and produce significant distortion, especially for amplitude modulation signals and, thereby, are not preferred options for audio applications.

7.2.1.5 Class D

A class D amplifier is used to operate as a voltage or current switching amplifier. Under ideal conditions, class D amplifiers are 100% efficient. One of its main applications is to convert DC into AC. The circuit configuration of a class D amplifier employs a resonant network (e.g., an LC low-pass filter network) in order to reject high-frequency noise imposed by the PA power supply. When used as a voltage switching amplifier, the voltage waveform across the switching network is a square wave, and the current passing through them is a half sine wave. Also, the current waveform through the resonant network is a sinusoidal wave. Using a class D amplifier as a current switching amplifier, the voltage waveform across the switching network is a half sine wave, the current passing through them is a square wave, and the voltage waveform across the resonant network is a sinusoidal wave.

In general, class D amplifiers are suitable for low megahertz (MHz) frequency applications because at such frequencies this class assumes an ideal switch with no output capacitance of the transistor. At high frequencies, power losses due to the transistor output capacitance are considerable. In addition, the switching time is another limiting factor of class D switching amplifiers operating at high frequencies.

7.2.1.6 Class E

Similar to class D, class E is used as a switching transistor. A class E amplifier exhibits very high power efficiency, very low switching losses, and can operate at relatively higher frequencies since the device output capacitance doesn't require to be small when compared to class D. Using a class E amplifier in a push–pull configuration provides a significant reduction on the even harmonics. Class D, nevertheless, has better output power capabilities than class E amplifiers. The realization of a class E amplifier requires an inductive load and one main drawback in this class is the large voltage swing across the device, nearly four times the transistor drain voltage ($4V_{DD}$).

7.2.1.7 Class F

Class F PAs can be categorized into two groups: odd harmonics and even harmonics (the even harmonics PA is also known as inverse class F, F^{-1}). A class F PA incorporates multiple harmonic resonators at the output of the amplifier. Odd harmonics class F PA has the output resonator impedances tuned to act as an open circuit at the odd harmonics and thereby no odd harmonic power is dissipated. Similarly, for even harmonics, the tuning of the resonant circuits is made so that they present an open circuit at even harmonics and a short circuit at odd harmonics; thereby no even harmonic power is dissipated. Hence that at each harmonics frequency there is no real power generated because there is either no current or no voltage. In theory, a class F amplifier has 100% efficiency assuming that one can control an infinite number of harmonics. In this class, there are few drawbacks regarded to the output capacitance of the device that requires an inductor to cancel out its effect. Output capacitance limits the PA efficiency at high frequencies. In addition, it is difficult to control more than five harmonics (i.e., harmonic resonators) as their loss contributions deteriorate the PA efficiency.

Some other classes such as class G, I, S, and T exist and are suitable for use as switching amplifiers operating at low frequency.

7.2.2 Common PA configurations

7.2.2.1 Doherty amplifier

The Doherty amplifier was invented by William H. Doherty in 1936 [2] during which broadcast radios and trans-oceanic radios were of rising importance that triggered the need for powerful transmitters. Its main purpose is to amplify signals exhibiting varying amplitudes while maintaining both linearity and a good level of power efficiency.

The block diagram of a Doherty PA is shown in Figure 7.3. It consists of two power amplifiers: the main power amplifier, also known as the carrier amplifier and an auxiliary PA called the peaking amplifier. The carrier amplifier could be constructed as a class AB or a class B and is used to amplify the average signal level, whereas the peaking amplifier is usually a class C that is aimed to handle high power ratio levels and therefore has better power handling capability than that of the carrier amplifier. Doherty configuration implies that the two amplifiers are biased differently, and also requires a splitter at its input and a combiner at its output.

Doherty PA can be classified into two main types, symmetric and asymmetric, in terms of the PA transistor used in the design configuration. The symmetric type is one of the earliest configurations of Doherty amplifier architectures. The advantages of using this type are their design simplicity that incorporates two identical devices for both the peaking and carrier transistors, and a simple 3-dB power splitter on the input of the amplifier. Although the design is simple, this

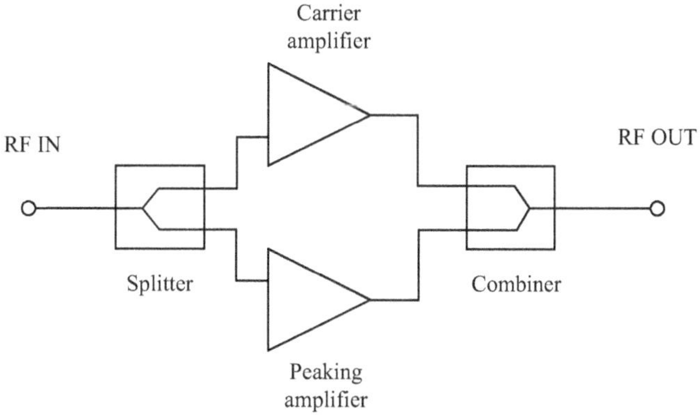

Figure 7.3 Basic block diagram of a Doherty PA

comes at the cost of performance. The asymmetrical Doherty type is the most commonly used type that incorporates transistors with different peak-power capabilities as discussed earlier in this section.

Doherty amplifier provides improved power efficiency when compared to balanced amplifiers, especially when operating below the PA saturation level. However, this configuration has a limited bandwidth due to the difficulty to maintain the phase shifts of the power splitter over a wide bandwidth. Employing two amplifiers, power splitter and power combiner, in this configuration instead of a single amplifier leads to a higher cost. In addition, the power splitter and combiner are selected and designed so that the carrier amplifier and the peaking amplifier can operate without loading each other.

7.2.2.2 Direct-coupled amplifier

A direct-coupled amplifier or a DC amplifier configuration is established by directly connecting the output of one stage transistor amplifier to the input of the next stage transistor in a way that permits zero frequency signals to pass from the input to the output. This type of configuration finds applications in computers, test instruments, and regulator circuits. Figure 7.4 demonstrates a direct connection between two transistors forming a direct-coupled amplifier. Hence, there is no capacitor in the connection path, and thereby DC signals can pass through. The voltage divider network R_1 and R_2 and the resistor R_{E1} form the biasing and stabilization network. As can be observed, the output voltage of the first transistor provides input bias to the second transistor.

This configuration provides a simple mean of connecting amplifiers with a minimum number of components. Additionally, the absence of coupling devices (i.e., reactive elements) leads to low-cost designs. Direct-coupled amplifiers are high gain amplifiers applicable for low-frequency applications, usually, from 0 Hz to 1 MHz. A feedback bias network is typically used to control the gain and bandwidth of the amplifier. This configuration can exhibit gain at and near zero frequency since the low-frequency response of the amplifier is not limited by the use of blocking or

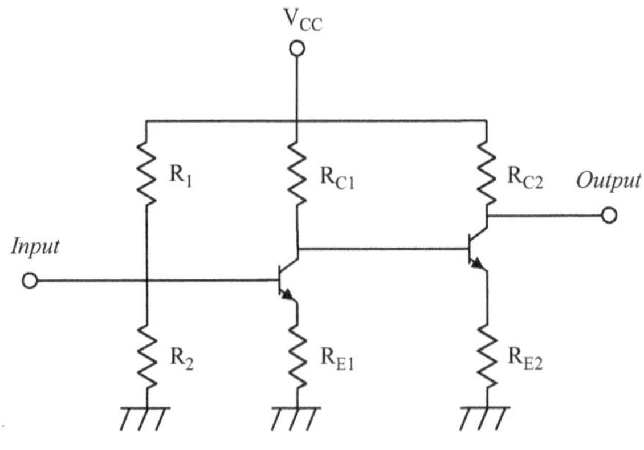

Figure 7.4 A two-stage direct-coupled amplifier

coupling capacitors which would reduce the amplification to zero at zero frequency. However, the direct-coupled configuration is incapable of amplifying high-frequency signals. Another problem associated with such configuration is that the operating point of the transistors can be drifted due to the device temperature variations. Small changes in the DC bias voltages due to the device temperature effects or power-supply variations are amplified by the subsequent stages and can result in a significant drift in the DC levels throughout the amplifier circuit. This unwanted change in the output signal without an intended change in the input signal limits the minimum input signal that can be amplified.

7.2.2.3 RC coupled multistage amplifier

The direct-coupled amplifier configuration suffers from limited output voltage swing. This can be addressed by enabling separate DC conditions in each stage. The transistor stages can then be coupled using capacitors and thereby more suitable for high-frequency applications. Figure 7.5 shows a two-stage RC coupled common emitter amplifier. The input capacitor C_{in} presented at the input of the amplifier circuit couples AC signal to the base of the first transistor. Similarly, C_{out} at the output of the amplifier circuit couples AC signal to the outer circuitry. The coupling capacitor prevents DC bias interference between the transistor stages and controls the shift of operating point while allowing AC signal to pass through with no or minimal attenuation.

In this configuration, the second transistor stage acts as a load to the first stage. This means that the output impedance of the first stage comes in parallel with the input impedance of the second stage. The fact that the input impedance of an amplifier is low, while its output impedance is high makes the effective load resistance to decrease which is considered as one of the main disadvantages of the RC coupled amplifier.

7.2.2.4 Transformer coupled amplifier

A transformer coupled amplifier circuit consists of two or more transistor stages connected using coupling transformers as illustrated in Figure 7.6. This configuration

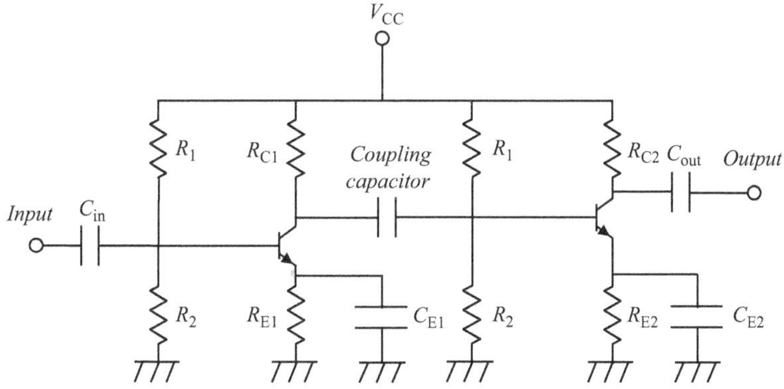

Figure 7.5 A two-stage RC coupled amplifier

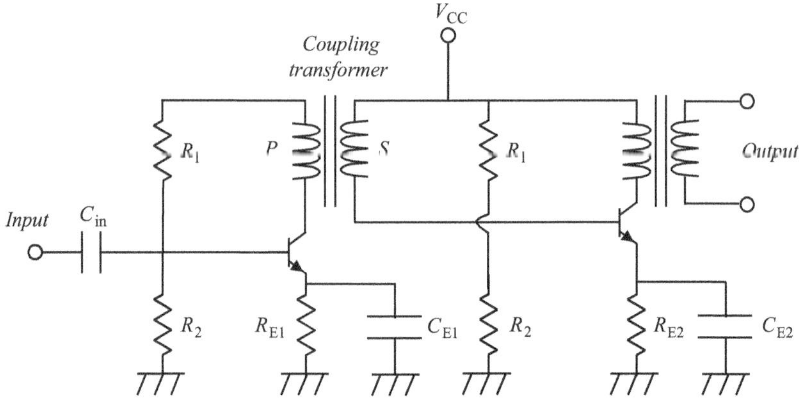

Figure 7.6 A two-stage transformer coupled amplifier

Table 7.2 Feature comparisons between different types of amplifier coupling methods

Semiconductor technology	Direct-coupled	RC coupled	Transformer coupled
Cost	Low	Intermediate	High
Impedance matching	Good	Not good	Excellent
Applications/suitability	Very low-frequency amplification	Voltage amplification	Power amplification
Form factor	Small	Intermediate	Large
Frequency response	Excellent	Excellent at low frequencies	Poor

overcomes the problem of decreased effective load resistance discussed in the RC coupled amplifiers. The impedance transformer in this circuit has the property that the low input impedance of the second stage (i.e., the load for the first stage) can be presented as a high impedance to the first stage. Transformers usually provide good impedance matching which is considered one of the advantages in this configuration. It is thus desirable in applications where maximum power transfer and broad impedance matching are required. In addition, transformer coupled amplifier provides high gain and efficient operation since there is no power loss in the collector and base resistors. The drawback of this configuration, however, is the considerable gain variation over the frequency of operation that results in poor frequency response.

The purpose of using coupling devices is to provide DC isolation when connecting one amplifier stage to the next stage and also to transfer AC output signal from one stage to the input of the next stage. Table 7.2 summarizes the common characteristics of each coupled amplifier configuration mentioned so far.

7.2.2.5 Differential coupled amplifier

The differential amplifier configuration is widely used in RF test instrumentation amplifier circuits since it provides low drift and is beneficial in DC applications. It is also the basic building block of an operational amplifier (op-amp). The basic configuration of a differential amplifier possesses two inputs and two outputs. When the two inputs are excited with an out of phase signals, the amplifier is said to be operating in the differential mode. Similarly, when both input signals exhibit the same phase and magnitude, the amplifier is then said to be operating in the common-mode. In this configuration, when the output signal is taken from one of the outputs with reference to ground, the amplifier is then called a single-ended output amplifier. Otherwise, when taking the output as the difference between the two output signals, the amplifier is called double-ended. Figure 7.7 demonstrates the configuration of a double-ended differential amplifier.

Usually, it is desirable to have a differential-mode gain to be greater than the common-mode gain so that the amplifier will respond only to the difference between the input signals. As a metric to quantify the ability of the amplifier to reject common-mode signals, the term common-mode rejection ratio (CMRR) is defined as the ratio of the differential-mode gain to the common-mode gain, in decibels, and is calculated as:

$$\mathrm{CMRR} = 20\log\left(\frac{A_d}{A_m}\right) \tag{7.1}$$

where A_d is the differential-mode gain and A_m is the common-mode gain.

Under an ideal common-mode rejection ratio condition and when both input signals to the amplifier have the same voltages, then the output of the amplifier is zero, and the amplifier is said to be rejecting the signal.

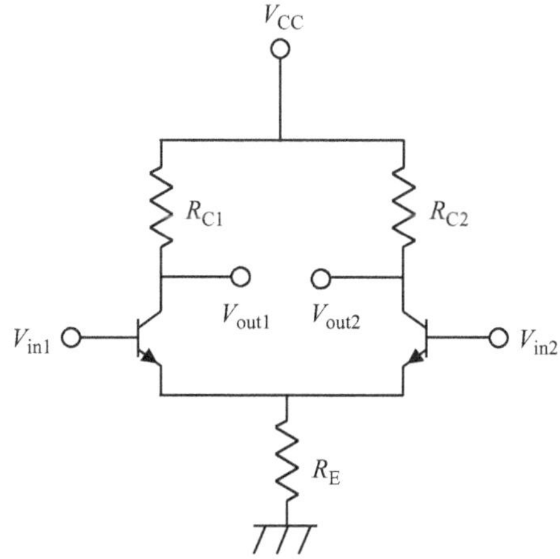

Figure 7.7 A differential coupled amplifier

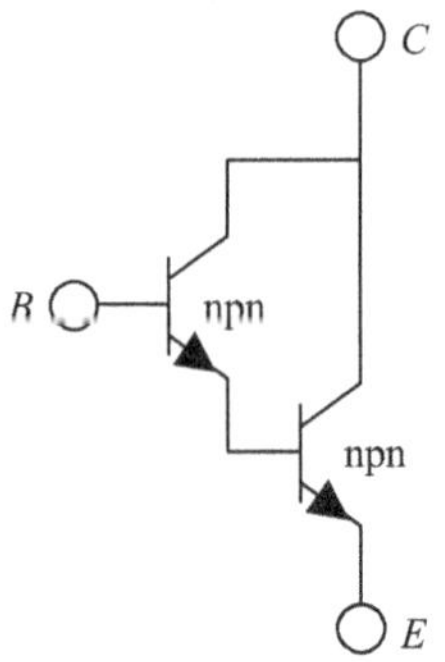

Figure 7.8 Darlington-connected transistors

7.2.2.6 Darlington amplifier

To achieve an increase in the current gain and the input impedance of an amplifier circuit, one can connect two transistors as shown in Figure 7.8 consisting of a pair of npn or pnp transistors connected together so that the emitter of the input transistor is connected to the base of the output transistor and both collectors of the transistor are directly connected together. This type of connection is known as Darlington configuration, named after its inventor, Sidney Darlington. The benefit of this connection is allowing the current to be amplified by the first transistor and to be further amplified by the second one from a low base current. The total current gain in this configuration can be simply calculated as the current gain of the first transistor multiplied by the current gain of the second transistor.

Darlington amplifier finds many applications in power regulators, sensors, and audio amplifiers.

7.2.3 PA stability

Unconditional stability of the PA must be ensured to avoid possible oscillation in the amplifier circuit. Unconditional stability means that the device is stable for any passive source and load impedances over the operating frequency range. In practical active circuits, oscillation may occur due to positive feedback parasitics from the output to the input of the circuit through external elements forming a positive feedback loop. In some transistor configurations where the performance could drift due to temperature variations, it is necessary to examine the stability of the transistor over an extended range of frequencies to ensure reliable PA operation. Any oscillation within the operating bandwidth would impact the PA gain performance. Generally, a two-port network is stable if the magnitude of both its input and output reflection coefficients are less than one. Using the k–Δ test, an unconditionally stable device implies that $k > 1$ and $|\Delta| < 1$ over the frequency range of interest. The stability factor (k) is calculated as:

$$k = \frac{1 - |S_{11}|^2 - |S_{22}|^2 + |\Delta_S|^2}{2|S_{12}S_{21}|} \tag{7.2}$$

and Δ is the determinant of the S-parameters matrix and is calculated as:

$$|\Delta| = S_{11}S_{22} - S_{21}S_{12} \tag{7.3}$$

Alternatively, a simplified single formula that can be utilized is the μ test which was defined to equivalently reach to the same conclusion [3] when $\mu > 1$:

$$\mu = \frac{1 - |S_{11}|^2}{\left|S_{22} - \Delta_S S_{11}^*\right| + |S_{12}S_{21}|} \tag{7.4}$$

The μ stability factor measures the distance from the center of the Smith chart to the nearest unstable location. Therefore, the larger the μ value, the more stable the amplifier is. Using the Smith chart to check for stability, under the unconditionally stable condition, both the input and output stability circles of a transistor must be completely outside the Smith chart and both the input (Γ_{IN}) and output (Γ_{OUT}) reflection coefficients of the two-port network are <1:

$$|\Gamma_{IN}| = \left|S_{11} + \frac{S_{12}S_{21}\Gamma_L}{1 - \Gamma_L S_{22}}\right| < 1 \tag{7.5}$$

$$|\Gamma_{OUT}| = \left|S_{22} + \frac{S_{12}S_{21}\Gamma_S}{1 - \Gamma_S S_{11}}\right| < 1 \tag{7.6}$$

Hence, for a unilateral amplifier (i.e., the reverse gain is zero), $\Gamma_{IN} = S_{11}$ and $\Gamma_{OUT} = S_{22}$.

In the case of combining multiple devices in order to meet a specific output power requirement, there will be possible modes of oscillation equivalent in their number to the number of the added devices. In such cases, stability cannot be ensured using the k–Δ or μ test which only addresses a single mode of operation [4]. Analyzing each individual amplifier in a multistage amplifier as a single two-port and their k–Δ factor is necessary, but it is not a sufficient test for ensuring the overall stability. This is because any of the two-port network amplifiers is not perfectly unilateral (i.e., $S_{12} > 0$), any change in the output (load) impedance will cause a change in the input impedance of the network and thereby unstable condition may be presented to the overall network. In a similar fashion, the input impedance of a two-port network influences the output impedance of the network. In summary, instability can occur when the terminating impedance on one side of the two-port amplifier is transformed through the network in a way that the amplifier presents a negative impedance at the other port.

Taking the case of two active devices connected in parallel, the stability is determined through the even- and odd-mode stability [4]. Similar to the discussion in Chapter 5, an odd-mode oscillation would occur when the circuits act as if there is a ground plane between them and satisfies the following impedance conditions:

$$\begin{aligned} \mathrm{Re}(Z_{io} + Z_{oo}) &< 0 \\ \mathrm{Im}(Z_{io} + Z_{oo}) &= 0 \end{aligned} \tag{7.7}$$

while an even-mode oscillation would occur when the circuits act as if there is an open-circuited plane between them and satisfy:

$$\begin{aligned} &\mathrm{Re}(Z_{eo} + Z_{eo}) < 0 \\ &\mathrm{Im}(Z_{eo} + Z_{eo}) = 0 \end{aligned} \tag{7.8}$$

where Z_e and Z_o are the impedance seen at each port under even- and odd-mode excitation, respectively. If the conditions in (7.7) and (7.8) are met for a mode at a specific frequency, then oscillations will occur at that mode of operation.

Generally, a resonant circuit will exhibit a stable oscillation if three conditions are met. For a series resonant circuit:

$$\begin{aligned} &R_L - R_S < 0 \\ &X_L + X_S = 0 \\ &\frac{d(X_L + X_S)}{df} > 0 \end{aligned} \tag{7.9}$$

Similarly, for a parallel resonant circuit:

$$\begin{aligned} &G_L - G_S < 0 \\ &B_L + B_S = 0 \\ &\frac{d(B_L + B_S)}{df} > 0 \end{aligned} \tag{7.10}$$

where R_L and R_S are the load and the source resistance, reactively, and G_L and G_S are their conductance. X_S and X_L are the two reactance elements in series or parallel configuration, and B_L and B_S are their respective susceptance.

7.2.4 PA linearization

With the complex modulation schemes used in mobile communication systems, linear amplification is necessary. Nonlinear amplification tends to produce unwanted intermodulation distortion (IMD) and adjacent channel spectral regrowth that leads to violations of the out-of-band (OOB) emissions standards mandated by regulatory bodies. In this contexts, there exist three main PA linearization techniques that can be used based on the targeted correction in the nonlinear amplification, bandwidth over which the linearization can be performed, and design complexity. The three techniques are the feedback, the feedforward, and the predistortion methods.

The feedback techniques can be further categorized into two methods: the negative analog feedback and envelope feedback. The negative analog feedback methods are efficient for use at low-frequency since at high-frequency such techniques can cause PA gain loss and may introduce instabilities. The envelope feedback as the name implies it compares the envelope of the output RF signal to the input and then modifies the output RF signal accordingly.

The feedforward amplifier consists of two loops (carrier cancellation loop and IMD cancellation loop) and exhibits relatively improved linearization performance over broader bandwidth when compared to the feedback techniques; however, they

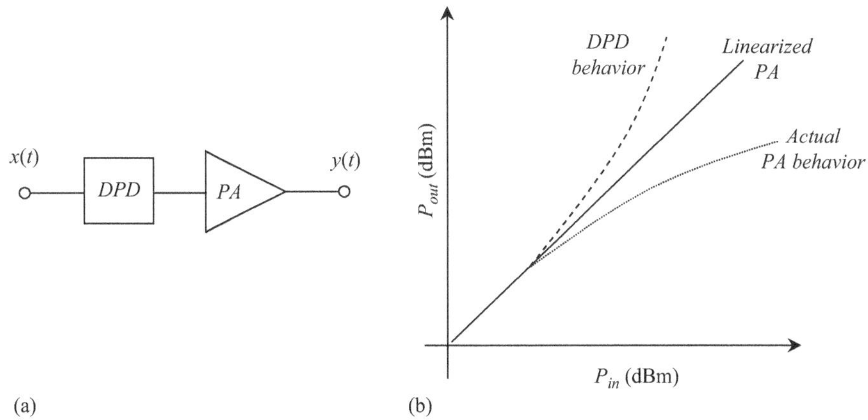

Figure 7.9 (a) A simplified block diagram of a PA in cascade with a DPD and (b) the input–output power behavior with and without a DPD

are relatively complex to design. The improvement that can be achieved on the IMD and adjacent channel power ratio (ACPR) can be in the range of 20 dB. In general, the feedforward methods have poor efficiency since there are essentially two PAs involved.

The predistortion linearization technique is one of the predominantly and most effectively used methods in commercial PA applications and consumer electronics due to its low complexity, maintaining high power efficiency, and providing an unconditionally stable solution. The underlying principle of the predistortion method is that a predistortion circuit can generate an inverse transfer function in the baseband signal to form an expanding nonlinear power curve of the PA that is complementary to the PA gain compression characteristics by virtue of today's capabilities of advanced digital signal processing and A/D converters [5]. In such technique, the distortion is corrected before it passes through the PA and thereby, the predistortion system produces a small level output power that doesn't impose a significant influence on the efficiency of the PA. Both digital and analog RF predistortion solutions are available. Figure 7.9 demonstrates an example of amplifier linearization utilizing digital predistortion (DPD) as an input circuit to the PA.

7.2.5 PA biasing

Biasing networks are one of the critical aspects in ensuring stable operation of any RF PA circuit design and require knowledge of the device behavior. PA biasing implies the application of DC voltage to establish a level of current and voltage to construct the desired operating point on the device IV characteristics which define the region that will be used for amplification. When the operating point on the device characteristics is a fixed point, it is also called the quiescent point or abbreviated as the *Q*-point. By using different transistor bias conditions, one can achieve different functions. A transistor is expected to maintain the operating point stable to achieve the intended amplification. Taking the example of a class A PA, the *Q*-point is selected so that it is located at the mid-point of the active region (also known as the linear region) of the transistor where a complete waveform can be reproduced without distortion. This is under the condition that the RF input drive signal will never exceed the allowed signal swing range around the *Q*-point.

To achieve an optimum PA efficiency, the optimum load resistance (R_{opt}) is constrained by the current and voltage swing as follows:

$$R_{\text{opt}} = \left(\frac{v_o}{i_o}\right) \tag{7.11}$$

where v_o and i_o are the voltage and current swings, respectively. A matching network is usually required to present the optimum load to the amplifier. This is because the impedance vector over frequency presented by the other circuitries at the output of the PA does not necessarily provide the optimum load to the PA.

A transistor can operate in three main regions of the device characteristics: linear, cutoff, and saturation regions. In applications that require signal amplification, the PA transistor is biased so that it operates in the linear region. Operating near the saturation or the cutoff regions is usually avoided as this will impose distortion on the amplitude of the output signal. A transistor operating in the saturation region acts as a short circuit as the current reaches its maximum value. In contrast, a transistor operating in the cutoff region acts as an open circuit since no current will be flowing through the PA output.

In PAs, the magnification of the output signal power is the result of the transfer of energy from the DC power supply (i.e., under a given bias condition). Generally, PAs with fixed biasing conditions provide the best power efficiency when operating at the maximum output power. Operating at relatively lower output power levels renders PAs to be inefficient. Therefore, the bias level needs to be a function of the input signal. In this context, several adaptive biasing techniques such as average power tracking (APT) and envelope tracking (ET) have been developed as discussed in detail in Chapter 8 to improve power efficiency across a wide range of operating power levels.

When biasing an RF active circuit there are some considerations that need to be taken into account, for instance, the change of the bias point of a particular quiescent current can change with changing temperature. Therefore, thermal compensation is a concern for most bias circuitries. This phenomenon, in some cases, can cause device failure if such thermal effects are not accommodated in the bias circuit. Different transistor technology will possess different bias point rate of change with temperature. In a bipolar transistor, the change occurs at a fixed rate of −2.4 mV/°C, while for a given FET it is inconsistent and thereby will require characterization over the expected range of operating temperature. Moreover, having a power supply driving different classes of amplifiers usually leads to large current variations and tends to have poor RF and envelope frequency response. In such case, proper isolation between the power supply lines is necessary. For example, a linear RF PA such as class A has a very low supply current equals almost zero, whereas, for class-B amplifiers, the full current capability of the supply is required. Isolation can be obtained by means of decoupling elements which could be placed close to the proper circuit point to ensure optimum decoupling. Also in BJTs, having bias resistor and capacitor on the emitter can cause low-frequency instability, increase the NF, and decreasing the gain of the amplifier. Thus it is desired to have the emitter terminal to be grounded so that no emitter feedback can be caused by the ground inductance.

7.2.6 PA distortion

Generating a large signal will naturally suffer from the nonlinearity of the used PA. Nonlinearity introduces signal distortion in the form of spectral regrowth (i.e., introduce new unwanted frequencies) that can spill over the neighboring bands. Usually constant envelope modulations such as frequency modulation (FM) are not impacted by the PA nonlinearity; however, it is not a preferred scheme due to its low spectral efficiency. Other modulation schemes with a strong variation of amplitude provide better spectral efficiency and demand high requirements for the linearity of the PAs.

Any electrical network including PA networks can be classified into memoryless or with memory. A memoryless network is an instantaneous circuit to the applied excitation, while a network with memory has an energy storage element (i.e., L or C) or a thermal storage unit that causes memory to the system. In other words, when an output of a system at a time t does not depend only on the input signal and relies as well on prior values of the signal, the system is then said to have memory effects. PA memory effects can be attributed to the dynamic effects such as electrical, electronic, and electro-thermal effects. Such effects represent an additional source of distortion and nonlinear behavior in PAs and usually, it is not accounted for in PA models. When the electrical characteristics of the PA are distorted by the temperature changes of the transistor, this leads to electro-thermal memory effects, also known as thermal power feedback (TPF). The electro-thermal effects are attributed to the dynamic temperature variation due to the device self-heating and its impact on the electrical properties of the transistor. Similarly, an undesirable type of electronic defect in semiconductor materials (known as semiconductor trapping effects), especially in GaAs-based devices, can cause potential memory effects and thereby PA distortion. In practice, any unintentional bias modulation is one of the significantly observed memory effects [6,7]. The electrical memory effect in PAs is seen via the relationship between the harmonic components in the output signal of the PA and how they are mixed into the fundamental band.

Memory effects cannot be described by steady-state analysis that is based on periodic stimulus such as a sinusoidal wave of the nonlinear transfer function of the device. In practice, modern RF wireless modules use digitally modulated waveforms and not steady-state signals. For PAs to improve the output power linearity and OOB emissions, they are designed to have strong memory effects. In addition, PA matching networks with harmonic trap functions contribute to the memory effects.

Total harmonic distortion (THD) is a measure of the device linearity by means of calculating the root sum of squares (RSS) of the voltage amplitudes of the first five or so harmonics of the fundamental frequency. Equivalently, the THD can be calculated in terms of the total power contained in all harmonics in watts and expressed in percentage as:

$$\mathrm{THD}(\%) = \frac{\sqrt{P_{H2} + P_{H3} + P_{H4} + \ldots + P_{Hn}}}{P_1} \times 100 \tag{7.12}$$

where P_{H2}, P_{H3}, ..., P_{Hn} are the power levels of the harmonic content of the PA output in watts and P_1 is the power level at the fundamental frequency. The THD of the PA is quantified by applying a single frequency (a pure sine wave) into the input of the PA, and measuring the output spectrum.

Depending on the modulation scheme used in the communication system and the type of the imposed distortion, changes in signal amplitude such as gain compression or gain expansion of a nonlinear device according to the level of the input power is referred to as amplitude-to-amplitude (AM-AM) distortion or conversion. Similarly, amplitude-to-phase (AM-PM) conversion is due to variation of the phase of the output signal according to changes in the amplitude of the input signal at a constant frequency. This deviation in phase is expressed in degrees/dB. Characterization of the AM-AM or AM-PM conversions is important in a PA system that relies on amplitude modulation or phase modulation schemes, respectively.

7.3 Low noise amplifiers

In wireless communication systems, LNAs are used to boost the amplitude of weak RF received signals to allow seamless signal processing, while maintaining minimal added distortion and noise. As discussed in Chapter 8, the first stage after the antenna in an RF receiver chain contributes the most to the system noise figure (NF). Therefore, RF LNAs can be used at this stage to reduce and control the overall NF of a receiver system. This is also one of the reasons why RF system engineers spend a considerable effort and time to optimize the first active device in the design of receiver front-ends. The design of an LNA entails understanding several critical parameters which if not designed properly can severely impact the receiver sensitivity. These parameters include NF, gain, bandwidth, linearity, and dynamic range. Understanding these parameters requires knowledge of the active device operation, types, impedance matching networks, and how they are integrated into a module environment in order to achieve optimal performance and accurately investigate the performance trade-offs. Measuring a NF level of less than 1-dB of an LNA necessitates careful attention to how the vector network analyzer is calibrated and how accurate are the measurement reference planes. In addition, it is important to correct impedance mismatch between the noise source and the device under test (DUT) for high measurement precision of NF. There are some considerations that need to be followed when integrating an LNA into an RF module. These include:

- The importance of considering the several variables of the intended electromagnetic (EM) environment when designing an LNA that could have a significant impact on the LNA performance. These variables include the laminate layout and signal trace routings, bias setting, selecting the proper impedance matching network/topology of the LNA input and output terminals as well as the proper lumped reactive components, and EM shielding.

- Isolate sensitive circuits by separating their ground planes, signal trace routing, and EM shielding.
- Avoid lossy elements such as reactive matching components along the signal path up to the LNA input. High-Q (i.e., minimal loss) matching elements provide better NF and gain performance. Moreover, NF performance can be improved by limiting the number of elements between the receiver's antenna and the LNA input, when possible.
- Mitigate aggressive digital signals lines with series resistance, when possible, to reduce their noise impact.
- Avoid routing traces carrying low-noise signals over planar voids which may significantly alter the presented parasitics to the signal carrying traces.
- To maintain control of the RF trace impedance, coplanar waveguides (CPW) traces are usually preferred over microstrip traces.

7.4 RF switches

RF switches are active elements employed in several RF transceiver modules and phase shifters. Their primary function is to control and direct the flow of RF energy from one circuit to another through the application of external control signals. For instance, frequency division duplex (FDD) transceiver modules that incorporate multiplexers with a common transmit and receive antenna require a switching mechanism to route the RF transmit signal from the PA to the antenna and simultaneously the received RF signal from the antenna to the LNA. As can be concluded, the switched signals are different in their frequency, amplitude, and phase. Therefore, isolation between the switching nodes is an important aspect in such applications to protect the sensitive receive circuits from the large RF transmit signals. In general, when designing an RF switch, isolation, insertion loss, and power handling are the main design parameters that need to be carefully considered.

The switching mechanism in RF switches can be achieved mechanically or electronically to switch from one state to another. In microelectronic modules, the module size is very limited and the required switching time is very short that cannot be handled mechanically. Thereby, electronic switching is more desirable. Electronically controlled RF switches can be realized using pin diodes or transistors. GaAs metal-semiconductor field-effect transistors (MESFETs) are suitable for use in RFIC applications with power handling capability that can reach up to 40 dBm. Using HEMT as an RF switching element is also possible for applications that require operation at high microwave frequency. However, HEMTs-based switches generate high distortion when compared to MESFETs. On the other hand, transistors as any other practical elements are not perfect switches. They comprise finite rise and fall time, finite on-state resistance, and finite output and input capacitances.

In terms of the switching topology in a particular application, RF switches can be realized in a series or a shunt connection to the transmission node of the circuit.

Both configurations are complementary in a sense that in the "on" state of the series switch, the switch will exhibit low impedance to allow the signal to flow, while a shunt switch will permit signal flow when the switch is in the "off" state possessing a high-impedance path to ground to maintain the signal flow in the circuit main path. Depending on the number of switching pins, a switch can be categorized as a single pole, single throw (SPST), single pole, double throw (SPDT), single pole, triple throw (SP3T), and so on.

Sections 7.4.1 and 7.4.2 will cover two types of commonly used RF switches: PIN diodes and FET switches.

7.4.1 PIN diode switches

A pin diode is one of the simplest semiconductor structures that consist of a two-terminal nonlinear device and behaves as a switch by allowing the current movement in one direction but not to the other. A diode is constructed of a pn junction device with IV characteristic behaves as shown in Figure 7.10. The diode will conduct current in the direction of the arrow defined in the symbol and can act as a short circuit, whereas reversing the voltage across the diode will result in no current flow in the opposite (nonconduction) direction and thus the diode acts as an open circuit with an effective impedance proportional to the junction capacitance. When there is no bias, the voltage across the pn junction (V_D) is zero, and therefore, the net current flowing through the junction is zero. The work principle relies on the fact that applying a forward bias ($V_D > 0$) as depicted in Figure 7.10, makes the

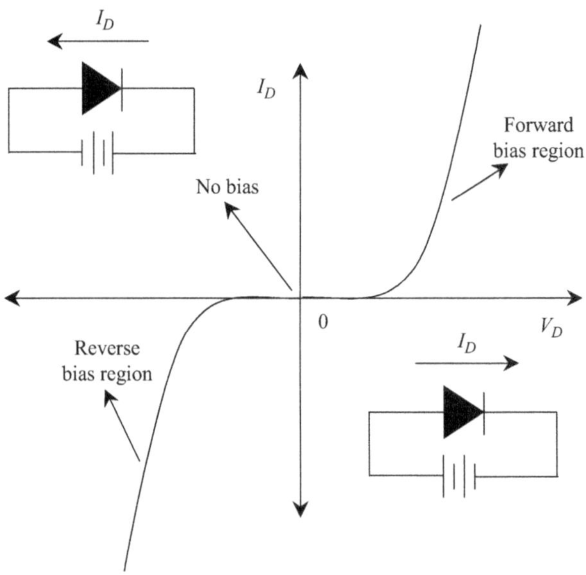

Figure 7.10 An IV characteristic curve of a forward and reverse bias pn junction diode

depletion layer in the pn junction to decrease as V_D increases, while with reverse biasing the pn junction ($V_D < 0$), the depletion layer increases as the voltage decreases and thus the impedance to the current increases. Under zero or reverse bias condition, the large impedance of the diode is dominated by the junction capacitance and at low-frequencies, this impedance is large. However, at microwave frequencies, the junction impedance may not be sufficiently large.

The pin diode under steady-state conditions follows the following IV relationship:

$$I_C = I_S\left(e^{V_D/V_T} - 1\right) \tag{7.13}$$

where I_C is the forward current, I_S is the temperature-dependent reverse saturation current, V_D is the forward voltage, $V_T = kT/q$ is the thermal voltage, k is the Boltzmann's constant, T is the diode temperature in Kelvin, and q is the electronic charge.

In general, PIN diodes and solid-state switches are widely used in modern phased array systems for their applicability as a planar solution at microwave frequencies and their low insertion loss. However, PIN devices consume more DC power than FET switches.

7.4.2 FET switches

A FET device can be thought of as a two-terminal device functioning as an impedance controller between its source and drain terminals. The amplitude of the RF signal (RF impedance) passing between the source and drain terminals is controlled by a DC signal applied to the gate terminal of the FET. When the gate terminal has no potential applied to it, the source to drain path will possess low impedance channel and thereby acts as a short circuit. Similarly, when the gate is biased with a negative potential, the drain to source impedance exhibits high impedance path, and with sufficient potential amplitude, the current flow can be prevented from flowing through the transistor and thereby an open circuit condition can be established. Hence, the low-impedance path of the FET switch is determined mainly by the open-channel resistance of the device. The equivalent circuit model of a FET switch can be represented by a low-impedance resistance for the transistor in the "on" state. In "off" state, the FET is dominated by the depletion layer and is represented by a high-impedance resistance (R_{off}) in parallel with a capacitor (C_{off}). The inclusion of the C_{off} element implies that the magnitude of the impedance has an inverse function of frequency (i.e., the off state impedance decreases as frequency increases).

FETs provide compact switching solutions to the RFIC modules and also compatible in terms of the semiconductor manufacturing and assembly processes. As the FET size increases to withstand high-power applications, substantial degradation on the switch isolation and insertion loss [8] could occur. In general, solid-state switches suffer from poor linearity at mm-wave frequencies and their low third-order intercept point (IP3) makes them highly susceptible to intermodulation distortion that significantly limits their dynamic range.

Other switching elements such as ferrite-based switches can handle high RF power levels with excellent performance but are very expensive to manufacture, consume significant DC biasing power, and usually require manual tuning. Also, RF micro-electro-mechanical system (MEMS) offer excellent performance compared to solid-state switches for only low- and medium-power applications [9]. The reliability and robustness of MEMS devices, however, are still questionable. Moreover, RF MEMS have very slow switching speed and uncertain lifetime [10].

7.5 Phase shifters

As steerable and switchable microwave circuits continue to take a significant role in the operations of modern technologies, phase shifters become increasingly important. Phase shifters are used in a myriad of applications that involve instrumentation, metrology, and communication. They are an essential part of beam forming networks, power combiners and dividers, linearization networks for PAs, and phase discriminators. The most prevalent use of phase shifters is in phased array antennas and radars used by many industries (e.g., autonomous vehicles, avionics) and military programs.

Phase shifters contribute to as much as 40% of the electronic equipment costs in phased array antenna systems owing to the manufacturing cost and the accuracy level of the employed RF switching elements. Current phase shifters are based on ferrite materials, PIN diodes, transistor/FET switches, or RF MEMS switches.

Existing phase shifter designs fall into three main classes [8], all compatible with the MMIC technology. These classes include loaded-line, switched delay line, and reflection-type phase shifters. With the loaded-line design, typically, $\lambda/4$ TL sections are periodically loaded with reactive shunt elements to ground. The implementation of a loaded-line phase shifter results in excellent response for small phase delays (i.e., up to 45°); however, a 90° phase shift is essentially narrowband and thereby is not practical. Furthermore, loaded-line phase shifters require considerable chip area. In the switched delay line design, RF switches are employed to route the RF signal into the appropriate length of matched TL. The time delay is proportional to the difference in the physical lengths of the selected line and the reference line. Such design can achieve a wide bandwidth, however, the phase shift varies with frequency (hence, networks/lines are true time-delays) and the amount of chip space increases exponentially with the increase of the required phase states, resulting in impractical size for the realization of integrated circuits. Also in switched-line designs, two single-pole-double-throw switches are required, and as a result, an advanced fabrication technology is needed to realize phase shifter at high frequencies with low insertion loss. Reflection-type phase shifters are advantageous in doubling the phase delay per unit length as compared to standard designs, and the loss of the reflective terminating line is dependent only on the delay bit used. Separating between the incident and reflected voltage waves is possible through the use of directional couplers (e.g., Lange couplers), where the reflected signals combine destructively at the input port (hence, the input match is preserved), and combine constructively at the output port (i.e., the isolation port of the coupler). Reflection-type phase shifters are capable of providing wide bandwidth and require minimal space on the wafer [8]. Furthermore, if the utilized couplers are well balanced, the interactions between the cascaded circuits are reduced, resulting in very good port matching (in other words, the coupler serves as an impedance isolator).

Figure 7.11 illustrates an N-bit phase shifter constructed using N 3-dB couplers and N 1-bit switchable terminations that create switchable reflection coefficients.

For maintaining a constant loss for all phase states, loading susceptance pairs must alternate between complex-conjugate values, $\pm B$ [11]. Thus, the reflection coefficient of any two phase states is given by:

$$\Gamma_{on} = \Gamma_{off}^{*} \tag{7.14}$$

Although having a coupler in the realization of the phase shifter circuit is taking a considerable circuit size, Lange couplers are usually preferred as they possess relatively compact size and provide wideband performance when compared to other possible couplers that can be employed. RF MEMS switches could be connected to the ground plane utilizing wideband shorts (e.g., radial stubs). The switch connection provides two loading states to the phase shifter representing an SPST switch, which is adopted in most pulsed radar systems and wireless communication systems. Furthermore, SPST switches require less circuit area in planar circuits. Assuming ideal switches, the electrical lengths of the stub loading sections (i.e., reflective terminating stubs) in the phase shifter that ensure complex-conjugate susceptance ($B_1 = -B_2$) switching is given by [11]:

$$\tan(\theta_2) = -\tan(2\theta_1) \tag{7.15}$$

where θ_1 and θ_2 are the loading stub electrical lengths corresponding to Load 1 and Load 2 in Figure 7.11, respectively.

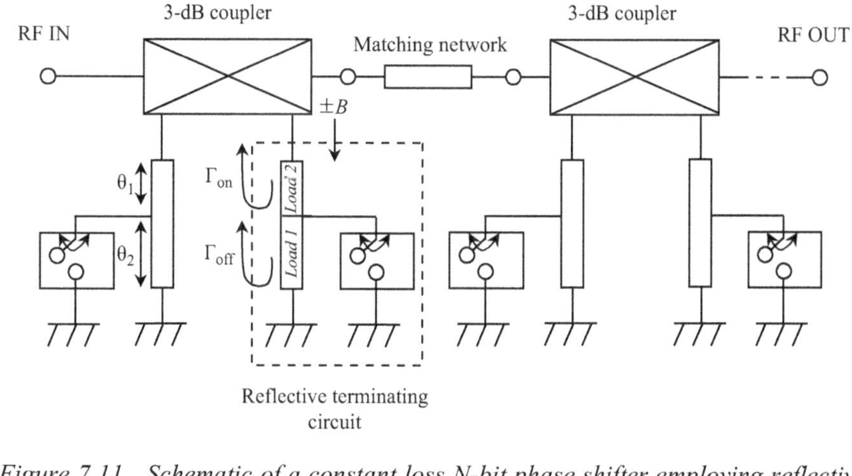

Figure 7.11 Schematic of a constant loss N-bit phase shifter employing reflective terminating lines as RF phase shifters

7.6 RF filters

RF and microwave filters are key building blocks in any modern RFFE systems. Nearly all RF and microwave components can be thought of, in one way or another, as a filtering element. In fact, components such as matching circuits, circulators, amplifiers, oscillators, and switches are filters that allow waves to transfer over their passbands and block or attenuate waves at their out-of-band. However, such components do not provide adequate filtering performance (e.g., selectivity, group delay, and insertion loss) to meet the stringent requirements by modern wireless systems when compared to the performance of dedicated filter designs. Filters can be classified depending on the used resonator technology into passive or active, lumped or distributed, linear or nonlinear, reciprocal or nonreciprocal, planar or three-dimensional (3D), and coupled or uncoupled resonator structures. The selection between the different filter realization technologies is dependable upon the intended application. Bandwidth, size, operating frequency, and the amount of loss that can be tolerated within the filter passband are all design factors that are used to determine which implementation technology can be used. Generally, narrowband filters exhibit a fractional bandwidth of 20% or less and this is also dependable on the used topology of the filter and the resonator technology.

Filters can be designed to exhibit several transfer function characteristics including lowpass, highpass, bandpass, bandstop, and all-pass filters. The reader may be familiar with the purpose of the first four types, but not the all-pass filter. An ideal all-pass filter characteristic does not alter the magnitude of the signal. However, it imposes phase delay as a function of frequency. Knowing that the derivative of phase is a time delay, therefore, one of the common application of all-pass filters is to provide a group delay equalization for a given network.

When compared to an ideal filter characteristic, one can observe that a realistic filter does not possess a sharp transition, from its passband to its stopband region, and its transmission within the filter's passband is not constant. Since an ideal filter characteristic requires an infinite number of elements or a network with an infinite group delay, using a realizable finite number of element filters, require an alteration on the characteristics of an ideal filter. This approximation can be described in the form of a polynomial approximation representing the transfer function of the filter. Common filter polynomial approximations and their main characteristics include:

- Butterworth or maximally flat response comprises a flat passband, however, with a slow transition to the stop-band;
- Chebyshev response which has a sharp transition at the cost of pass-band ripples;
- Inverse Chebyshev response exhibits a sharp transition at the cost of stop-band ripples;
- Bessel response comprises a maximally flat group delay; and
- Elliptic and quasi-elliptic response has the fastest transition at the cost of ripples in both the passband and the stopband. In addition, most filters exhibiting elliptic response characteristics are difficult to realize due to the requirement of extreme impedances (i.e., very low and very high impedance) of the resonant elements.

The filter selectivity can be improved by introducing a finite-frequency transmission zero (TZ), (i.e., the placement of a highly sharp attenuation through the transmission coefficient (S_{21}) of the filter at a specific frequency usually near its passband edges). The importance of placing a TZ is to achieve a filter with high selectivity on one side of the passband than the other (as is the case with diplexers or multiplexers) or on both sides if there are two or more TZs placed on the lower and upper stopband sides. The steepness of the filter or any resonant circuit, in general, can be measured by the shape factor which is defined as the ratio of the 60-dB bandwidth to the 3-dB bandwidth of the circuit. An ideal filter will result in a shape factor of one. A shape factor smaller than one is physically unrealizable and therefore impossible.

Moreover, reconfigurable filters are required in many multiband wireless communications and wideband radar systems. In response to this need, many tuning technologies and elements have been proposed and incorporated in the design of filters such as MEMS and mechanical tuning with electrically controlled methods. However, there is always a need for new tuning technique which leads to a dramatic reduction in the complexity of the external control system (e.g., biasing circuits).

As discussed in Chapter 3, the synthesis of all filter types based on any of the polynomial approximations mentioned above can be derived from a low-pass filter prototype. The low-pass filter prototype is a normalization or standardization tool for the RF filter designers to reference their designs and to be able to compare performance and usefulness of one topology over another.

Sections 7.6.1 and 7.6.2 will discuss some of the commonly used filters in RFIC and MMIC RFFE modules.

7.6.1 RF passive filters

7.6.1.1 Lumped-element filters

At microwave frequencies, lumped elements-based devices suffer from high insertion loss, limited power handling capability, and the influence of parasitic elements such as parallel resonances of inductors and series resonances of capacitors are pronounced. Thereby, the use of lumped elements as the primary resonant components to form a microwave filter is not a preferable option. At low frequencies, lumped elements have the advantage of small component size and reasonable electrical performance [12].

Hybrid filters that comprise both lumped and distributed elements are also a common practice in the field of microwave filters. In many applications, lumped elements are used as tuning or loading elements in cascade with the distributed resonant elements to form the resonant structure. Tuning can be performed using lumped reactive elements for controlling the inter-resonator coupling coefficient as in coupled-resonator filters and thereby adjusting the achievable bandwidth, or for tuning the location of the transmission zeros. Lumped reactive elements when used as loading components serve a primary purpose which is reducing the effective length of the distributed element.

7.6.1.2 Distributed microwave filters

Distributed resonators, transmission line or waveguide-based resonators, exhibit high order modes at multiples of the fundamental frequency. A distributed filter designed with such resonators will comprise spurious resonances at the OOB response which in many cases limits its OOB rejection performance from multiplexing with other filter bands located at any of the spurious frequencies. Some filter topologies allow the inclusion of capacitive loading elements that serve two purposes: reducing the physical length of a distributed resonator and at the same time extending the upper OOB, stopband, response to a higher frequency. Generally, at microwave frequencies, the *Q*-factor of distributed elements is higher than their counterparts of lumped elements.

Lumped inductors and capacitors can be realized using transmission line sections. By using Richard's transformation [13], it is possible to convert a lumped element to its equivalent distributed element, where the reactance of an inductor can be calculated as:

$$jX_L = jL\tan(\beta l) \tag{7.16}$$

and the susceptance of a capacitor is given by:

$$jB_C = jC\tan(\beta l) \tag{7.17}$$

where L and C are the normalized lumped element values, $\beta = 2\pi/\lambda$, λ corresponds to the cut-off frequency of the LPF, and X_L, B_C are the reactance and the susceptance of the lumped element components. The equations in (7.16) and (7.17) indicate that an inductor can be replaced with a short-circuited stub of length βl and characteristic impedance L, while a capacitor can be replaced with an open-circuited stub of length βl and characteristic impedance $1/C$ as depicted in Figure 7.12.

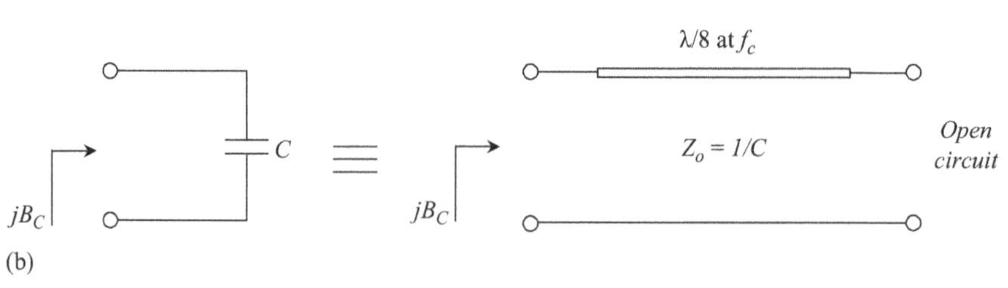

Figure 7.12 Richard's transformation for (a) an inductor and (b) a capacitor

Equation (7.18) maps the ω plane to the Ω plane to realize a two-port *LC* network using open or short-circuited stubs:

$$\Omega = \tan(\beta l) = \tan\left(\frac{\omega l}{v_p}\right) = 1 \tag{7.18}$$

where $\beta = \omega/v_p$ and v_p is the phase velocity and l is the stub length.

Kuroda's identities [13] are used to separate filter elements using redundant transmission line sections and also to transform short-circuited stubs into open-circuited stubs for more practical realization of microwave filters. In PCB technology, open-circuited stubs are easier to fabricate than short-circuited stubs because a via hole through the substrate is not required. The Kuroda's identities illustrated in Figure 7.13 are used to convert all series stubs to shunt stubs and to convert the resulting short-circuited stubs into open-circuited stubs. Here, $n^2 = 1 + Z_2/Z_1$.

Both Richard's and Kuroda's identities use transmission line sections of length equal to $\lambda/8$, for which tan $(\beta l) = 1$ in (7.16)–(7.18). These $\lambda/8$ transmission line sections forming a filter are called commensurate lines since they all possess the same length.

A duality between a distributed 3D waveguide (WG) section and a planar form called substrate integrated waveguide (SIW) has been reported in 2001 [14]. The development of SIW technology has opened new perspectives for circuits and systems in the microwave and millimeter-wave frequency range. This new guided wave structure comprises a waveguide section implemented on a planar dielectric substrate with two rows of metallic vias [14] equivalent to a magnetic wall that guides the transmitted waves. The separation between the two rows of vias (i.e., the SIW width) determines the cut-off frequency (in a similar fashion to that in WG resonators) depending on the mode of operation. This low-cost realization alternative of the traditional WG section inherits the merits from both easy integration of microstrips and low radiation loss of WGs. Among the different classes of SIW

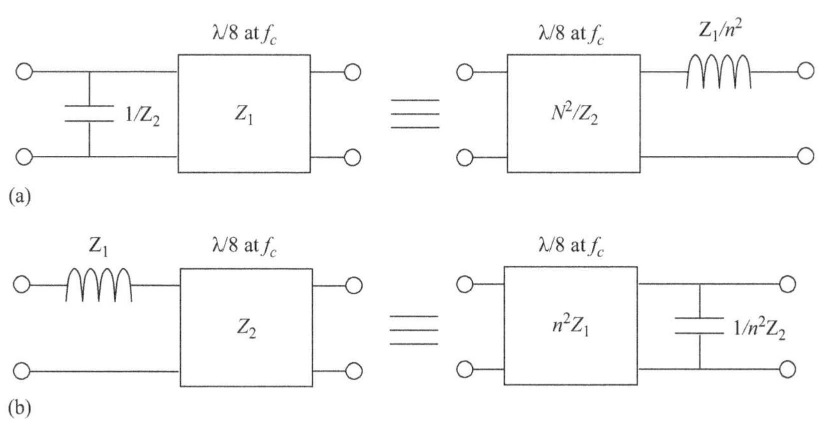

Figure 7.13 (a) Kuroda's 1st identity and (b) Kuroda's 2nd identity

components proposed in the literature, SIW filters have received particular attention, due to the possibility of achieving high *Q*-factor and better selectivity when compared to classical planar filters realized using microstrip and CPW technology.

7.6.1.3 Coupled resonator filters

In coupled resonator filters, many coupling structures are possible to couple energy from one resonator to another resonator and from a resonator to the termination port. In a coupled structure (either lumped or distributed), the coupling strength (also referred to as the coupling coefficient) between the resonators is an important parameter that must be in tune to achieve a specific frequency characteristics and bandwidth. Different coupling structures usually result in significantly different filter performance in terms of selectivity, insertion loss, group delay over the passband frequencies, and the tunability of the filter. As an example, several cavity filters and their compact alternatives planar SIW filters comprise inter-resonator coupling structures (e.g., iris coupling elements) that are used to control the energy coupled between the resonator modes and to control the frequency separation between the adjacent resonators with the possibility to achieve advance frequency characteristics such as introducing additional transmission zeroes at the OOB frequencies. The coupling structures include inductive or capacitive elements which may be adjustable.

Asymmetric filter characteristics are regularly used in the design of diplexers and multiplexers to achieve compact and low-loss filtering solutions for mobile wireless systems. The realization of an asymmetric characteristic using coupled resonator filter structure usually requires cross-coupling between the nonadjacent resonators. In most of the commonly used filter implementation technologies, it is physically and mechanically challenging to realize the cross-coupling between the resonators. Microwave filters that rely on cross-coupled resonators are usually more sensitive and difficult to tune in the lab after the part is built when compared to filters with straight couplings. The benefits that asymmetric cross-coupled filters provide in terms of electrical performance justify their usage despite the realization challenges.

7.6.1.4 Acoustic-wave filters

Piezoelectric acoustic-wave (AW) resonators offer a remarkable filtering solution when compared to their counterparts lumped and distributed resonators because of their high unloaded *Q*-factor, precise frequency control, reasonable low-frequency variation over temperature, small size at gigahertz frequencies, compatibility with conventional photolithography and silicon IC fabrication processes, suitability for electro-mechanical tuning techniques, low-cost, and applicability for mass production. In addition, AW transducers which are used to convert mechanical or acoustical energy into electrical energy find several other applications in devices such as microphones, phonographs, and acoustic sensors. However, AW-based resonators have limited power handling capabilities when compared to lumped and distributed filters due to their miniature size.

Two main types of piezoelectric acoustic-wave resonators exist, namely, surface acoustic wave (SAW) and bulk acoustic wave (BAW). Filters based on SAW resonators are one of the most commonly used filters in commercial RF telecommunications and consumer products. Recent advances in BAW technology and BAW-based filters have also gained unprecedented adoption in applications where power handling, high selectivity, and high-performance filtering solutions are of major concern. Both technologies, however, have a similar principle of operation which is converting an electrical signal into acoustic vibrations and then back to an electrical signal.

The structure of a SAW resonator consists of a pair of interdigital electrodes placed onto a piezoelectric substrate (e.g., lithium niobate and lithium tantalate) allowing waves to travel along the surface of the piezoelectric material. These waves are most frequently referred to as Rayleigh waves, named after their discoverer. Each electrode structure consists of a thin metal layer forming an interdigitated electrode fingers attached to a bus bar. When RF energy is applied across the input terminal of the resonator, the input transducer converts the electrical signal into surface acoustic waves that propagate in a direction perpendicular to the alignment of the interdigital fingers and these waves are picked up by the output transducer. The output transducer then converts the acoustic-waves back to an electrical signal which can be utilized and characterized. In SAW resonators, the resonant frequency is established when the spacing between the interdigital fingers equals half-wavelength of the acoustic wave.

The BAW resonator relies on the longitudinal wave propagation. The structure of a BAW resonator consists of a piezoelectric material (e.g., film of aluminum nitride) sandwiched between two electrodes. BAW resonators used in constructing BAW filters are commercially available in two forms: the thin film bulk acoustic resonator (TFBAR) and solidly mounted resonators (SMR).

The two technologies differ between them in terms of how it approaches the problem of confining the resonator energy. Generally, energy confinement in a BAW resonator is achieved by introducing a large acoustic impedance mismatch at the resonator electrode interface which also defines the type of the BAW resonator. The TFBAR is formed by developing an air cavity interface under the bottom electrode that acts as an acoustic short. Whereas in an SMR, energy confinement is achieved by introducing a sacrificial substrate layer under the bottom electrode of the resonator so that most of the acoustic energy is reflected back to the resonator active region. The SMR layers are deposited onto a stack of quarter-wavelength thick layers that alternate between high and low impedance to form what is called a Bragg mirror or a Bragg reflector. The reflector usually consists of alternating layers of silicon dioxide (SiO_2) and tungsten (W) placed underneath of the bottom electrode of the BAW resonator. Both technologies have their own advantages and disadvantages. For instance, the SMR possesses better mechanical/structural stability and thermal conductivity than the air cavity TFBAR. In contrast, the SMR may exhibit lower resonator Q-factor if the Bragg mirror is not designed properly to confine the acoustic energy constructively. BAW resonators surpass SAW resonators in terms of performance; however, the SAW fabrication process is simpler and cheaper.

The electrical performance of an AW resonator possesses two resonant frequencies: a series and a parallel (or antiresonant) resonant frequency. The series resonance exhibits an impedance minimum, while the parallel resonance shows an impedance maximum. The resonator performance can be represented by the modified Butterworth-Van-Dyke (mBVD) circuit model [15] illustrated in Figure 7.14(a) with an electrical characteristic shown in Figure 7.14(b). In this lumped element model, L_m and C_m are the motional (dynamic) inductance and capacitance, respectively. R_m is motional resistance and is used to represent the mechanical loss during the transformation between the electrical and acoustical domains. C_0 is the static capacitance across the resonator. R_1 and R_2 are to account for the resistive loss of the electrodes of the resonator. R_o describes the dielectric loss tangent of the piezoelectric film. This model is capable of providing an accurate fit of the real performance of an actual resonator. The region of the resonator's input impedance between the series and parallel resonances exhibits an inductive behavior, while the performance above the parallel resonance and below the series resonance are capacitive. The impedance peak above the series resonance occurs when part of the piezo-electric motional inductance parallel resonates with the acoustic resonator static capacitance due to the terminal electrodes of the resonator which are necessary for providing the electrical contact. Whereas at series resonance, the acoustic-wave resonator has a low input impedance which occurs due

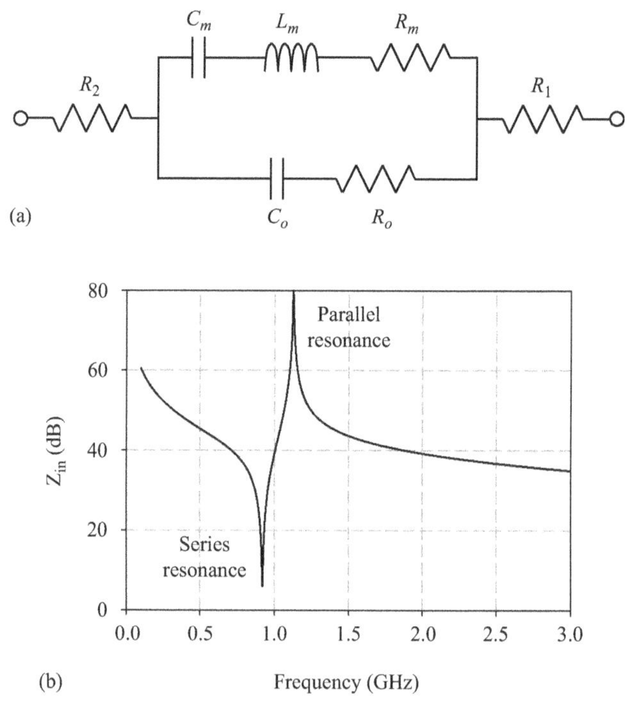

Figure 7.14 (a) A schematic diagram of an mBVD model, and (b) an input impedance response of an AW resonator or its equivalent mBVD model

to the piezo-electric motional inductance series resonating with the piezo-electric motional capacitance.

In many cases of AW filters design, it is desirable to have a means of slightly adjusting the resonant frequency so that a given design specifications can be met. For example, to extend the filter's bandwidth or to position a transmission zero in a specific location at the filter's OOB region. This can be accomplished by adding a tuning capacitor or inductor in series to the AW resonator to adjust the series-resonant frequency or in parallel to the AW resonator to shift the parallel-resonant frequency.

At the resonant frequency, the capacitive and inductive reactances are equal in magnitude and opposite in phase and only the resistance of the resonator remains to determine its *Q*-factor. A high resonator resistance leads to a resistive loss and makes its *Q* to be low. While most of the losses in a purely electrical-resonant circuit are caused by the incorporation of high resistive (lossy) coils, in an AW resonator, the loss is mainly introduced by the electrodes that provide the electrical connection when compared to the *Q* of the purely mechanical resonance measured in isolation of the electrical interface. Moreover, AW resonators have, in general, poor linearity performance.

The difference between the series-resonant frequency of the AW resonator and its parallel resonance determines the achievable bandwidth in AW filters. The ratio between the parallel and the series resonance is proportional to the effective electro-acoustic coupling coefficient (k_e^2) and is approximated by the second order Taylor series expansion as:

$$k_e^2 \approx \frac{\pi^2}{4}\frac{f_s}{f_p}\left(1 - \frac{f_s}{f_p}\right) \tag{7.19}$$

From the mBVD model response illustrated in Figure 7.14(b), it can be concluded that the out-of-band (OOB) impedance characteristic of an AW resonator and consequently an AW filter is capacitive [16].

The ladder topology illustrated in Figure 7.15(a) has been one of the most widely used configurations in AW filters comprising the high-performance AW resonators. This is attributed to the low sensitivity of ladder networks if compared to its counterparts of lattice topologies. In many instances, a ladder-type AW filter will incorporate inductor or capacitor elements connected in series or parallel across the shunt or series resonators. These elements provide the following benefits:

- Extending the filter bandwidth and thereby enhancing the passband insertion loss;
- Placing transmission zeroes at the desired OOB rejection frequencies; and
- Allows the use of a lower coupling coefficient piezoelectric material which implies using a thinner material possessing higher static capacitance. Having a high static capacitance means that we can obtain and a smaller size resonator for a given impedance.

Another configuration that is commonly used in low power applications and has the advantage of realizing filters with small form factor is the double mode

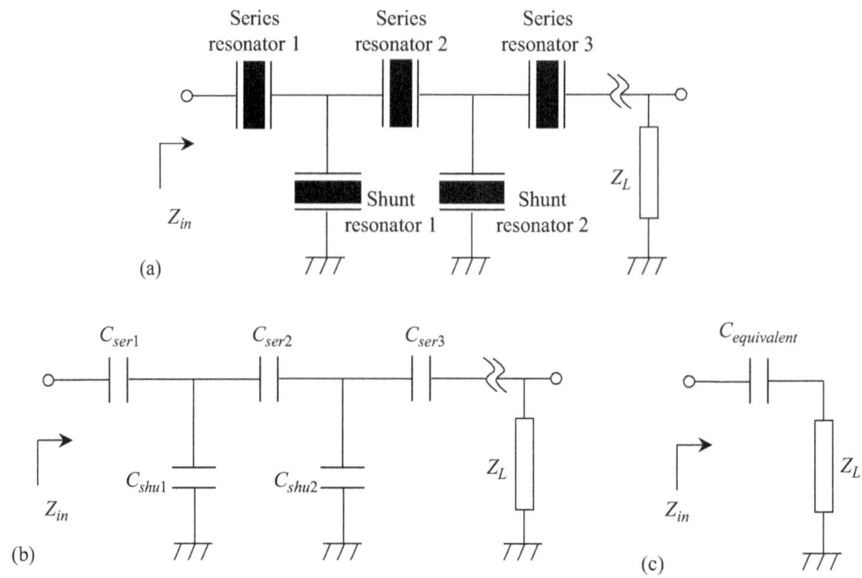

Figure 7.15 (a) A schematic representing a ladder-type AW filter comprising series and shunt crystal resonators; (b) an equivalent circuit model of the circuit in (a) showing that a capacitor can substitute each of the AW resonators at the OOB frequencies; and (c) an equivalent circuit that shows the effective capacitance seen at one of the filter terminals

surface acoustic wave (DMS) filters where acoustically coupled resonators in the longitudinal direction are used to form coupled SAW resonator filters. The performance of a DMS SAW filter suffers from poor stopband rejection levels at the upper side of the filter's passband [17] when compared to a ladder-type filter which also has a broader bandwidth and a better power handling. However, DMS filters have relatively better skirt selectivity and higher attenuation at the lower side of the passband.

The emerging 5G communication technologies are crafting new domains in the wireless communication market and major RFFE players are competing to provide devices that could be integrated in the next 5G smartphones. In terms of filtering solutions, it is expected that the low cost SAW filters will keep its market share for low band communications such as global system for mobile communication (GSM), second generation (2G), or third-generation (3G), whereas high performance filtering solutions such as BAW filters will focus on the 4G and 5G market. 4G and 5G technologies are the impetus behind the better integration of highly complex RFFE communication devices under a single module package. One form of integrated filter solutions in RFFE modules takes the form of multiplexers as will be discussed next.

7.6.1.5 Multiplexers

A multiplexer is a set of interconnected filters that can combine two or more channels carrying different frequency bands into a single broadband channel or split a single broadband channel into two or more channels. A multiplexer configuration provides several advantages in terms of cost and performance. Efficient realization of advanced communication technology concepts such as carrier aggregations (CA) was possible by the virtue of filter multiplexers. The concept of combining and splitting frequency bands to a common node, usually the system antenna node, has been investigated for over the past half-century, and significant advances have been made in this field at beginning of the development of satellite communication systems in the 1970s [18]. Aside from the fact that RF filters can be switched in and combined through an RF switch to form a multiplexer, in the literature there exist several multiplexing topologies where RF filters are directly connected to the system antenna node. The four most commonly used multiplexer configurations are [18]:

- circulator-coupled multiplexer;
- hybrid-coupled multiplexer;
- directional filter multiplexer; and
- star-junction or manifold multiplexer.

In a circulator-coupled multiplexer, the interaction between the filters is isolated by virtue of the unidirectional property of the incorporated circulators. This property enables seamless design and integration of a complex multiplexer. The use of magnetically biased ferrite material to construct an RF circulator renders realizing circulator-coupled multiplexers suitable only for large RF modules (e.g., those used in base stations) and not convenient for wireless microelectronic RF modules.

In a hybrid-coupled multiplexer, each channel consists of two identical filters and two identical 90° hybrid couplers. Hence, this configuration requires a significant increase in the size allocated per each channel filter. Size in RF microelectronic modules is usually a limited resource. Another critical aspect of this configuration is that in order for the two filter signals paths to add constructively at the output of the filter, there must be no or minimal phase deviation between the two routes. Therefore, tight fabrication tolerance is needed to maintain low phase deviation that also increases the manufacturing cost. On the positive side, the directional property of the hybrid couplers allows minimal interaction between the multiplexed filters. Additionally, having two filters per band entails that half of the input power is only passing through each filter and thereby resulting in relaxed power handling requirements, especially for transmit filters that carry high power levels.

A directional filter multiplexer usually comprises directive four-port filters cascaded in a modular concept and contains fewer components when compared to a hybrid-coupled multiplexer. The coupling nature of the directional filters requires them to be realized using either waveguide or multilayer coupled transmission line structures, which are large in size and are not applicable for microelectronic RFIC modules.

The star-junction multiplexer topology avoids the use of circulators or hybrid couplers by simply connecting the filters at their output node. However, the individual filters need to be carefully designed to minimize the interaction between them once connected. This topology explained in detail because it is the most commonly used topology in the design of AW multiplexers used in commercial wireless applications and consumer electronics due to its compact size and excellent filtering performance. In a star-junction or manifold multiplexer configuration, the typical structure consists of a number of filters connected to a short-circuited length of a transmission line (i.e., the manifold) or simply to a matching inductor as depicted in Figure 7.16.

The self-isolation and cross-isolation between the channel filters depend on the rejection floor provided by each filter at the other bands' frequency of interest. In general, multiplexers are classified into contiguous and noncontiguous multiplexers in terms of the proximity of the multiplexed bands in frequency. Contiguous bands require filters with high selectivity to allow rapid transition from the filter passband to its rejection band. The sharpness of this transition is proportional to filter's resonators Q.

Since AW filters can be modeled as capacitor networks at their OOB frequencies, the basic matching network of a star-junction AW multiplexer consists of a single shunt inductor placed at the antenna side of the multiplexer.

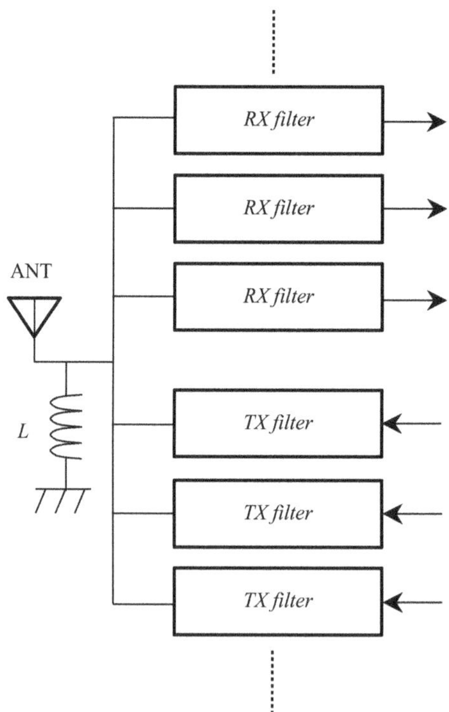

Figure 7.16 A star-junction multiplexers

In general, a filter in a multiplexer needs to present high OOB impedance at all other bands that are being multiplexed, while maintaining the desired in-band impedance matching. However, it is challenging to obtain high impedance at the filter-antenna interface over a wide span of frequencies while preserving minimum in-band loss. This is due to the impedance loading imposed by the other filters in the multiplexer (i.e., MUX loading) as it is impossible that a practical filter topology with a finite number of resonant elements can land the OOB impedance perfectly at the open side of the Smith chart for all other bands in the multiplexer. Therefore, the OOB performance of the multiplexed filters will deviate around the open circuit point on the Smith chart with an acceptable MUX loading given that a suitable phasing configuration is used. The general approach in realizing an optimum multiplexing point has been for many decades until today relying on iterative methods. Although several multiplexer synthesis techniques have been developed [19], they can serve as a good starting point and thereby achieving an optimal and competitive multiplexing performance can still be obtained through an iterative approach.

7.6.2 RF active filters

Active filters are constructed using active or amplifying devices such as MESFETs for filters operating at the microwave frequencies, complementary metal–oxide semiconductor (CMOS) for filters operating at RF, or OPAMP at low frequencies. The main motivations behind using active filters are the possibility to realize filters without inductors, without large capacitors at low frequency, and the possibility of designing electronically programmable and tunable filters. In addition, active filters are capable of covering applications ranging from few hundreds of megahertz to over ten gigahertz.

A high selectivity RF passive filter requires high-order filter topology in order to meet the design specifications. Having a high degree filter accompanies several challenges involving trade-offs between the filter performance and size. This includes high passband insertion loss and the need to precisely place several poles and zeros due to the several resonant elements involved in the filter topology in order to achieve the desired filter selectivity. In active filters, however, one can achieve the required selectivity with low loss at a reduced number of resonant elements and therefore yielding a less complex design [20]. In fact, an active filter design is considered to some extent as one of the RF filter miniaturization techniques where a reduced number of small resonators are used, and the consequential losses of the passive circuits are compensated by the incorporation of an active device. This is however at the cost of the additional active device and the need for a DC power supply to control the filter operation.

The use of lumped-element inductors in RF passive filters causes several concerns, especially, when operating at high frequencies, unlike operating at low frequencies where lumped inductor values are quite large and in most of the cases can be eliminated when placed in shunt configuration. To list some of the challenges, inductors are lossy elements and thereby can influence significantly high-Q filters. The self-resonant frequency of the inductor elements, in general, limits their usage in high-frequency applications. Inductors implemented either as embedded spirals in the laminate or as chip inductors require a considerable placement space which is a major challenge in microelectronic RFICs. Also, inductors are aggressive elements that can cause significant electromagnetic interference to nearby circuitries if not placed properly. In active filters, however, a feedback network and capacitors can only be used to realize an inductor-like characteristic [21]. An active filter comprises a transistor and a feedback network to achieve gain stability. The feedback network and the gain produced by the transistor magnify the effective value of capacitors and thereby the size of capacitors can be an issue. When a feedback network is used, the cutoff frequency of the transistors must be much larger than the operating frequency of the filters. Typically, a transistor cutoff frequency should be approximately ten times the operating frequency of the filter. In applications that require low noise levels and high power handling, active filters are not a preferred option due to their poor linearity and noise figure. An active filter has an associated noise figure equal to the insertion loss of the passive filter in addition to the noise contribution from the feedback network and the active devices.

The design of active filters can be demonstrated using the negative resistance concept which has been reported by Chang and Itoh [21] in order to cancel the losses in low Q resonant circuits. A simple and low sensitive configuration of an active filter is a cascade connection of a passive filter and an active gain block as depicted in Figure 7.17. This configuration, however, offers compensation for loss effects of the passive elements on average across the filter's passband without effectively improving the selectivity of the filter. Alternatively, one of the most common approaches in the field of RF active filter designs is to rely on transistors and their regenerative feedback networks (i.e., when a portion of the output signal is returned to the input and is in phase with the input signal) to provide Q-factor enhancement [22], especially, at the filter's band edges and therefore, high filter selectivity can be obtained. The incorporation of a regenerative feedback network to active filters may raise circuit stability concerns. However, the tolerance of such feedback networks is becoming more feasible with the continual advances in circuit fabrication and adaptive gain methods. Another approach to deal with the Q-enhancement at the filter's band edges and to improve the filter selectivity is to use active transversal and recursive filter techniques where an input RF signal is divided into multiple subcomponents that are magnified individually, time delayed, and then combined at the output as a composite signal [23].

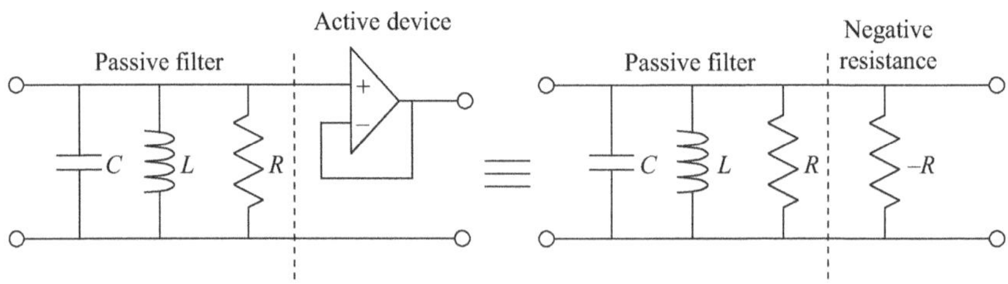

Figure 7.17 An active resonant circuit cascaded in series with an active device represented schematically by a negative resistance

The design of RF active filters relies on simulation oriented approaches as it is difficult to find comprehensive models that account for the active device parasitics.

Active filter configurations are mostly named after their inventors. Some common classes of active filters include the Sallen–Key filter comprising a finite amplifier gain, the state variable filters that incorporate several amplifiers, the Deliyannis–Friend filters that assume an infinite amplifier gain, and the Tow–Thomas biquadratic filters that comprise several amplifiers that can exhibit several filter transfer functions with the possibility to electronically program them.

7.7 CMOS controllers

The CMOS circuit is utilized as a digital logic circuitry that controls the communication buses between the integrated components in a module. In a complex RFFE module environment, CMOS provides a power management control to one or more peripheral chips such as an amplifier power levels and mode control, power regulators, DC-to-DC converter, electrostatic discharge handling, and bias management including complex biasing techniques such as envelope tracking [24].

For such applications, CMOS is a preferred technology for its high performance, low latency, and low cost. The CMOS circuit is formed by *n*-channel and *p*-channel metal-oxide-semiconductor-field-effect transistors (MOSFET). The realization of CMOS as a silicon on insulator (SOI) device offers low loss frequency characteristics and high isolation performance when compared to standard bulk CMOS.

The mobile industry processor interface (MIPI®)-based application processors for power management is a serial interface that uses CMOS I/Os to connect the integrated power controller points of the integrated devices with the power management IC. MIPIs are used to monitor and control device performance in a given application by dynamically adjusting various supply voltages in a real time to maintain the targeted performance levels. MIPI also helps in simplifying the interconnection of devices from different manufacturers and thereby reduces the production cost and time to market. For an optimal performance of such interface, it must exhibit a short response time, and the ability to support multiple processor devices.

7.8 RF circulators and isolators

Ferrimagnetic materials (or ferrites) are the fundamental component in the design of RF circulator or isolator. Depending on the crystal structure of the ferrite materials, they can be classified into three main categories: spinels, hexagonal, and garnets [25]. Materials with hexagonal crystallization are used to create ceramic magnets that are mainly used at millimeter wave applications. At RF and microwave frequencies, the spinels and garnets materials are commonly used.

Both circulators and isolators are ferrimagnetic nonreciprocal passive devices that usually consist of three ports and used to circulate the flow of energy from one port to its adjacent port, however, not to the other port. Wave circulation can be made to circulate either clockwise (CW) or counterclockwise (CCW). In fact, circulators can comprise any number of ports. For example, a one port circulator can be thought of as a short circuit. A three-port circulator when one of its ports terminated in a matched load, it can be used as an isolator, which also can be considered as a two-port circulator and performs the same function as a section of transmission line. The use of a load in an isolator prevents the available power to be reflected from the load back to the source.

RF circulators and isolators can be implemented using different technologies such as waveguide, coaxial, and microstrip structures. One of the primary uses of circulators is the transceiver duplexers allowing the transmission of two signals over one channel. Their applications extend to several RF systems such as radar, wireless communications, distributed antenna systems (DAS), and laboratory test setups and instruments where sensitive RF sources require protection from reflected high power RF signals.

There are common electrical specifications and performance trade-offs that are important to identify for any intended application. These specifications include insertion loss, return loss, power handling, linearity, performance dependency on temperature, termination rating for isolators, magnetic shielding, and isolation level. The isolation level over a given bandwidth quantifies the separation of signal levels on the adjacent ports of the device for waves traveling in the reverse direction and measured in dB. The nonreciprocity property of the ferrite material ensures low transmission loss in one direction and high loss in the reverse direction. The greater the isolation level, the less the influence from one port to the others. Maximum input power that a circulator or an isolator can sustain is dependable upon the source type (i.e., CW or pulsed). Operating at high peak power levels can cause breakdown or arcing and exacerbate the nonlinearity effects of the ferrite material that can cause an increase in the insertion loss when designed below the ferrite resonance. A device with high insertion loss leads to a dissipated power resulting in a device temperature rise. The inclusion of magnets in RF circulators and isolators to control the signal flow make them susceptible to placement in close proximity to another source of magnetic fields that may degrade its electrical performance. Therefore, magnetic shielding is required in some cases and needs to be designed during the initial design phase. Also, care should be taken when storing such devices or integrating them in a system. When operating at high power levels with poor matching conditions or poor return loss at the output port of an isolator, part of the power will be reflected causing an additional rise in the device temperature and thereby causing the performance to degrade and possibly permanent damage can occur before even reaching the rated input power. As can be concluded, designing a circulator requires considerations of several interrelated parameters.

The selection of the ferrite material is one of the important aspects in the design of circulators and depends upon the specified operating frequency, bandwidth, insertion loss, ambient temperature, and RF input power level. Ferrite resonators with any 3-fold symmetry shapes could be used (i.e., disk or triangle shape) [26,27]. Most of RF circulators and isolators are designed above the ferrite resonant frequency. The ferrite magnet operating above-resonance in the saturation mode is preferred since much higher magnetic field, and low field loss is not a problem when compared to that at resonance and below resonance cases [25]. To achieve this condition, the saturation magnetization should satisfy:

$$4\pi M_S < \frac{\omega}{\gamma} - H_a \tag{7.20}$$

where $4\pi M_S$ is the ferrite saturation magnetization, ω is the radian frequency of the ferromagnetic resonance, γ is the gyromagnetic ratio, and H_a is the anisotropy field associated with the selected material. There are various types of magnetic materials which include paramagnets, ferromagnets, antiferromagnets, ferrimagnets, and diamagnets. All magnetic materials contain magnetic moments. In a paramagnet material, the magnetic moments tend to be randomly orientated due to the thermal effects and when there is no magnetic field. Applying a magnetic field makes these random magnetic moments to align in parallel to the applied field, and also their magnetization is proportional to the applied field. In a ferromagnetic material, the magnetic moments tend to align parallel to each other under the application of magnetic field. Unlike the moments in a paramagnet material, the moments in ferromagnets will remain parallel when a magnetic field is not applied. In an antiferromagnetic material, the magnetic moments adjacent to each other tend to align antiparallel to each other without the application of a magnetic field. When these magnetic moments are equal in magnitude and opposite in direction, they result in no magnetization. In a ferrimagnetic material, the magnetic moments are not of the same size and the adjacent magnetic moments of different magnitudes aligned antiparallel. This implies that more than one type of magnetic ion is forming the ferromagnetic material. In this type of material not all the magnetic moments contribute constructively to the overall magnetization. In a diamagnetic material, the moments are repelled by the application of a magnetic field that creates an induced magnetic field in them which oppose the external applied magnetic field. This is in contrast to paramagnetic and ferromagnetic materials which are attracted by the applied magnetic field.

When the ferrite material is magnetized, the magnetic moments of the electrons spin at a frequency proportional to the magnetic field biasing. The ferromagnetic resonance occurs when a rotating RF magnetic field has the same direction and frequency as the spin of electrons in the ferrite material. At ferromagnetic resonance, the maximum coupling of the energy from the RF signal to the ferrite material occurs. In circulators, the ferromagnetic resonance is not desirable because the circulator would be extremely lossy as depicted in Figure 7.18. Another region of loss also occurs at very low magnetic field bias. This occurs due to the fact that the applied magnetic field is not sufficient to fully saturate or align the individual magnetic moments of the ferrite material. With the knowledge that high insertion loss occurs at both the low field and ferromagnetic resonance regions, circulators typically operate at the low loss regions in the below and above resonance, as shown in Figure 7.18.

The following discussion demonstrates the basic and important properties of three-port networks which are the basis for many RF and microwave devices including circulators, isolators, power splitters, and combiners. As is well-known, it

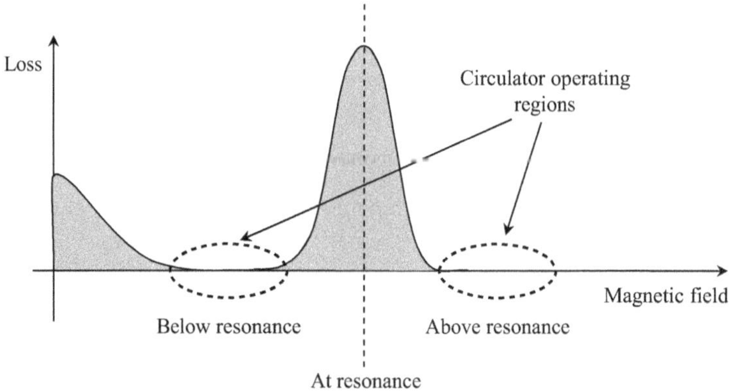

Figure 7.18 Ferromagnetic resonance curve showing the main regions of circulator operation based on the relationship between the magnetic field strength and the field loss

is impossible to obtain a three-port network that is lossless, reciprocal, and matched at all ports. As will be shown next, by relaxing one of these three conditions, the other two can be satisfied. To explain whether a lossless and reciprocal three-port network can be perfectly matched at all ports or not, consider a three-port reciprocal network that has its transmission parameters satisfy $S_{21} = S_{12}$, $S_{32} = S_{23}$, and $S_{31} = S_{13}$. A matched network will also satisfy $S_{11} = S_{22} = S_{33} = 0$. Since a reciprocal network is also symmetric, we obtain:

$$[S] = \begin{bmatrix} 0 & S_{21} & S_{31} \\ S_{21} & 0 & S_{32} \\ S_{31} & S_{32} & 0 \end{bmatrix} \tag{7.21}$$

A lossless network must satisfy $\sum_{k=1}^{3} S_{ki} S_{kj}^* = 0, \ \ i \neq j$. Thus, if $S_{32} S_{31}^* = 0$, this implies that $S_{31} = 0$ or $S_{32} = 0$. If $S_{31} = 0$, the S-matrix is written as:

$$[S] = \begin{bmatrix} 0 & S_{21} & 0 \\ S_{21} & 0 & S_{32} \\ 0 & S_{32} & 0 \end{bmatrix} \tag{7.22}$$

From (7.22), when $|S_{21}| = 1$ and $|S_{32}| = 1$, thus $|S_{21}|^2 + |S_{32}|^2 \neq 1$. Similarly, if $S_{32} = 0$, the S-matrix is written as:

$$[S] = \begin{bmatrix} 0 & S_{21} & S_{31} \\ S_{21} & 0 & 0 \\ S_{31} & 0 & 0 \end{bmatrix} \tag{7.23}$$

From (7.23), when $|S_{21}| = 1$ and $|S_{31}| = 1$, thus $|S_{21}|^2 + |S_{31}|^2 \neq 1$. In conclusion, a lossless and reciprocal three-port network cannot be perfectly matched at all its ports.

By relaxing one of these three conditions (i.e., the reciprocity condition as in circulators and isolators), a lossless and matched network can be obtained. For a nonreciprocal network (i.e., $S_{ij} \neq S_{ji}$), the S-matrix for a clockwise wave circulation in an ideal circulator is given by:

$$[S] = \begin{bmatrix} S_{11} & S_{12} & S_{13} \\ S_{21} & S_{22} & S_{23} \\ S_{31} & S_{32} & S_{33} \end{bmatrix} = \begin{bmatrix} 0 & 0 & 1 \\ 1 & 0 & 0 \\ 0 & 1 & 0 \end{bmatrix} \tag{7.24}$$

This implies that the incident energy at port 3 is delivered to the load at port 1, the incident energy at port 1 is delivered to the load at port 2, and the incident energy at port 2 is delivered to the load at port 3. From (7.24) this propagation of energy in a circular fashion is expressed as:

$$\begin{bmatrix} b_1 \\ b_2 \\ b_3 \end{bmatrix} = \begin{bmatrix} a_3 \\ a_1 \\ a_2 \end{bmatrix} \tag{7.25}$$

Similarly, the S-matrix for a counterclockwise wave circulation in an ideal circulator is given by:

$$[S] = \begin{bmatrix} S_{11} & S_{12} & S_{13} \\ S_{21} & S_{22} & S_{23} \\ S_{31} & S_{32} & S_{33} \end{bmatrix} = \begin{bmatrix} 0 & 1 & 0 \\ 0 & 0 & 1 \\ 1 & 0 & 0 \end{bmatrix} \tag{7.26}$$

In practice, RF design engineers evaluate the circulator performance through two performance parameters, the IL and isolation factor. The IL is defined as $-20 \log|S_{21}|$, and the isolation factor is defined as $-20 \log|S_{31}|$, assuming the wave circulates from port 1 to port 2.

7.9 RF mixers

Almost all RF receivers in addition to many other electronic systems employ frequency conversion or heterodyning circuits. This is also the reason why receiver architectures are called superheterodyne receivers. RF mixers are three-port passive or active devices that are designed to mix the frequencies of two input signals [28]. These two input signals are usually the local oscillator (LO) signal, and the receiver or transmitter RF signal. The signals mixing process produce sum and difference frequency products of the two distinct inputs of the RF and LO frequencies and present the resultant frequency at the third port. An RF mixer can also be used as a phase detector or as a demodulator. A simple passive mixer can consist of a single Schottky diode [29], or a field-effect transistor [30] that can operate in a way capable of producing the required frequency products. Hence, passive mixers have no conversion gain, unlike active mixers that also contributes to the overall system noise.

An ideal mixer is expected to convert a signal from one frequency to another while preserving the properties of the input signal in terms of phase and amplitude. Therefore, mixers are expected to perform a linear operation without affecting the modulation properties of the signals at its input ports. Practically, mixing two frequencies to provide a new frequency requires the use of a nonlinear device. This in return makes mixers exhibit intermodulation distortion and other spurious responses due to the device linearity. When the resultant frequency is greater than the input signal frequency, then the process is called up-conversion and vice versa. In RF transceiver systems, converting an RF signal down to a lower intermediate frequency (IF) is essential for seamless signal processing in the receive RF modules by mixing the received RF signal from the low-noise amplifier (LNA) with the LO signal. This allows frequency selection and gain control at lower frequencies where high Q-factor filters (RF image-rejection filters) can be realized. Whereas converting an IF signal into a higher RF frequency is also necessary for efficient transmission as in transmitting modules.

Three common types of active RF mixer include unbalanced, single-balanced, and double-balanced designs. Each of these mixer topologies has its own advantages and disadvantages as explained in detail in [28]. A single-balanced mixer can be realized using two diodes with inputs and outputs phased in a way that one of the inputs (e.g., LO) cancels at the output port (i.e., IF port). A double-balanced mixer as depicted in Figure 7.19 is advantageous and it can cancel both inputs (i.e., LO and RF) at the output port (i.e., IF port). In addition, harmonics of the balanced signals are also canceled.

To determine the quality of an RF mixer operation, several performance metrics need to be considered. These metrics include:

- Conversion gain or loss which represents the ratio of IF to RF power in dB, and also called single-sideband conversion gain or loss since the output power of the mixer includes only the sum or the difference frequency of the two input signals;
- Image noise suppression to improve the system noise figure;
- Isolation level between the mixer ports (i.e., LO, RF, and IF ports) to ensure low interference with the IF output port;
- Linearity performance and input intercept point; and
- Frequency ranges of the RF, LO, and IF signals for which the mixer is designed.

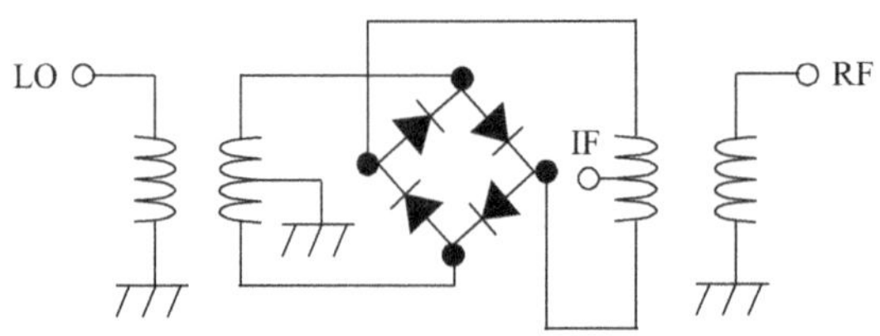

Figure 7.19 A basic configuration of a three-port double-balanced RF mixer

In general, active mixers (i.e., FET mixers) are advantageous over diode mixers due to their superior gain and noise figure performance and because high-quality diodes are difficult to produce in monolithically integrated circuit technologies.

7.10 RF oscillators

An RF oscillator is basically a signal generator that produces a periodic AC output signal from its DC supply voltage without any input signal. Oscillators are an important part of any communication systems and are used primarily to generate the carrier or local oscillator signals. The basic operation of RF oscillators relies on amplifying the circuit generated noise once a DC power is applied to the circuit and then fed to the input of a feedback network. The feedback network is a resonant circuit with a filtering characteristic that allows the signal frequency to pass the desired frequency and rejects other frequencies. The feedback network is usually constructed using reactive components in the form of *PI*-, *T*-network, transformer, or a combination of them.

Crystal oscillators were invented in the 1920s. The early oscillators used vacuum tubes, and the emphasis back then was on using a low number of tube-oscillator circuits, preferably one, while being able to achieve the maximum power out of the oscillator circuit. The reason for using a fewer number of power amplifier tubes in the transmitter circuits was mainly to reduce the manufacturing cost due to the oscillator circuit. The advent of the IC transistor circuits with their very small size, the number of the used transistors in an oscillator circuit is no longer a concern. This, in fact, resulted in better frequency stability and allowed feeding the transistor circuits with lower power supply voltage to mitigate the oscillator's crystal overdrive problem since oscillators with IC transistors can operate at low power output and then adding an amplifier stage to boost the power come at a low cost.

In a two-port network, the selection of the terminations with a negative real input or output emittances will cause oscillations. For designing a low phase-noise oscillator, the biasing circuit should be carefully regulated to avoid an unwanted noise injection. Significant variations on the supply voltages or currents can cause undesirable output power fluctuations. For this reason, some oscillator topologies use noise feedback bias networks to achieve phase noise cancellation which can provide a significant noise improvement within the bandwidth of the used circuit.

The commonly used biasing condition in an oscillator's transistor is to have a similar biasing as a Class A operation. In the case when stability and noise are of less concern, and high output power is a requirement, the oscillator may be biased for class C operation.

RF oscillators are generally classified into two types:

- Relaxation, multivibrator, or nonlinear oscillators that can generally produce nonsinusoidal output signal such as square, sawtooth, or triangle waveforms switching between two unstable states.
- Harmonic or sinusoidal oscillators that can produce a stable near sinusoidal output waveform. Their design is based on a positive feedback circuit approach.

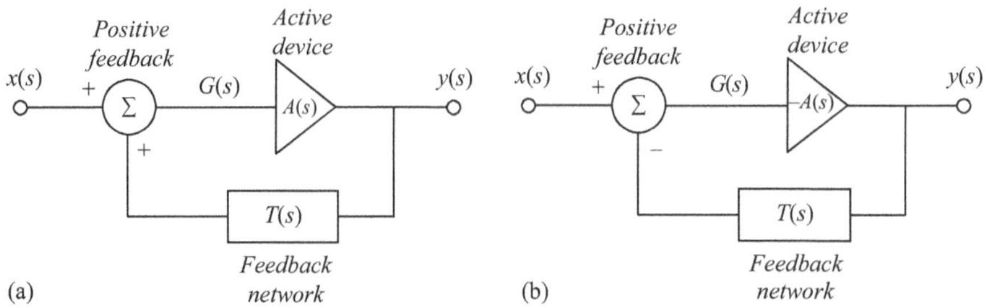

Figure 7.20 A basic RF oscillator circuit using a positive feedback network along with (a) a noninverting amplifier and (b) an inverting amplifier

Figure 7.20 shows the basic block diagram of an oscillator consisting of an amplifier and a feedback network. Here, the output power of an oscillator can be achieved by controlling the load termination presented to the transistors. In oscillators, we need slight instability to achieve proper phase noise. Good oscillators are usually low gain.

By assuming that the positive feedback network and the amplifier transistor do not load each other, we can describe the closed-loop transfer function in Figure 7.20 as:

$$\mathrm{y}(s) = \mathrm{x}(s)\frac{A(s)}{1 - A(s)T(s)} \tag{7.27}$$

The term $A(s)T(s)$ represents the gain of the network around the feedback loop and is also referred to as the loop gain. Oscillation or nonzero output can occur without input signal ($x(s) = 0$) when the term $1–A(s)T(s) = 0$. Having a proper sustainable oscillation condition using a positive feedback network requires meeting the Barkhausen criterion. The Barkhausen criterion is the relationship between the amplifier's gain ($A(s)$) and the oscillator's feedback factor (T(s)) and is expressed as:

$$1 - A(s)T(s) = 0 \tag{7.28}$$

And for the oscillation to start, the following condition must be met and is given by:

$$|A(s)T(s)| > 1 \tag{7.29}$$

Since the oscillator circuit comprises nonlinear devices, upon startup, (7.29) can seamlessly be accomplished. Once the currents and voltages magnitudes increase and the amplifier reaches saturation, the loop gain condition in (7.28) can be achieved, and therefore a steady-state condition is obtained. The Barkhausen criterion in (7.28) should be established at one frequency which determines the oscillation frequency. The transistor amplifier usually exhibits a relatively wideband frequency response. At high-frequency feedback networks, it is difficult to define the loop gain of the feedback oscillator due to the loading impact of the feedback network on the amplifier at such frequencies. Given that an amplifier input impedance decreases with the increase of frequency and that the output impedance of an amplifier is not zero due to the parasitic inductive and capacitive couplings between the devices of the surrounding components, the loop gain does not have a straightforward relation between the feedback network and the amplifier gains. This is one of the reasons why it is difficult to realize a precise feedback oscillator circuit at gigahertz frequencies.

Basic oscillator's circuit configurations include Miller, Colpitts, low capacitance load, high resistance load, Pierce, modified Meacham, harmonic-Butler common base, and harmonic-Butler emitter follower [31].

The other type of oscillators is the multivibrator which consists essentially of a two-stage oscillator and can perform a variety of functions such as the generation of square waveforms, frequency division, the voltage to frequency conversion, generation of time delays, and storage of digital data. Multivibrators can be classified into three types depending on their stability conditions: astable, monostable, and bistable. The astable multivibrator circuit is used to generate a square waveform of a known period without an external trigger signal. It switches continuously between two unstable states where one oscillator stage operates in the on-state while the other stage remains in the off-state until a reversed condition of the two stages is reached. The monostable multivibrator has one of its stages is permanently stable, while the other is quasi-stable. When an external trigger signal is applied, the stable state changes to a quasi-state for a predefined period of time and then returns to the stable state. In the bistable multivibrator, both stages are stable and the circuit requires two trigger signals (one for each stage) to switch from one state to the other. The bistable circuit maintains its output state for an indefinite period unless an external trigger pulse is applied forcing it to change state. This type of multivibrators with two stable states are also known as Latches and Flip-flops and used in sequential type logic circuits.

7.11 Microwave baluns

A microwave balun is a component used to switch the terminals of a circuit from balanced to unbalanced and vice versa by providing the required impedance transformation as the input/output impedances of the balanced and unbalanced ports are usually unmatched. A balanced output is established when two conductors have equal amplitude and opposite phase.

Microwave baluns play an important role in many classical RF applications [32] such as balanced-mixers, balanced frequency multipliers, push–pull amplifiers, phase shifters, impedance transformers, and dipole antennas. In addition to the primary purpose of incorporating baluns that is to connect a balanced load to unbalanced source in RF circuit designs, baluns are also used to provide wide matching bandwidth to cope with the multiband integrated systems and the emerging broadband technologies. Conventional baluns are generally narrowband circuits due to the limitation imposed by the even mode to odd mode impedance ratio as in coupled balun transformers or due to the use of a section of transmission line that is a fraction of a wavelength as in transmission line transformers. Several attempts have been conducted to realize different wideband designs, such as Marchand, multisection coupled-lines, composite right/left-handed transmission lines, and coplanar waveguide baluns. Each, however, has its own pros and cons in terms of the achievable bandwidth, operating frequency, and physical realization. Active baluns using transistors are also used as in RF integrated circuits.

Designing a balun requires having an amplitude balance (i.e., having both amplitudes of the two output signals matched), and phase balance (i.e., How well the phase of one of the output signals could maintain a correlation of the phase of the second output signal).

Lumped element balun can be realized using surface mount devices (SMDs) to form a center-tapped transformer that can provide a 180° phase difference between its balanced ports when its center tap nodes are grounded. The realization of a balun structure using one of the planar implementation technologies is desirable for simple low-cost fabrication; however, it is challenging to have a simple balun structure that also possesses a good electrical performance. In this context, distributed baluns have been developed to provide a simple means of transition between a microstrip line (unbalanced line) and a stripline (balanced line). Coupled microstrip lines have been also utilized to realize a balun function. One configuration can make a use of a Wilkinson power divider (WPD) to divides the signal into two signals with equal amplitude and phase and then followed by the second section of coupled transmission lines that can provide −90° and +90° phase shift for the output signals to achieve the 180° output phase difference. Two possible coupled line configurations to achieve the 180° phase shift are illustrated in Figure 7.21(a) and their equivalent S-matrix is expressed as:

$$[S]_{\text{Short}} = \begin{bmatrix} 0 & j \\ j & 0 \end{bmatrix} \tag{7.30}$$

$$[S]_{\text{Open}} = \begin{bmatrix} 0 & -j \\ -j & 0 \end{bmatrix} \tag{7.31}$$

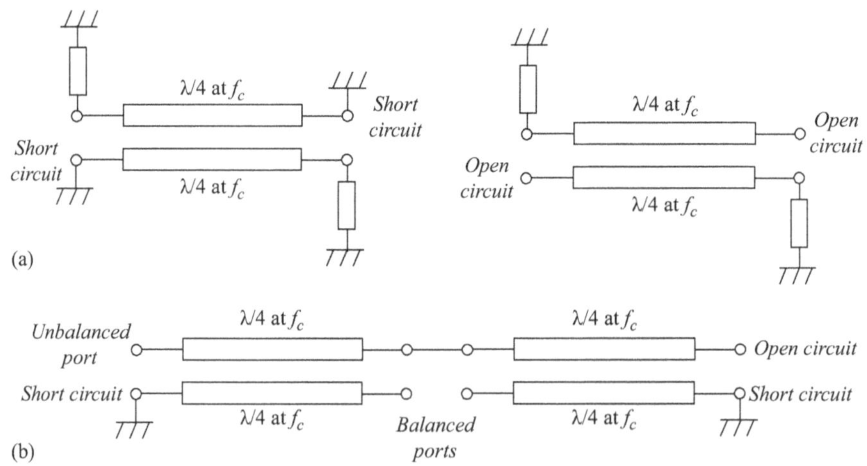

Figure 7.21 (a) A 180° phase shift using a quarter wavelength coupled lines with open/short terminations and (b) a conventional Marchand balun with a uniform coupled line sections

A wideband CPW balun with the aid of a multistage out-of-phase WPD was proposed in [33]. However, this configuration is limited to the number of consecutive stages that can be used in a WPD that may degrade the overall electrical performance due to the resulting crossovers.

Another common balun configuration is the Marchand balun which can be implemented using a planar fabrication technology while providing a broad bandwidth and good amplitude balance performance. The Marchand balun comprises two pairs of quarter wavelength coupled lines as illustrated in Figure 7.21(b). Three nodes are left open (one for the unbalanced port and two for the balanced port), and the remaining nodes are shorted. Marchand baluns can also be realized using lumped elements, and other nonplanar implementation technologies.

Attempts were also carried out to increase the even-mode impedance between the coupled lines and thereby increasing the operational bandwidth. Given that an even-mode impedance depends on the capacitance between the coupled lines, and the coupled lines and the ground plane, the bandwidth enhancement methods relied on modifying such capacitances by, for example, removing the ground plane underneath the coupled lines.

7.12 RF power limiters

RF Power limiters find applications in several areas that are sensitive to harsh EM environments and potentially exposed to high power microwave (HBM) including high-power radar systems, electronic warfare (EW) systems. Sensitive microelectronics RFIC transceiver modules also require the use of power limiters. The main purpose of RF limiters is to pass low-power signals within a predefined threshold and to protect against high-power signals that could damage the successive stages in a system by limiting or attenuating their high-power amplitudes. For instance, RF limiters are used to protect low noise amplifiers from excessive power at the antenna of a receiver system because LNAs are very sensitive devices designed to receive weak signal levels. This is the reason why RF limiters are also called receiver protectors.

Important performance parameters that need to be taken into consideration when selecting an RF limiter for a particular application include insertion loss, maximum input power for both CW and pulsed waveforms, turn on time, spike leakage, flat leakage, recovery time, and isolation. RF limiters as a protective circuit supposed to have the lowest loss possible when the signal passes through it. Low loss limiter can be achieved when placed in a shunt configuration with respect to the signal path. The maximum input power of a CW waveform is the maximum power a limiter can handle under normal operating conditions. Of course, as any semiconductor device, excessive RF power can cause permanent damage to solid-state-based limiters and thus, limiters can operate safely within a finite range of CW amplitudes or a given pulse-width and duty cycle as the case with pulsed waveforms. Maximum input peak power is the maximum pulse power a limiter can handle without breaking down. The turn-on time of an RF limiter is the time at which the limiter starts to operate and limit the input signal amplitude. The faster the turn-on time of the limiter, the better in handling fast rise time of pulsed signals.

Usually, short-pulse, high power input signals can cause significant damage to the sensitive components in a system. Other limiter technologies are based on vacuum discharge tubes or plasma components which are designed to withstand short pulse signals with a greater amount of power. However, they have larger turn-on times (<10 ns) when compared to solid-state-based limiters (<1 ps). This is because of the relatively more delay in the breakdown of gas into a low-impedance plasma where the limiter starts its function. The spike leakage occurs when the RF input pulse possesses a faster rise time than that of the turn-on time of the RF limiter. This causes a spike in RF output power until the limiter turns on. There will be no concern about the spike leakage when the limiter turns on faster than the rise time of the input pulsed signal. Plasma-based limiters usually exhibit wider spike leakage (i.e., passes more energy) compared to the solid-state limiters. Flat leakage is the CW output signal of the limiter that occurs under high input power conditions and depends on the designed output power level of the RF limiter. Flat leakage occurs after the spike leakage is complete where the limiter continues to transmit a constant power in an ideal case. The recovery time after the incidence of the high power signal is the time needed for the free charge carriers in a solid-state limiter (e.g., diode) to settle into its steady state. This parameter is important in radar and EW transceiver systems where exposure incidents to high power microwaves render them blind for an interval of time (i.e., the recovery time). Isolation of a limiter, usually given in decibels, is described by the fixed fraction between the limiter's output and input power. Generally, limiters have maximum isolation level that can reach up to 40 dB.

The implementation technology of RF limiters relies mostly on the use of solid-state devices such as PIN diodes, Schottky diodes, and FETs. Limiters comprised of Si PIN diodes are extensively used in many systems because they are compatible with MIC fabrication technologies, possess low capacitance that translates into better thermal performance, and do not require an external power supply and biasing network.

The working principle of a PIN diode as a limiter placed in a shunt to ground configuration is that when an input signal with a peak voltage exceeds the forward bias voltage of the PIN diode (i.e., 0.7 volts), the signal amplitude will start to clip. Below the threshold voltage, the diodes appear as high impedance element to the current passing through it and thus minimal impact on the main path of the input signal. Unbiased diodes with high capacitance limit the operation at high frequencies and also results in impedance mismatch that requires impedance tuning to maintain the proper operation of the following stages connected to the RF limiter.

The power limiting mechanism is achieved by reflecting or absorbing the incident power through directional devices (i.e., circulators, isolators, or couplers) to direct the high-power signal path into a lossy element to absorb or terminate them.

In some cases, multistage diode limiters with different threshold values are used for applications that require additional isolation to protect sensitive components in a system when compared to a single-stage limiter that can typically protect against 20–30 dB of excess incident power levels depending on the characteristics

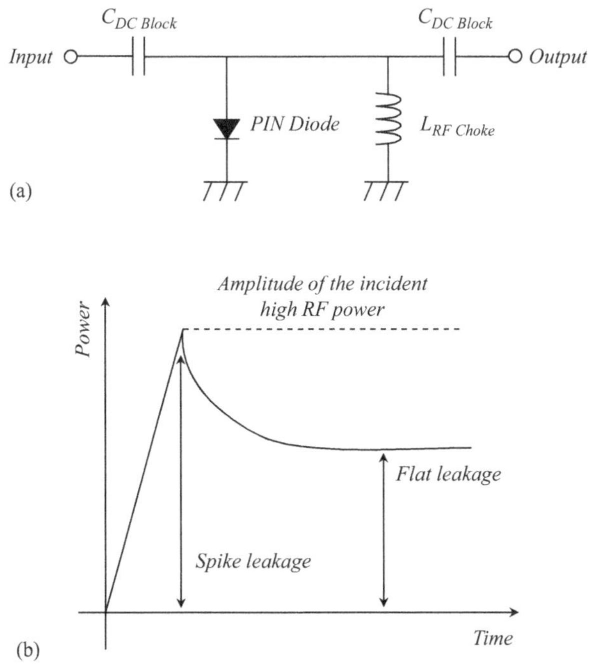

Figure 7.22 *(a) A schematic of a single stage RF limiter comprising a pin diode shunted to ground, and (b) output power response over time of a solid-state RF limiter showing the occurrence of the spike leakage as initial transient power passing through the limiter and the flat leakage after the limiter is turned on*

of the input signal such as the operating frequency, and the characteristics of the diode [34]. Careful attention must be paid to the placement of each stage in a multistage configuration. Each stage is required to be spaced by a quarter wavelength or odd multiples of quarter wavelength to obtain a good impedance matching at the input/output ports. This helps in canceling the effect of the reflected power of the first diode near the incident power source. The PIN diodes in a multistage limiter possess high impedance for incident power below the threshold level. Once a high input power above the threshold level is applied, the second diode is selected so that it has faster turn-on time than the first which causes a standing wave along the quarter wavelength line with low voltage at the second diode and a high voltage at the first diode that in return turns on the first diode. Then, the multistage power limiter starts its function.

7.13 Diversity antennas

Unlike single-input and single-output (SISO) antenna transceiver systems, MIMO systems comprised of multiple antennas at the transmitter and receiver provide efficient spectral usage by introducing additional degrees of freedom such as spatial multiplexing and diversity [35].

A diversity system employs multiple antennas to receive multiple copies of the same signal sent across independently multipath fading channels and is intended to improve the signal-to-noise ratio and to reduce multipath effects. Signal multipath interference occurs when RF waves reflect off different objects through the transmission channel and once they arrive at the receiver side, they interfere with each other, in most cases destructively (i.e., with different phase delays). The fact that the received signal on the multiple antennas will comprise uncorrelated noise characteristics between the antennas, the signals when combined and reconstructed will provide a better quality signal. Similarly, diversity can also be obtained using multiple transmit antennas by using space-time coding (STC) techniques which can be used to transmit over multiple antennas and achieve similar performance gains as can be obtained by multiple antenna diversity receptions. As the number of transmitting or receiving antennas (i.e., the number of independent fading channels) increases, the overall transmission performance improves its reliability (i.e., decrease the probability of error) linearly. For a system with (N_T) transmitter antennas and (N_R) receiver antennas, the maximum number of diversity paths is N_T times N_R. With the many advantages that antenna diversity has, it comes also with some drawbacks associated with the increased number of antennas, requiring extensive signal processing and thereby an increase in the power consumption of the circuits. Hardware overhead also increases, but a fair amount of circuits and building blocks can be shared among multiple signal paths.

The spatial multiplexing enables a MIMO transmitter/receiver to increase its throughput effectively. Here, each spatial channel carries independent bit streams from each of the transmit antennas and as a result, increases the data rate of the system, but not necessarily reliability, by allowing each receiver antenna to receive a linear combination of the transmitted signals.

References

[1] Yole Développement, *RF Front End Modules and Components for Cellphones*, Market & Technology report, Mar 2017

[2] Doherty W.H. "A New High efficiency power amplifier for modulated waves." *Proceedings of the IRE*. 1936;**24**:1163–82

[3] Rollett J. "Stability and power-gain invariants of linear two ports." *IRE Transactions on Circuit Theory*. 1962;**9**(1):29–32

[4] Freitag R.G. "A unified analysis of MMIC power amplifier stability." *IEEE MTT-S Microwave Symposium Digest*; Albuquerque, NM, 1992. pp. 297–300

[5] Ding L. "Digital predistortion of power amplifiers for wireless applications." *PhD thesis*. Georgia Institute of Technology, Atlanta, GA, Mar 2004

[6] Gharaibeh K.M. *Nonlinear Distortion in Wireless Systems: Modeling and Simulation with MATLAB*. Hoboken, NJ: Wiley; 2011

[7] Vuolevi J., and Rahkonen T. *Distortion in RF Power Amplifiers*. Norwood, MA: Artech House; 2003

[8] Rebeiz, G.M. *RF MEMS Theory Design and Technology*. 1st edn. Hoboken, NJ: John Wiley & Sons; 2004

[9] Rebeiz G.M., Patel C.D., Han S.K., Ko C.-H., and Ho K.M.J. "The search for a reliable MEMS switch." *IEEE Microwave Magazine*. 2013;**14**(1):57–67

[10] Oberhammer J. "Novel RF MEMS switch and packaging concepts." *PhD dissertation*. Royal Institute of Technology, Stockholm, Sweden, 2004

[11] Murphy T.A., Gipprich J.W., Hines M.E., and Kryger D. "A six-bit GMIC phase shifter for 6-18 GHz." *IEEE MTT-S International Microwave Symposium Digest*; Albuquerque, NM, Jun 1992. pp. 1171–74.

[12] Levy R., Snyder R.V., and Matthaei G. "Design of microwave filters." *IEEE Transactions on Microwave Theory and Techniques*. 2002;**50**(3):783–93

[13] Pozar D. *Microwave Engineering*. 3rd edn. New York, NY: Wiley; 2005. pp. 412-416

[14] Deslandes D., and Wu K. "Integrated microstrip and rectangular waveguide in planar form." *IEEE Microwave Wireless and Component Letters*. 2001;**11** (2):68–70

[15] Larson J.D., Bradley P.D., Wartenberg S., and Ruby R.C. "Modified Butterworth-Van Dyke circuit for FBAR resonators and automated measurement system." *2000 IEEE Ultrasonics Symposium. Proceedings: An International Symposium (Cat. No.00CH37121)*; vol. 1, San Juan, Puerto Rico, 2000. pp. 863–68

[16] Jamneala T., Bradley P., Feld D.A., Clark D., and Chien A. "A generic passive matching network for multiplexers." *IEEE Transactions on Ultrasonics, Ferroelectrics, and Frequency Control*. 2007;**54**(6):1272–75

[17] Morita T., Watanabe Y., Tanaka N., and Nakazawa Y. "Wideband low loss double mode SAW filters." *IEEE 1992 Ultrasonics Symposium Proceedings*; Tucson, AZ, 1992. pp. 95–104

[18] Cameron R.I., and Yu M. "Design of manifold-coupled multiplexers." *IEEE Microwave Magazine*. 2007;**8**(5):46–59

[19] Macchiarella G., and Tamiazzo S. "Synthesis of star-junction multiplexers." *IEEE Transactions on Microwave Theory and Techniques*. 2010;**58**(12): 3732–41

[20] Steer M. *Microwave and RF Design: A Systems Approach*. 2nd edn. Raleigh, NC: Sci-Tech Publishing; 2013

[21] Chang, C.Y., and Itoh, T. "Microwave active filters base on a coupled negative resistance method." *IEEE Transactions on Microwave Theory and Techniques*. 1990;**30**(12):1879–84

[22] Katzin, P., Bedard B., and Ayasli Y. "Narrow-band MMIC filters with automatic tuning and Q-factor control." *IEEE MTT-S International Microwave Symposium Digest*; Atlanta, GA, 1993. pp. 403–06

[23] Schindler M.J., and Tajima Y. "A novel MMIC active filter with lump and transversal Elements." *IEEE Transactions on Microwave Theory and Techniques*. 1989;**37**(12);2148–53

[24] Lee C.F., and Mok P.K.T. "A monolithic current-mode CMOS DC-DC converter with on-chip current-sensing technique." *IEEE Journal of Solid-State Circuits*. 2004;**39**(1):3–14

[25] Linkhart D.K. *Microwave Circulator Design*. 1st edn. Norwood, MA: Artech House; 1989

[26] Helszajn J., and Sharp J. "Verification of first circulation conditions of turnstile waveguide circulators using a finite-element solver." *IEEE Transactions on Microwave Theory and Techniques*. 2005;**53**:2309–16

[27] Tsankov M.A., and Milenova L.G. "Accurate design of partial-height ferrite resonators for waveguide circulators." *Journal of Magnetism and Magnetic Materials*. 1990;**83**:435–36

[28] Fong K.L., and Meyer R.G. "Monolithic RF active mixer design." *IEEE Transactions on Circuits and Systems II: Analog and Digital Signal Processing*. 1999;**46**(3):231–39

[29] Vendelin G.D., Pavio A.M., and Rohde U.L. *Microwave Circuit Design Using Linear and Nonlinear Techniques*. New York, NY: Wiley; 1990

[30] Maas S.A. "A low distortion GaAs MESFET resistive mixer." *Microwave Journal*. 1988;**31**(3):213–20

[31] Matthys R.J. *Crystal Oscillator Circuits*. Malabar, FL: Krieger Publishing Company; 1992

[32] Sturdivant R. "Balun designs for wireless, mixers, amplifiers and antennas." *Applied Microwaves*. 1993;**5**:34–44

[33] Lim J., Park U., Oh S., Koo J., Jeong Y., and Ahn D. "A 800- to 3200-MHz wideband CPW balun using multistage Wilkinson structure." *IEEE MTT-S International Microwave Symposium Digest*; San Francisco, CA, 2006. pp. 1141–44

[34] Garver R.V. *Microwave Diode Control Devices*. Norwood, MA: Artech House, Inc.; 1976

[35] Lozano A., and Jindal N. "Transmit diversity vs. spatial multiplexing in modern MIMO systems." *IEEE Transactions on Wireless Communications*. 2010;**9**(1):186–97

Chapter 8
Component- and module-level measurements

8.1 Introduction

Wireless service providers are the driving force of today's advances in the mobile communication sectors. The growing demand for more bandwidth and the increasing number of mobile wireless users has led to the rapid development of next-generation broad- and multiband cellular technologies. Long-term evolution (LTE)-advanced which is one of the third generation partnership project (3GPP) technology standards was introduced to accommodate the demand on high-speed wireless communications and data terminals by extending the data capacity via the enablement of a new service capability known as carrier aggregation (CA). In CA, the number of aggregated carriers can be different in the uplink (UL) than the downlink (DL); however, the number of UL carrier components must not exceed the number of the DL components. Furthermore, CA allows up to five component carriers to be aggregated with bandwidths ranging from 1.4 to 20 MHz. These possibilities proclaim different CA configurations and therefore complex radio frequency (RF) front-end transceiver module design and testing procedures. Consequently, new challenges have arisen in terms of designing and testing RF modules to support multiple frequency band combinations, managing nonlinear effects, and blocking spurious modes. An RF module is a basic building block when we look into the top-down design flow of an RF product implementation. The development and testing of RF modules are receiving lots of attention from the industry today due to the rapid growth in module business coupled with growth in module content and complexity.

RF and microwave characterization is an immensely important part of validating theoretical concepts and verifying design qualities. In many cases, at the final development stages of a product, fine performance tuning based on accurate model and simulation results may not be adequate to yield a competitive performance. At such circumstance, design tuning decisions can rely instead on high fidelity measured data of the RF module prototype to determine the final performance. In today's highly complex RF modules, it is important to understand that RF testing can impact the product time-to-market when the correlation between one test equipment to another, and a fixture to another are not taken into consideration. It is then important to apply the appropriate fudge factor for parts measured by different equipment of the same type. Testing large volume parts usually need to be carried out in a short period of time using the shortest test procedure as possible at a

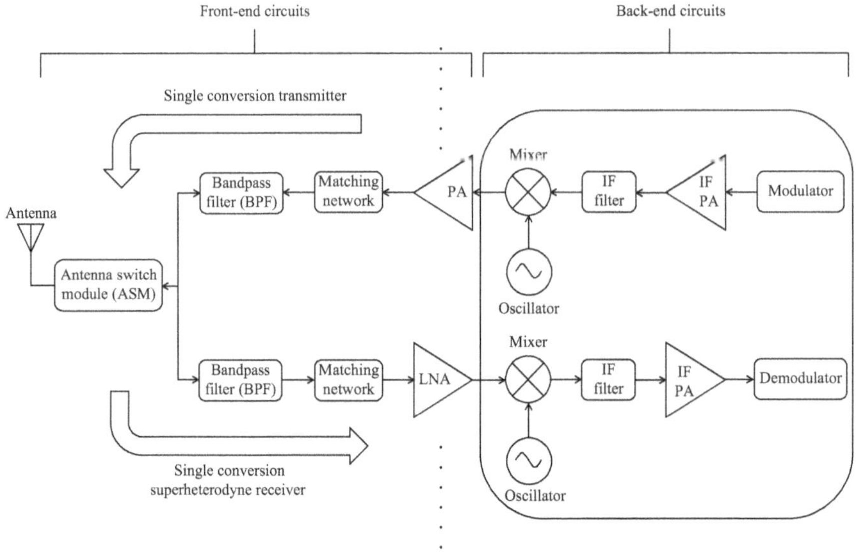

Figure 8.1 Typical wireless transceiver block diagram illustrating the microwave components in both: (a) a single conversion transmitter and (b) a single conversion superheterodyne receiver

low cost while being able to determine the pass/fail condition of the device under test (DUT). High volume testing is usually a less comprehensive procedure than characterizing a device during the prototype stage. Therefore, careful attention must be paid when identifying the short test procedure to ensure it can meet the product quality requirements.

Regardless of the different transceiver architectures that an RFIC module can have, this chapter aims to outline common measures and test setups used in the characterization of RF modules—large and small signal measurements.

Figure 8.1 illustrates a generic RF transceiver block diagram demonstrating the primary passive and active components in an RF transmit and receive chains. Different characterization techniques are available to test the individual performance of passive and active components in an RF module as well as the performance of the full RF chain. RF and microwave measurements can be classified under several categories: time and frequency domain, small and large signal, discrete component and system level, receiver and transmitter-related measurements, etc.

8.2 Small-signal measurements

A small-signal is a terminology accompanied with the analysis and characterization of electronic circuits operating at relatively a small magnitude alternating current and voltage when compared to the direct bias current and voltage of the device. As discussed in Chapter 3, at microwave frequencies, power and phase are preferable quantities that can be easily and accurately measured as opposed to voltages or currents. Therefore, traveling-wave-based parameters such as *S*-parameters are one of the commonly used linear parameters in the characterization of an RF device. Several important measures can be derived from the two main wave parameters, the reflection coefficient (Γ), and the transmission coefficient (T). These measures include return loss (RL), insertion loss (IL), mismatch loss (ML), the voltage standing-wave ratio (VSWR), and group delay. Small-signal characterizations of RF and microwave circuits are carried-out typically using a vector network analyzer (VNA).

The general phasor form of the reflection coefficient can be described as:

$$\Gamma = \frac{Z_l - Z_s^*}{Z_l + Z_s} = |\Gamma| e^{j\theta} \tag{8.1}$$

The reflection coefficient is usually used to measure how well the impedance matching is between two successive RF stages and is often expressed in dB and referred to as RL:

$$\mathrm{RL(dB)} = -20 \log_{10} |\Gamma| \tag{8.2}$$

Another dimensionless quantity that measures impedance matching is the VSWR and is related to the reflection coefficient by:

$$\mathrm{VSWR} = \frac{1 + |\Gamma|}{1 - |\Gamma|} \tag{8.3}$$

The transmission coefficient of a lossless network in terms of Γ is given by:

$$T = 1 + \Gamma = \frac{2Z_l}{Z_l + Z_s} \tag{8.4}$$

and Z_S and Z_l are the source and load impedances of the network, respectively. The transmission coefficient when expressed in dB, yields to a commonly used quantity which is the IL. The same parameter in dB is referred to as gain in the case of evaluating active devices:

$$\mathrm{IL(dB)} = -20 \log_{10} |T| = -20 \log_{10} |1 + \Gamma| \tag{8.5}$$

Another performance parameter that is typically used to quantify loss due to impedance matching is the mismatch loss and is given by:

$$\mathrm{ML(dB)} = -10 \log_{10} \left(1 - |\Gamma|^2\right) \tag{8.6}$$

The reflectivity and transmissivity of EM waves are real quantities for a lossless media and complex quantities for a lossy media. Knowing that in a lossless media the incident energy is the sum of the reflected energy and the transmitted energy at the media interface, the transmitted power (P_T) can be written in terms of the wave reflection coefficient and the incident power (P_{in}) as:

$$P_T = P_{\mathrm{in}} \left(1 - |\Gamma|^2\right) \tag{8.7}$$

where $|\Gamma|^2$ is the fraction of incident power that is reflected.

For many applications, the knowledge of the signal phase variations over frequency, especially, in broadband circuits is an important parameter that can significantly impact the RF system performance. Therefore, group delay (Gd) is used to calculate the rate of change of phase over frequency. ϕ in (8.8) is the phase in radians and ω is the angular frequency in radians per second:

$$\mathrm{Gd} = -\frac{d\phi}{d\omega} \tag{8.8}$$

8.3 Large-signal measurements

A device or a transceiver module behavior under small- and large-signal conditions is different. Under large-signal conditions, the time-variant signals possess significantly high magnitudes where the nonlinear effects of an active device start to play a role in the overall system performance. Load selections under large-signal are constrained by the IV characteristic of the active device whereupon load behavior over a broad frequency range must be taken into account to cover not only the fundamental design frequency but the harmonic frequencies as well. Understanding the nonlinear phenomena of a given device by carrying out the proper characterization techniques enables system engineers and circuit designers to efficiently communicate design tradeoffs to meet the overall system specifications. A thorough understanding of the design tradeoffs is often gleaned from several laboratory experiments, where for example, power amplifier (PA) and module engineers usually examine small and large signal measurements and try to make a correlation between both tests to identify the proper offset in performance and eventually reduce the redesign cycle time of complex modules. Different module designs may end up with different offset values due to the different EM parasitics in each case. Therefore, every new module design needs to be dealt with as a new case having new offset values.

Several characterization methods and techniques have been developed to improve our understanding of the device and system level performance under large signal conditions. Linearity is one of the critical measures in RFFE transceiver modules. In general, solid-state-based devices such as PAs, low-noise amplifiers (LNAs), and mixers exhibit nonlinear characteristics that dominate the system-level performance. In this context, several characterization methods and definitions have been developed to gauge the nonlinearity characteristics accompanying such devices including harmonics, gain compression, intermodulation distortion (IMD), phase distortion, and adjacent channel power (ACP), and several others. Sections 8.3.1 and 8.3.2 will discuss some of the important concepts and measures used to characterize nonlinear distortion in transceiver modules.

8.3.1 Transmitter-related measurements

8.3.1.1 Harmonics

Harmonics are usually unwanted signals generated by a nonlinear device that produce signal components at integer multiples of the fundamental operating frequency as depicted in Figure 8.2.

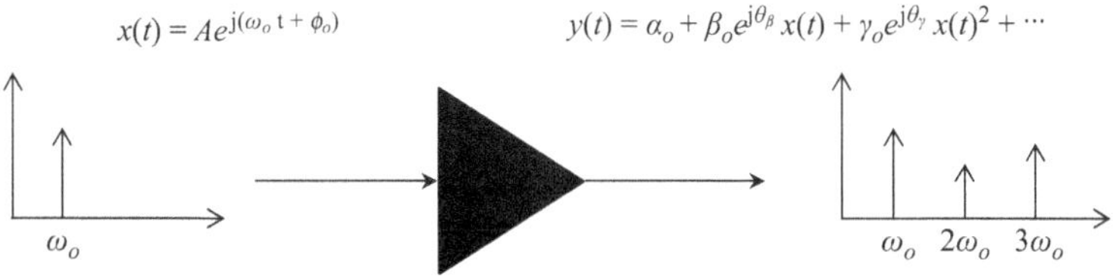

Figure 8.2 Harmonic frequency components produced by a nonlinear device

Considering an input signal $X(\mathrm{t}) = A_{\mathrm{in}} \cos \omega_1 t$ applied to a linear device with a linear gain of α_1, the output signal then can be written as $Y(\mathrm{t}) = \alpha_1 X(t) = \alpha_1 A_{\mathrm{in}} \cos \omega_1 \mathrm{t}$. On the other hand, a nonlinear device generates an output signal that can be mathematically represented by a Taylor series expansion in the following power series form:

$$Y(\mathrm{t}) = \alpha_1 A_{\mathrm{in}} \cos \omega_1 \mathrm{t} + \alpha_2 A_{\mathrm{in}}^2 \cos^2 \omega_1 t + \alpha_3 A_{\mathrm{in}}^3 \cos^3 \omega_1 t + \cdots \tag{8.9}$$

The coefficients, $\alpha_{1,2,3,..}$, in (8.9) are the corresponding nth partial derivative coefficients of Y in terms of the input signal (X) evaluated at a given bias point, temperature, and other factors. α_1 is the small signal gain. Depending on how large the input signal is, one can accurately determine how many terms are needed to represent the device nonlinear behavior using a Taylor series. Using trigonometric identities, one can rewrite (8.9) in the following form:

$$\begin{aligned} Y(\mathrm{t}) &= \alpha_1 A_{\mathrm{in}} \cos \omega_1 t + \frac{\alpha_2 A_{\mathrm{in}}^2}{2}(1 + \cos 2\omega_1 t) \\ &+ \frac{\alpha_3 A_{\mathrm{in}}^3}{4}(\cos 3\omega_1 t + 3 \cos \omega_1 t) + \cdots \end{aligned} \tag{8.10}$$

Equation (8.10) can be further rearranged to demonstrate the harmonic frequency contents and their amplitudes as:

$$Y(\mathrm{t}) = \underbrace{\frac{\alpha_2 A_{\mathrm{in}}^2}{2}}_{\text{DC}} + \underbrace{\left(\alpha_1 A_{\mathrm{in}} + \frac{3\alpha_3 A_{\mathrm{in}}^3}{4}\right) \cos \omega_1 t}_{\text{Fundamental}} + \underbrace{\frac{\alpha_2 A_{\mathrm{in}}^2}{2} \cos 2\omega_1 t}_{\text{2nd harmonic}} + \underbrace{\frac{\alpha_3 A_{\mathrm{in}}^3}{4} \cos 3\omega_1 t}_{\text{3rd harmonic}} + \cdots \tag{8.11}$$

Equation (8.11) shows that as the power of the input signal (A_{in}) increases, the amplitude ratio of the harmonics to the fundamental increases as well. Although the harmonics amplitudes are less than the fundamental amplitude, their rate increases in a rate faster than the amplitude increase of the fundamental frequency. It is therefore obvious that harmonics can cause drastic distortion to a system operating at these harmonic frequencies and, thus, must be filtered out. Figure 8.3 shows the measurement setup used in evaluating the harmonic distortion. Harmonics are straightforward to measure using a spectrum or vector network analyzer as their frequency locations are known a priori.

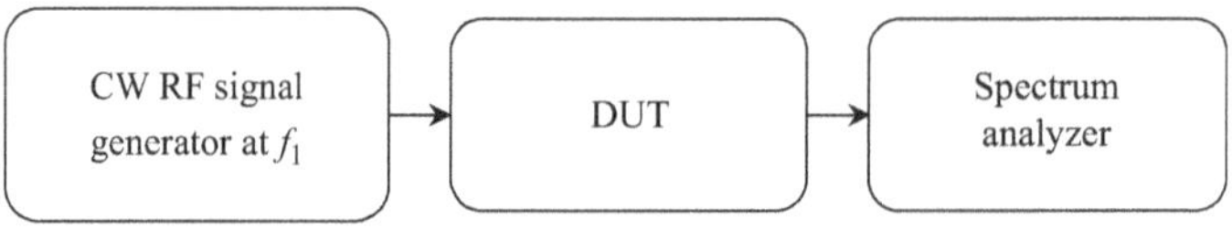

Figure 8.3 A diagram shows the setup of measuring a device harmonics

8.3.1.2 Intermodulation distortion

In addition to harmonic distortion caused by active circuits in a transceiver module, intermodulation distortion (IMD) is another critical nonlinear parameter that must be quantified and measured. As can be seen in (8.11), harmonics of a single tone input signal are well spaced from the fundamental frequency, whereas having a second or multitone input signal with a small offset of the fundamental frequency results in unwanted frequency products (spurious frequencies) that may fall nearby the fundamentals. IMD is a customary measurement in transceiver systems and is typically performed by applying two continuous wave (CW) signals with a small offset in frequency to a power combiner circuit that feeds the transmitting side of the module. Another possible scenario could be a transmitted fundamental frequency is mixed with a jammer frequency received by the system antenna. Considering a two-tone input signal $X(\text{t}) = A_1 \cos\, \omega_1 t + A_2 \cos\, \omega_2 t$, the output of a nonlinear device given by a Taylor series model would be in the following form:

$$Y(\text{t}) = \alpha_1(A_1 \cos\omega_1 t + A_2 \cos\omega_2 t) + \alpha_2(A_1 \cos\omega_1 t + A_2 \cos\omega_2 t)^2 + \alpha_3(A_1 \cos\omega_1 t + A_2 \cos\omega_2 t)^3 + \cdots \tag{8.12}$$

Equation (8.12) can be expanded to illustrate the generated IMD frequencies and their amplitudes as in the following form:

$$\begin{aligned} Y(t) = {} & \underbrace{\frac{1}{2}\alpha_2\left(A_1^2 + A_2^2\right)}_{DC} + \alpha_1 A_1 \cos\,\omega_1 t + \alpha_1 A_2 \cos\,\omega_2 t + \alpha_2 A_1 A_2 \cos(\omega_1 + \omega_2)t \\ & + \alpha_2 A_1 A_2 \cos(\omega_1 - \omega_2)t \\ & + \frac{1}{2}\alpha_2 A_1^2 \cos 2\omega_1 t + \frac{1}{2}\alpha_2 A_2^2 \cos 2\omega_2 t + \frac{1}{4}\alpha_3 A_1^3 \cos 3\omega_1 t + \frac{1}{4}\alpha_3 A_2^3 \cos 3\omega_2 t \\ & + \left[\frac{3}{4}\alpha_3 A_1^2 + \frac{3}{2}\alpha_3 A_2^2\right] A_1 \cos\omega_1 t + \left[\frac{3}{4}\alpha_3 A_2^2 + \frac{3}{2}\alpha_3 A_1^2\right] A_2 \cos\omega_2 t \\ & + \frac{3}{4}\alpha_3 A_1^2 A_2 \cos(2\omega_1 + \omega_2)t + \frac{3}{4}\alpha_3 A_2^2 A_1 \cos(2\omega_2 + \omega_1)t \\ & + \frac{3}{4}\alpha_3 A_1^2 A_2 \cos(2\omega_1 - \omega_2)t + \frac{3}{4}\alpha_3 A_2^2 A_1 \cos(2\omega_2 - \omega_1)t \end{aligned} \tag{8.13}$$

Scrutiny of (8.13), one can classify the IMD products by looking at the amplitude order of each term and then finding the corresponding IMD frequencies.

The following IMD spectrum components will result by the sum and difference products of each of the fundamental input tones (f_1 and f_2) and their harmonics:

- Second order: $2f_1$, $2f_2, f_1 + f_2, f_1 - f_2$,
- Third order: $3f_1$, $3f_2$, $2f_1 \pm f_2$, $2f_2 \pm f_1$,
- Fourth order: $4f_1$, $4f_2$, $2f_2 \pm 2f_1$, etc.

Figure 8.4 depicts the IMD spectrum with second and third order IMD products caused by the sum and difference products of each of the fundamental input tones (f_1 and f_2). As can be seen, the challenging intermodulation products are the second and third terms as these terms fall in close proximity to the fundamental tones.

As can be observed from Figure 8.4, IMD3 products are critical distortion located near the two transmitter tones, and they may affect adjacent channels which could be receiving channels. In general, all odd order IMDs are close to the two fundamental tone frequencies when compared to the even order IMDs and care should be taken to ensure that no spectral leakage occurs at the adjacent bands. Measurement setup of IMD is illustrated in Figure 8.5.

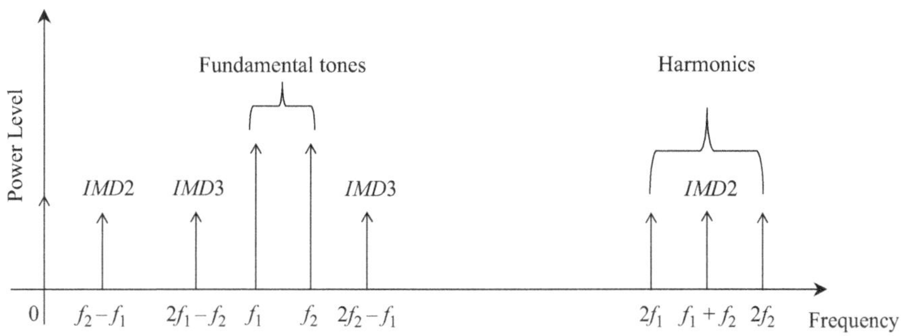

Figure 8.4 A diagram shows the spectra of fundamental tones and intermodulation products using the two-tone method

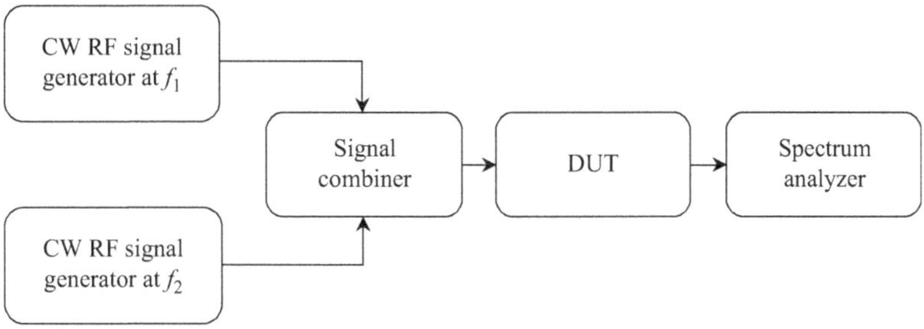

Figure 8.5 A diagram shows a two-tone measurement setup for measuring IMD products

8.3.1.3 Third-order intercept point

As will be shown next, increasing the power level of the fundamental input signal can lead the device to gain compression where part of the output power goes to the harmonic frequencies. Similarly, third-order intercept point (TOI), also called the intercept point of the third order (IP3), is an important device characteristic related to the IMD products. In principle, TOI is a theoretical point at which every 1 dB increase in the tones input power will drive the power of the third order IMD product to increase by 3 dB as illustrated in Figure 8.6. In other words, as the input power increase to a level where the first-order and third-order products are equal in power, this point indicates a TOI point.

Additionally, the IIP3 in logarithmic scale can be calculated in terms of IMD3 power level using:

$$\mathrm{IIP3}(\mathrm{dB}_m) = \frac{\left(P_{\mathrm{out,fundamental\ tone}} - P_{\mathrm{out,IMD3}}\right)}{2} + P_{\mathrm{out,fundamental\ tone}} \tag{8.14}$$

Similarly, the IMD3 power level can be calculated from the IIP3.

8.3.1.4 Gain compression point

A nonlinear device, for example, a PA, maintains a constant small signal gain (i.e., linear gain with a slope of 1); however, at high input power levels the device reaches saturation and causes the small-signal gain to drop or compress. Most often this output to input power characteristic is referred to as AM–AM distortion. A 1-dB compression point is a common measurement that refers to the point at which the amplifier gain drops by 1 dB when compared to the small signal gain. The peak output power of a PA is often specified at the 1-dB compression point.

The gain decrease at compression is related to the device linearity, and at this level of input power, the device starts to deviate from the linear region as depicted in Figure 8.7. As noted in (8.13), part of the output power appears at the harmonics

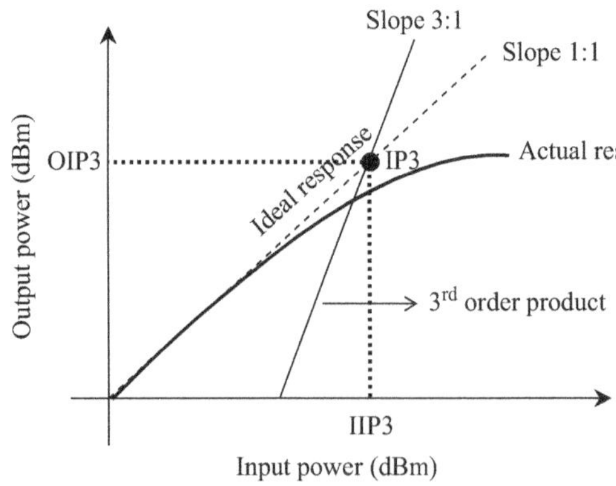

Figure 8.6 Third-order intercept point where first-order and third-order lines intersect

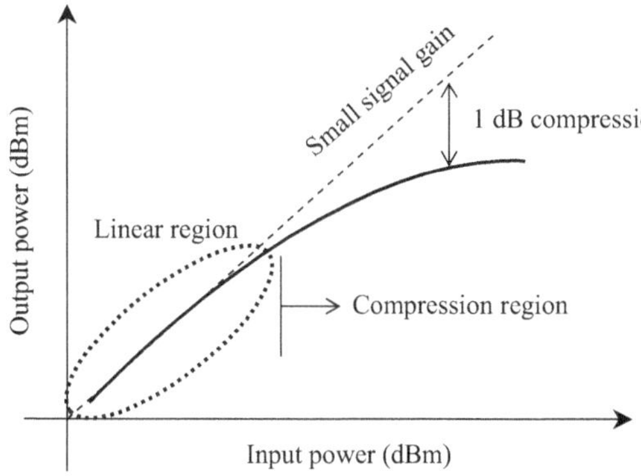

Figure 8.7 Input–output power relation of a nonlinear system showing the 1-dB gain compression point

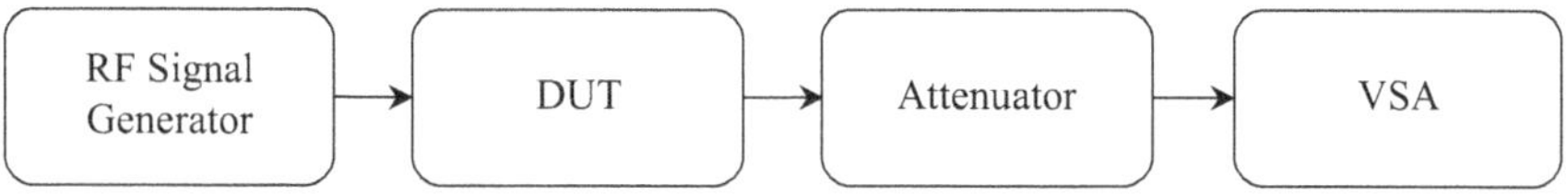

Figure 8.8 Measurement setup of gain compression

rather than the fundamental signal. Further increase of the input power drives the device to saturate, and the output power becomes constant. It is worth mentioning that the vast majority of nonlinear devices follows the same behavior. However, there are some cases where gain may be negative at higher input power levels.

The measurement setup in Figure 8.8 shows how a 1-dB compression point of a nonlinear device can be evaluated. An attenuator is used before the vector signal analyzer (VSA) to avoid driving the VSA into compression which may damage the analyzer receiver due to the power used from the signal generator to compress the DUT.

By knowing the nonlinear characteristic of a device, it would be possible to find the input power at which the 1 dB gain compression occurs. Recall from (8.13) the terms related to the input power of the fundamental frequency include:

$$G_{\text{Fundamental}} = \left(\alpha_1 A_{\text{in}} + \frac{3\alpha_3 A_{\text{in}}^3}{4}\right) / A_{\text{in}} = \alpha_1 + \frac{3\alpha_3 A_{\text{in}}^2}{4} \tag{8.15}$$

Here, a 1-dB compression means that the amplifier gain has decreased by one decibel from the small signal gain and thus from (8.15) by excluding α_1 which is the linear gain, the remaining part is related the 1 dB reduction in gain. In terms of logarithmic power, (8.15) translates to:

$$20 \log\left(1 + \frac{3\alpha_3 A_{\text{in}}^2}{4\alpha_1}\right) = -1 \tag{8.16}$$

Solving (8.16), yields:

$$A_{\text{in}} = \sqrt{0.145\frac{\alpha_1}{\alpha_3}} \tag{8.17}$$

At high compression levels, the PA begins to saturate where an increase in the input power will have no increase in output power.

8.3.1.5 Transmitter leakage and spurious emissions

In frequency-division duplexing (FDD) applications, both the transmitter and receiver operate at different carrier frequencies, thus simultaneous two-way communication is possible. Transmit signal leakage, also known as TX-leakage power, is a common term used to note the effect on the receiver sensitivity performance during the operation of the TX signal. In this case, having sufficient isolation between the transmitter and receiver terminals is crucial to maintain high receiver sensitivity or low noise floor. In other words, when a transmitter generates out-of-band phase noise that falls within the receiver band, the receiver noise floor will degrade leading to poor receiver sensitivity. Although transceiver modules are designed with sufficient isolation levels between its terminals, the integration with other parts for additional functionality may degrade isolation.

Transmitter spurious emissions include emissions due to the harmonics, parasitics, intermodulation products, and frequency conversion products that are generated by the nonlinear components in the transmitting devices when mixed, modulated, or amplified.

8.3.1.6 Envelope and average power tracking

Before proceeding to the subject of envelope and average power tracking, it is helpful to define the PA efficiency as it represents the main point of discussion in this section. The power efficiency of a PA is a measure of how much of the DC power is converted into RF output power. The output power of a PA is always less than the DC supplied power. The ratio of the output to the DC input power is known as the PA efficiency (η). The drain or collector efficiency of PA is defined as:

$$\eta = \frac{P_{\text{RFout}}}{P_{\text{DC}}} \tag{8.18}$$

A more prevalent and meaningful definition of PA efficiency is the power-added efficiency (PAE) that takes into account the input RF power added to the PA as in (8.19):

$$\text{PAE} = \frac{P_{\text{RFout}} - P_{\text{RFin}}}{P_{\text{DC}}} \tag{8.19}$$

PAE in (8.19) can be approximated to that in (8.18) if the value of P_{RFin} is small. In a single stage PA with a small gain, P_{RFin} is substantial, and in this case, PAE is a suitable measure for efficiency, while in a multistage PA and high total gain the value of P_{RFin} is much less than P_{RFout} and here the drain or collector efficiency becomes a good approximation for measuring the power efficiency.

PA power efficiency significantly impacts battery life, recharge cycles, and operating temperature of transceiver modules. Transmitting systems that support high data rate technologies such as W-CDMA and LTE are accompanied with modulation schemes that generate variable envelope signals (e.g., OFDM-LTE) or more specifically high peak-to-average power ratio (PAPR) signals. Traditional PAs powered by fixed bias voltage are most efficient when they operate at peak input signal power or at constant amplitude signal (e.g., FM signal) as they operate at the edge of the compression point. PAs, however, do not constantly operate at peak power levels and hence the amount of energy not being transformed into useful RF power is dissipated as heat. This in return may require cooling considerations as in the case of base stations. Moreover, having a high PAPR as in LTE technology can drive the PA into far compression region where PA output becomes highly nonlinear; therefore, PA output power is typically lowered to avoid high compression and this results in average output power below the optimum efficiency point. Both low average input power (i.e., high PAPR) and not operating at peak input power most of the time leads to poor PA efficiency.

As a result, a power management approach has become an essential requirement of almost any modern transceiver system. Among the different amplifier efficiency improvement techniques (e.g., Doherty amplifier, digital predistortion, DC tracking, etc.), envelope tracking (ET) and average power tracking (APT) are more prevalent and quickly gaining traction amongst commercial PA vendors. This is because ET and APT are more linear than other existing techniques which require the PA to operate in the high compression region.

ET and APT are conceptually similar and both are means of improving PA efficiency, and to some extent linearity through continually adjusting the supply source in accordance to the RF input power level. Although ET and APT technologies seem new, the concept has been around for many years [1]. Early adoption of ET technology was in cellular base stations and broadcast systems and then recently in mobile phone handsets.

APT as can be concluded from its name operates only when the PA's average input power level changes, which occur over time slots and not instantaneously as in ET. Hence, APT works on a slower timescale than ET with power supply adjustment at a rate of 1–10 ms and this may not be preferable for high PAPR signals because at signal peaks the average power cannot be reduced without deteriorating linearity. Thus for low PA input power levels, APT is necessary.

Each method has its peculiar advantages and drawbacks. It is understandable that PAs with ET capabilities are relatively more efficient than those with APT envelope tracker. However, ET requires a high bandwidth power supply (i.e., at least two times the modulation bandwidth), so that accurate tracking is maintained between the modulated supply voltage and the envelope of the transmitted RF signal. Preserving synchronization between the two waveforms is another challenge that is necessary for having maximum PAE and good linearity. Reliable operation of an ET system can lead to improved linearity (e.g., ACLR) by operating at the edge of the compression region. In the case of systems with sufficient linearity margin, a tradeoff between system efficiency and linearity is possible. In a sense, operating in farther compression region will deteriorate linearity, but PAE efficiency will improve.

PAs designed to support ET mode usually can't be used to support APT. The PA design dependence on the power supply mode is arising from the performance degradation due to how the PA supply network is designed. ET-PAs operate in compression most of the time to increase PAE and this, on the other hand, makes PAs more susceptible to power supply feed-through noise. In order to maintain reasonable linearity conditions or ACLR, low capacitance supply network is needed to ensure very low supply noise fed-through the PA output. Otherwise, tighter specifications on the TX-RX isolation levels will be required in a transceiver module. Ideally, a dual-mode envelope tracker capable of effectively operating with high PAPR and low average power levels is mandatory for best overall system efficiency. The challenge underlies in the switching mechanism between the two modes without distorting the signal.

Figure 8.9 demonstrates three different PA biasing cases. As can be seen, ET provides relatively better efficiency than APT, and both are more efficient than PAs with a fixed power source. The supply modulator block in Figure 8.9(c) is a simplified representation of the envelope tracker unit in a system. An actual ET unit usually consists of several blocks that start with an envelope detector from the I-Q signals of the baseband input power signal. Then, an envelope shaping table or lookup table is generated and stored in a memory unit (e.g., RAM or ROM) that can be used to map the envelope signal or baseband input signal into the appropriate power supply voltage according to the PA preferred mode of operation (e.g., maximum efficiency, constant gain, or constant gain-compression). To this end, the signal is in its digital form and need to be converted into analog format through a digital-to-analog converter (DAC) in order to prepare the signal to be supplied to the power supply control terminal. Additionally, a delay balance unit may be used to synchronize the time misalignment between the RF input signal and the supply envelope signal. Shaping table for designing a constant gain as an example will correlate the PA output power to the transducer power gain at different bias voltages. By specifying a constant gain value, the bias voltage can be read at different PA output powers as illustrated in Figure 8.10.

8.3.1.7 Error vector magnitude

Error vector magnitude (EVM), also called receive constellation error (RCE), is a time domain measurement that evaluates gain and phase accuracy of digitally modulated carriers of a communication system [2,3]. For high spectral efficiency, modern communication systems adopt high levels of M-ary modulation schemes which make their constellation diagrams densely packed and consequently symbols are more susceptible to signal impairments (i.e., power efficiency is decreased). Generally, modulation errors are a result of differences in phase, frequency, or

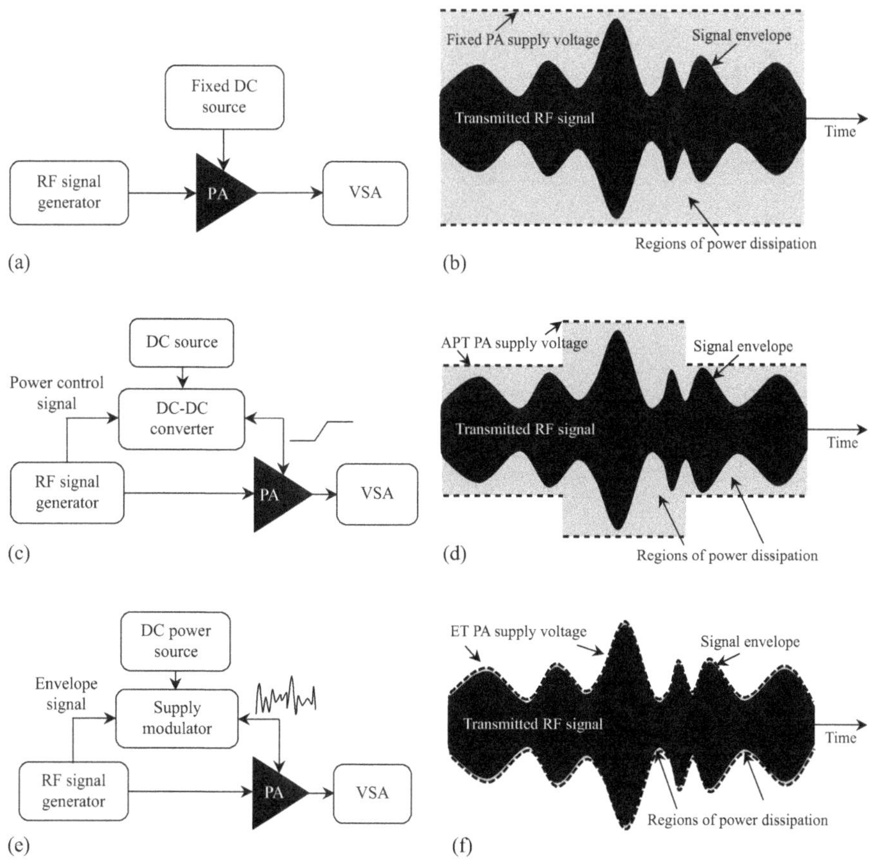

Figure 8.9 Three scenarios of PA bias voltage showing regions of dissipated power (gray regions): (a) conventional PA with fixed bias; (b) PA with APT operation illustrating the adjustment of bias voltage according to the average input power level changes; and (c) PA with ET operation illustrating the adjustment of bias voltage according to the envelope of the input RF signal

amplitude of the signal being modulated. These differences can be due to I-Q mismatch in the modulator or demodulator circuits, multipath fading, PA nonlinearity, and RF interferers that cause EVM to degrade.

As an exemplary illustration, Figure 8.11 depicts an EVM of a QPSK signal where all phasors ideally have the same amplitude. An added noise would misplace the received signal at a different phase and/or magnitude location than the ideal location causing a symbol error. EVM is then the difference vector on the constellation diagram between the ideal and received symbol locations.

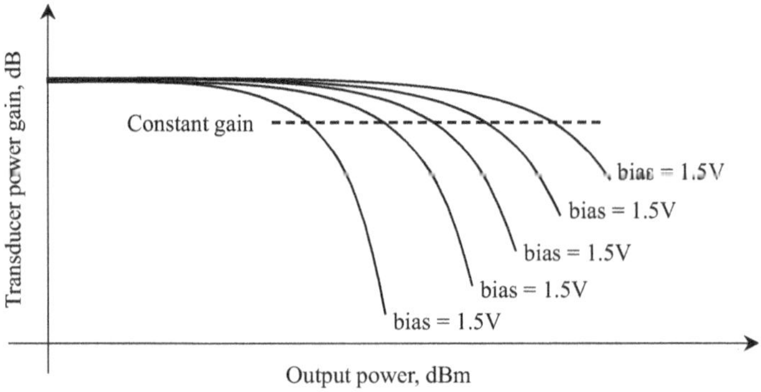

Figure 8.10 PA output power in dBm versus transducer gain used in creating ET shaping table

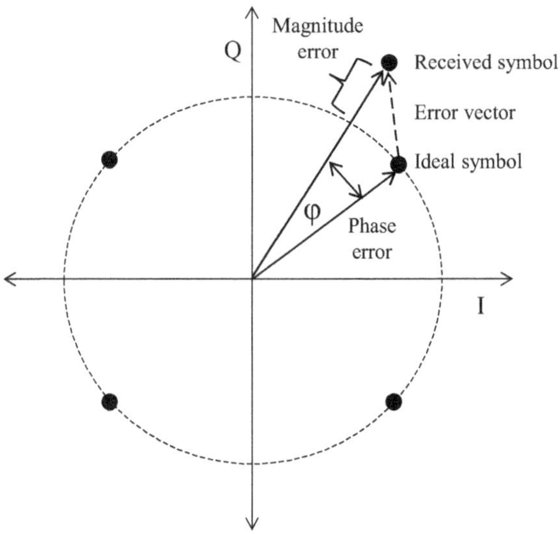

Figure 8.11 Illustration of EVM measurements on a constellation (or I-Q) diagram for a PSK signal

EVM has already been part of several system standards including 3GPP specifications for LTE-advanced technology. In an LTE carrier, EVM is measured at intervals called resource blocks (RBs). The number of RBs is determined by the LTE carrier bandwidth. Although EVM is relatively equivalent to bit error rate (BER), signal-to-noise ratio (SNR), and adjacent channel power (ACP), its importance comes from the inclusion of several system impairments into one quantity which is beneficial in determining the overall quality of an RF system.

EVM is typically measured in dB or percentage and mathematically represented by the power ratio of the root-mean-square (RMS) of the error vector (P_{error})

to the RMS of the reference signal ($P_{avg,ref}$) as in (8.20). Averages are calculated at multiple symbol periods as:

$$\mathrm{EVM}(\%) = \frac{\frac{1}{N}\sum_{i=1}^{N}\left|S_i - S_{\mathrm{ref},i}\right|^2}{\frac{1}{N}\sum_{i=1}^{N}\left|S_{\mathrm{ref},i}\right|^2} \mathrm{X100} = \sqrt{\frac{P_{\mathrm{error}}}{P_{\mathrm{avg,ref}}}} \mathrm{X100}, \mathrm{EVM(dB)}$$

$$= 10\log_{10}\left(\frac{P_{\mathrm{error}}}{P_{\mathrm{avg,ref}}}\right) \tag{8.20}$$

where $S_{\mathrm{ref},i}$ is the normalized *i*th symbol in a stream of measured symbols, S_{i} is the ideal normalized constellation point for the *i*th symbol, and N is the number of unique symbols in the constellation. As discussed in Section 8.3.1.6, PAs in transmitting systems employ different techniques to enhance the system power efficiency. These techniques require fast adjustment to the power supply switching unit as in PAs with envelope trackers or powering and disabling the PA in fast switching times as in Doherty PAs. Delays in switching the PA can cause undesirable transient effects that degrade transmitter performance. Therefore, dynamic EVM is another metric that uses a square waveform to emulate the actual operating conditions of the transmitter transients.

EVM can be directly measured using a VSA which is an instrument that can measure magnitude and phase of an input RF signal at specific frequency and bandwidth. Modern instruments come with ready application program interface (API) for demodulation signal analysis.

8.3.1.8 Adjacent channel leakage ratio

Adjacent channel leakage ratio (ACRL) measurement is one of the universal mobile telecommunications system (UMTS)—3GPP required specification [4] that is used to estimate how much power is leaking from one channel (e.g., a transmitter) to an adjacent and alternate channel (e.g., a receiver). ACLR is an important system parameter that determines the performance quality of co-existing systems operating at adjacent frequency bands. Power leakage occurs because of the imperfection of the transmitter spectrum mask due to the spectral regrowth resulting from the nonlinear effects of the transmitter and the nonideal filter selectivity of the adjacent receiver. ACLR is calculated by taking the ratio of the transmitter integrated power to the integrated power of the adjacent channels. ACLR measurement is generally carried out at two adjacent channel offsets at ±5 MHz and ±10 MHz from the TX channel [5]. It is worth pointing out that ACLR can be defined for any frequency offset and not necessarily those specified above.

The 3GPP specifications require both TX and RX powers to be measured within a filter response. In wideband code division multiple access (W-CDMA) channel carriers, there is a set of test models used to test the transmitter performance known as universal terrestrial radio access (UTRA). Similarly, long-term evolution (LTE) employs a series of signal configurations known as evolved

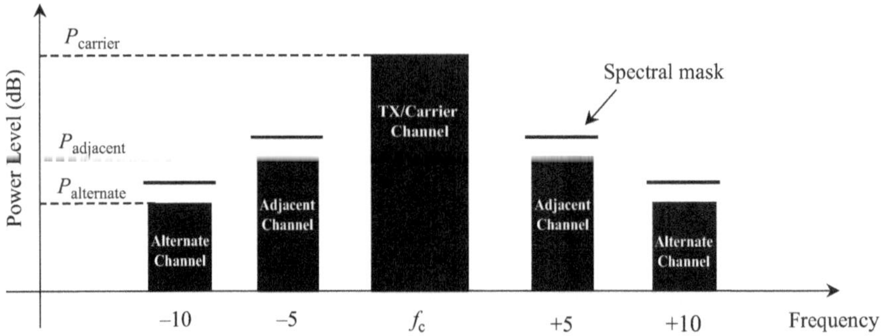

Figure 8.12 Illustrative ACLR response showing the frequency offsets from the carrier frequency (f_c) and power attenuation levels at the adjacent and the alternate channels

universal terrestrial radio access (E-UTRA) to test the transmitter performance. While ACLR of E-UTRA channels are measured using a rectangular shape filter with adjacent channels of the same BW of the carrier, UTRA are measured using root-raised cosine (RRC) filter to measure the power in the respective channel with a roll-off factor of 0.22 and a bandwidth equal to the chip rate of the radio access mode, that is, 3.84 MHz for the carrier BW and the adjacent channels.

Pass and fail criteria of an ACLR measurement is defined by a spectral mask that sets power limit levels at each of the adjacent channels according to the 3GPP specifications. Figure 8.12 shows a general graphical illustration of an ACLR measurement that can be carried out using a spectrum analyzer. ACLR is measured relative to the carrier power and expressed in dBc as:

$$\mathrm{ACLR}(\mathrm{dB}_C) = 10\log_{10}\left(\frac{P_{\mathrm{adjacent}}}{P_{\mathrm{carrier}}}\right) \tag{8.21}$$

8.3.2 Receiver-related measurements

Generally, receiver architecture can be distributed into three main building blocks: a front-end block followed by an intermediate frequency (IF) block and then a back-end block. The front-end block typically includes all of the following components operating at the carrier frequency: receiving antenna, LNA to improve the receiver sensitivity and also has a significant influence on the overall performance of the receiver chain, and RF filters to eliminate spurious signals. The IF stage contains a mixer that down-converts the carrier frequency to the IF frequency through a local oscillator (LO) and an IF filter to eliminate near interferers. The back-end block includes baseband processing and demodulator circuits operating below the IF frequency.

In Sections 8.3.2.1–8.3.2.4, the focus will be on common receiver measurements that are used to evaluate the system performance between the detector input at the back-end module as well as the preceding stages to the input of the receiver antenna of the front-end module.

Noise in communication systems can be defined as any undesired signals that interfere with the signal of interest. Thereby, noise sets a limit on the smallest signal that can be detected by a receiver system. The main sources of noise can be classified into two main categories: external and internal noise. External noise can further be classified as man-made noise such as noise from power lines, electric switches, and several industrial sources or natural sources (e.g., lightning and cosmic rays). The internal noise sources, also referred to as white noise, or thermal noise, are generated by the thermal agitation of electrons in the conductive parts of an RF component. The noise power spectral density of white noise is considered constant over a wide frequency range. The integrated noise power (N_d) of the thermal noise generated in all stages preceding the detector over the bandwidth of the receiver is given by:

$$N_d = KTB \tag{8.22}$$

where K is Boltzmann constant in MKS units and equals to 1.38×10^{-23}, T is the room temperature in Kelvin and equals to 290°, and B is the receiver signal bandwidth. The thermal noise power (dBm) in a 1−Hz bandwidth at room temperature is calculated as:

$$N_o(\text{dBm}) = KT(1\,\text{Hz}) = -174 \tag{8.23}$$

Equation 8.23 demonstrates that doubling either the bandwidth or the absolute temperature will add 3 dB to the available noise power. Moreover, noise power is a linear function of temperature.

8.3.2.1 Noise figure

The noise figure of a component is a unit-less number that quantifies the degradation from the SNR at the input of a device, which can be a single stage or a complete receiver [6] system, to the SNR at the output of the same device as illustrated in Figure 8.13.

The thermal agitation of electrons and their random movement causes thermal noise which in return degrades the device SNR. Assuming an additive noise power (N_{device}), two noise terms defined by the IEEE standard are used in the noise analysis of an RF component or a system. The two terms are: the noise factor (F) which is the linear term for measuring noise power and the noise figure (NF) which

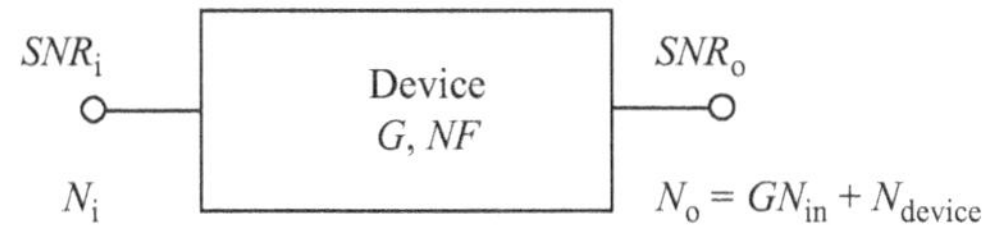

Figure 8.13 A two-port device showing signal-to-noise ratios at its input and output terminals

refers to the logarithmic term of the noise factor and expressed by:

$$F = \frac{\mathrm{SNR}_i}{\mathrm{SNR}_o} = \frac{N_o}{G \times N_i} = \frac{G \times N_i + N_{\mathrm{device}}}{G \times N_i} \tag{8.24}$$

$$\mathrm{NF(dB)} = 10\log_{10}(F) \tag{8.25}$$

Thermal noise has a broadband characteristic that can affect the detection of weak signals by a given receiver bandwidth. Therefore, to integrate different RF and microwave components within a receiver system (i.e., between the antenna and the IF stage), two key parameters need to be considered for each of the integrated elements. These parameters are the power gain (G) and the NF. Here, the power gain is a factor because signal losses introduced by the added circuitry correspond to a rise in the receiver noise figure. Additionally, parasitics due to RF components' packages increase as the frequency increases. Most physical structures of the packages are often dominated by parasitic capacitances that can influence the receiver noise figure.

Figure 8.14 shows a two-port network formed by cascading N components connected in series. The power gain is defined as the ratio of power delivered to the load to that available from the source, that is, the magnitude of the forward scattering parameter $|S_{21}|^2$. The term power gain is associated with active RF devices that need external power source to operate, while in passive components the term insertion loss is commonly used.

The total power gain (G_{Total}) of a system can be calculated using the following equation:

$$G_{\mathrm{Total}} = G_1 \cdot G_2 \cdot ...G_N \tag{8.26}$$

where $G_1, G_2, ..., G_N$ are the linear terms for power gain of the individual components in a system.

From (8.24) and (8.26) the total noise factor of a system of N components can be given by:

$$F_{\mathrm{Total}} = F_1 + \frac{F_2 - 1}{G_1} + \frac{F_3 - 1}{G_1 \cdot G_2} + ... + \frac{F_N - 1}{G_1 \cdot GT_2 \cdot ...G_{N-1}} \tag{8.27}$$

Equation (8.27) implies that the higher the gain and/or the lower the noise factor of the first stage, the lower the contribution of the subsequent stages. Hence, the system NF is controlled by the order of the system components. This is also the reason why an LNA is commonly used as one of the first stages in the design of a receiver system. A straightforward way to measure the NF of a device is through a

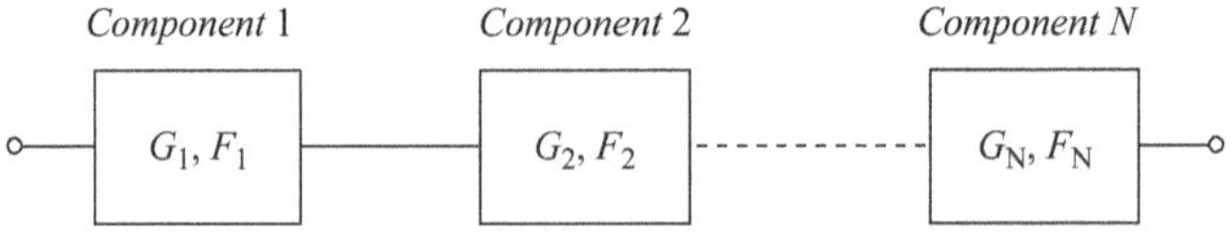

Figure 8.14 System noise by cascading N system components in series

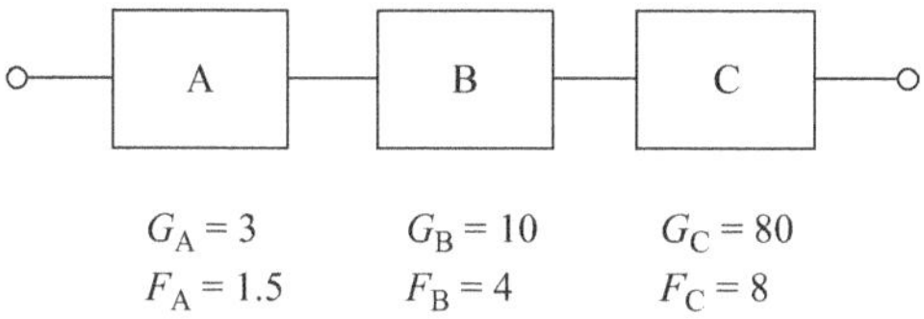

Figure 8.15 Numerical example of system integration by cascading three components, each with its gain and noise figure values

noise figure analyzer where a noise source is used to generate noise at the RF input of the DUT and then precisely measure the SNR at the device output. Since the input SNR ratio due to the analyzer noise source is known a priori, the NF of the device can be calculated.

To demonstrate the importance of identifying the NF for each of the system components besides their order in the system, considering the following three cascaded sub-blocks as illustrated in Figure 8.15 and their gain and noise factor. We will now calculate the system NF for different cascade arrangements of the sub-blocks.

Assuming A-B-C configuration of the three sub-blocks, the overall noise figure is

$$F_{\text{Total}} = 2.733 \quad \text{and} \quad \text{NF}_{\text{Total}} = 4.37\,\text{dB}$$

and assuming B-C-A configuration results in

$$F_{\text{Total}} = 4.7 \quad \text{and} \quad \text{NF}_{\text{Total}} = 6.72\,\text{dB}$$

while assuming C-A-B configuration will result in

$$F_{\text{Total}} = 8.0187 \quad \text{and} \quad \text{NF}_{\text{Total}} = 9.04\,\text{dB}$$

As can be observed, the role of the first stage in an RF system can significantly impact the overall system performance.

8.3.2.2 Receiver sensitivity

The receiver sensitivity is the smallest RF signal level at the receiver antenna input that can produce a detectable signal. Two factors can affect the receiver sensitivity in a wireless transceiver module. First, the TX leakage into the receiver input. Second, the surrounding noise captured by the receiver antenna. Degradation in sensitivity due to TX leakage can be mitigated through the redesign iterations of the module by improving the TX-to-RX isolation level. The picked up noise by the receiver is an uncontrollable factor and largely dependable on the test environment.

The receiver sensitivity can be calculated from:

$$\text{SNR}_d = \frac{G_{Rx} \times \text{Sensitivity}}{N_d} \tag{8.28}$$

where G_{Rx} is the power gain and N_d is the noise power of the receiver system including all stages from the antenna to the receiver input within the receiver bandwidth. SNR_d is the power SNR at the receiver input. Equation (8.29) expresses the NF in decibel of the receiver system in terms of the SNR_d:

$$\mathrm{NF(dB)} = 10\log_{10}\left(\frac{\mathrm{SNR}_i}{\mathrm{SNR}_d}\right) \tag{8.29}$$

$$\mathrm{Sensitivity(dBm)} = -174 + \mathrm{NF(dB)} + \mathrm{BW(dB)} + \mathrm{SNR}_d\mathrm{(dB)} \tag{8.30}$$

The receiver sensitivity in (8.30) is used by many of the RF system engineers for several spread spectrum systems and cellular receivers. It allows designers to choose the proper design tradeoffs in terms of link budget and the receiver parameters (such as LNA gain, linearity, NF, and bandwidth). Systems with high linearity typically produce more noise than those with less linearity. Therefore, finding an optimal balance between noise and linearity performance of a receiving system is the key for meeting the various requirements of a specific application.

Considering the example in Figure 8.16, we would like to calculate the noise threshold at the input of the receiver so that one can determine the minimum detectable transmitter power level. Assuming an effective radiated power (ERP) of the transmitter equals to +85 dBm that is encountering a −220 dB of path loss (PL) in its propagation toward the receiving system. At the receive side, assume a receive antenna has a gain of a 50 dB over a 100 MHz bandwidth and a receiver NF

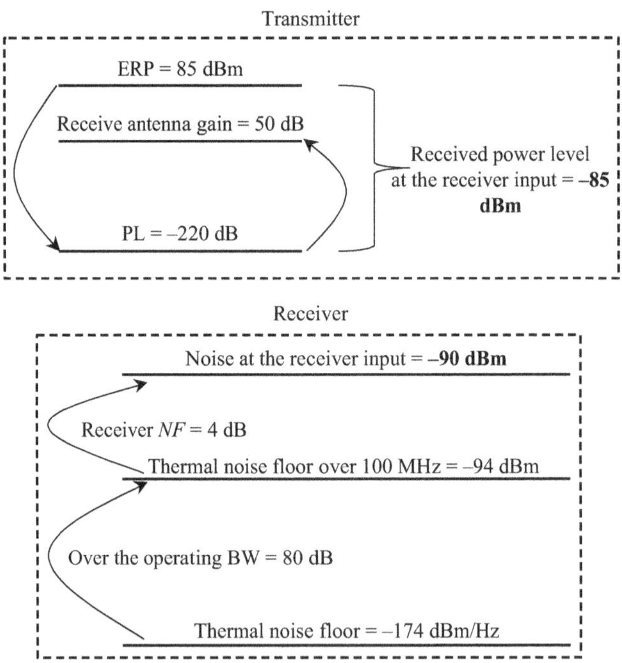

Figure 8.16 Numerical example illustrating the power and noise levels at a receiver input

of 4 dB. Starting from the thermal noise floor at room temperature as calculated in (8.23), we want to find the noise level at the receiver input using (8.30).

As can be seen, the power level at the input of the receiver exceeds the noise floor of the receiver by a margin of 5 dB, which implies that the receiver can properly sense the signal. From this example, one can conclude that the critical parameters that affect the receiver sensitivity are the NF of the receiver, the bandwidth, and the received SNR at the receiver input.

8.3.2.3 Desense

In an FDD system, both transmitter and receiver operate at different frequencies and can operate simultaneously. In such systems, two sources of noise may affect the sensitivity performance of the receiver, external noise directly captured by the system's antenna and internal noise (duplexer feed-thru) due to the simultaneous operation of the transmitter and receiver causing the transmitter noise to spill into the RX frequencies and therefore causing degradation in the SNR at the receiver input. This is also referred to as receiver desensitization or shortened as desense (DS). A strong transmitter noise will affect the receiver ability to sense weak signals and thereby an improvement in the isolation level will be required to recover the receiver sensitivity performance. The duplex desense is defined as:

$$\mathrm{DS(dB)} = 10\log_{10}\left(\frac{\text{Sensitivity when TX is enabled}}{\text{Sensitivity when TX is disabled}}\right) \tag{8.31}$$

In the case of one antenna shared between the TX and RX paths in a system, the mechanism is called duplexer/TX leakage into the RX band, and when separate antennas are used, the DS is due to direct EM coupling between the TX and RX antennas.

For a given isolation level between the TX output and the RX input at the RX frequencies, the noise level that reaches the antenna input (N_a) is calculated by:

$$N_a = 10^{\frac{N_{\mathrm{SSB}}}{10}} P_{Tx} \cdot \mathrm{BW}_{Rx} \cdot \frac{G_{Rx}}{\mathrm{ISO}_{TR}} \tag{8.32}$$

where P_{Tx} is the TX power in watt, N_{SSB} is the single sideband phase noise of the TX synthesizer at RX frequencies in dBc/Hz, BW_{Rx} is the receiver bandwidth in Hz, G_{Rx} is the power gain from the antenna to the RX input, $\mathrm{ISO}_{\mathrm{TR}}$ is the isolation level between the TX output to the RX input at the RX frequencies.

Receiver DS can be mitigated by having highly selective front-end filters, PAs with high IP3, and/or high antenna attenuation between the TX and RX bands.

The measurement setup illustrated in Figure 8.17 is for a receiver DS in a duplexer configuration that has a shared antenna for the TX and RX paths.

The measurement can be carried out by increasing the RF power until the RX sensitivity performance is reached according to a given value that assumes the signal is usable at this level. In a digitally modulated system, the sensitivity value will be given in terms of BER or SNR in analog systems. Typically, desense values of 1 dB or less are considered acceptable.

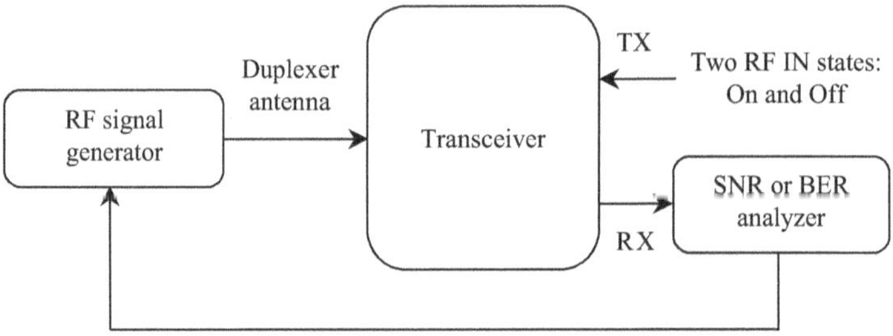

Figure 8.17 Desense measurement setup of a duplex system due to TX leakage into the receiver band

8.3.2.4 Dynamic range

As discussed in Section 8.3.2.2, a receiver has a lower limit to the RF signal that can be sensed. Similarly, a receiver cannot operate with unlimited large RF signal due to the fact that a high signal level will drive the receiver's active circuits out of the usual operating condition or biasing point causing signal distortion. The dynamic range (DR) as specified in (8.33) is the ratio between the largest RF power at the antenna input (saturation power calculated at the RX center frequency) and the minimum detectable power (receiver sensitivity power). Hence, the DR of a receiver is dependent on the system architecture and signal modulation type:

$$\mathrm{DR}(\mathrm{dB}) = 10\log_{10}\left(\frac{\mathrm{Saturation}}{\mathrm{Sensitivity}}\right) \tag{8.33}$$

In order to control the high signal power level, adaptive attenuators may be used for proper operation. The measurement setup for a receiver DR is done in two steps: first the sensitivity is measured and then the saturation is evaluated by increasing the input signal level until the receiver output performance degrades to the sensitivity value. The signal generator power at this stage is the value of saturation [7].

8.4 Radiated immunity and emission measurement

As contemporary RF modules continue to deploy an ever-increasing number of circuitry and components to transmit and receive RF energy for different purposes, an increasing interest in evaluating RF modules for possible susceptibility to electromagnetic interference (EMI) and to ensure their electromagnetic compatibility (EMC) with nearby systems sharing an electromagnetic (EM) environment has become evident. For example, oscillators and electronic switching devices act as potential sources of EMI and such devices can affect the proper operation of nearby systems, for instance, radio receivers, and sensors. Having a successful EM compliance testing at the final product stage requires performing several pre-compliance testing throughout the product development cycles.

One way to test the radiated immunity and radiated emissions of a device is by using a reverberation chamber (RC) as depicted in Figure 8.18 [8]. RC is an enclosure consisting of metal walls with a metallic field stirrer capable of creating plane-wave excitations in almost all angles with different wave polarization. Therefore, an RC can represent a good model for a complex EM environment and is relatively low in cost when compared to anechoic or semianechoic chambers because no absorbers are involved. An RC can be thought as a high-Q metallic cavity with a large number of modes that is capable of continuously varying its boundary conditions. One limiting factor of RCs is that they need to be large so that the fundamental mode resonance of the chamber is well below the frequency range of interest. Typically, large mode mixer or stirrer controlled by a motor spanning the height of the chamber is used to stir the field and to create a statistically uniform field inside the chamber. Hence, this is a deterministic problem where randomness for repeatability is maintained.

In the case of immunity test, the DUT is immersed in an RF field due to a transmitter inside the chamber, and the device performance is monitored to conclude its susceptibility profile. In the case of an emission test, emissions from any part of the DUT are measured through a receive antenna and maximum radiated power is calculated. Most EMC standards measure radiated emissions at a distance of 10 m; however, it is also common for precompliance purposes to measure at 3 m instead and by adjusting the test conditions accordingly. Measuring at a closer distance allows the use of smaller test sites and helps in obtaining an accurate reading of the SNR. Radiated EMI emissions measure the electric field strength which is measured in volts/meter. Most often the units of radiated emission are

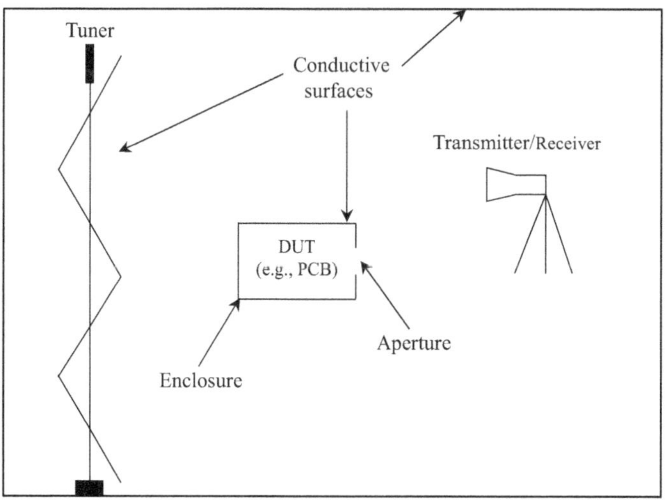

Figure 8.18 Reverberation chamber measurement setup for radiated immunity and radiated emissions

converted into dBμV to compare with the limits of the EMC emissions standards conveniently.

Radiated emissions measurement can be performed as follows:

- A DUT is set to the off mode and a tuner scan is performed to search for signals over the band of interest. Signals amplitude data are then recorded simultaneously at each tuner step.
- A DUT is set to the on mode and a tuner scan is performed to search of the signals which are also recorded at each tuner step.
- Emission data of the DUT when off and on are compared against each other to mark the signals that were not present when the DUT was in the off mode.
- Data are then checked for compliance with the EMC standards to determine if it passes or fails the test.

The large number of field data resulted from the RC test is usually characterized by statistical means such as probability density functions (PDFs) of the real and imaginary components of the electric and magnetic fields.

An alternative, less comprehensive, and less expensive way to identify suspected locations and sources of *E*- and *H*-field emissions from a device can be done through *E*- and *H*-near field probes. While loop shape probes are used to detect *H*-fields, stub probes are used to measure the *E*-fields. Some of the *H*-probes are designed so that they are shielded against capturing the *E*-fields. Probes can be connected to an RF measuring instrument such as a spectrum analyzer to view the field amplitude against frequency.

8.5 Wafer-level RF measurement

When cellular phones became widely and commercially available in the late 1990s, it was inevitable for RFIC manufacturers to cope with the vast demand on delivering large chip/module quantities in a short time span and ensuring performance quality, while maintaining a low operation and test cost. Since then, advanced RF testing methods have been developed and automated to serve the increasing demand.

While meeting the specifications of a design is considered the most challenging task, logistics preparation for mass production and planning for device and system characterization are critical for design wins and surviving a competitive market. This has led the device manufacturers to pursue a very extensive testing that might require up to 100% of the components manufactured to be tested. It is then obvious that for high throughput, tests must be made very fast and accurate.

During the late 1980s, the followed procedure was to dice a wafer and then test a few samples from each quadrant on a wafer. Understanding the inefficiency of this method has led to a vastly improved way of testing devices that are fabricated on wafers using photolithography fabrication methods which can be created in large volumes at a low cost. In this context, the advent of microwave probes was significant to the development of microelectronic RF integrated circuits that enabled precise calibration and thus accurate device characterization. Moreover, high pin-count probe cards are available in the form of ground-signal-ground (G-S-G) probes which are easy to characterize and use at microwave frequencies. With G-S-G probe configuration, probe stations can offer a stable 50 Ohms match to the DUT, non-destructive and highly repeatable microwave measurements conducted at the wafer level. Probe stations are typically calibrated using calibration standards (i.e., short, open, and through) fabricated on the actual wafers.

Probe automation of probe stations with probe manipulators has become prevalent in large volume measurements accompanied with fast measurement instruments (e.g., VNAs). The on-wafer test illustrated in Figure 8.19 can be performed to characterize several important parameters, such as *S*-parameters, power, noise figure, etc. RF on-wafer tests are advantageous over using test fixtures in several aspects. On-wafer measurements are nondestructive and repeatable with excellent accuracy every time a test is performed. On-wafer tests can be easily automated to yield high throughput in a short time span. Such measurement capability can give an early prediction of what a module level performance would look like as well as enabling accurate model extraction for the development of design kit libraries. Calibration of the test equipment is performed at the probe tips, and this eliminates uncertainties that may arise from the use of test fixtures or other added circuitry used for evaluation purposes and not part of the actual device being characterized.

Some considerations need to be taken into account when using this method. For example, probe tips are required to be cleaned up after few thousands of touchdowns. Otherwise, it affects the contact resistance causing an increase in resistance reading and eventually erroneous data. The lack of good grounds near the DUT pins can alter the device performance; therefore correlations have to be made between the actual packaged device performances versus the wafer environment where grounding usually are not as good.

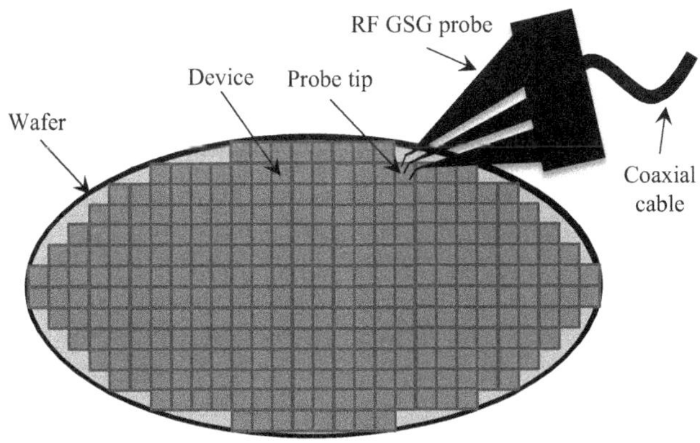

Figure 8.19 Illustration of an on-wafer probe measurement using an RF probe with three points of contacts

8.6 Phase stability in RF test cables

Phase stability of RF coaxial test cables is an important factor that ensures the accuracy of the measured data, especially when measuring high-frequency analog signals or high-speed digital signals, as well as applications that are sensitive to phase variation such as precision phase shifters, phase cancelation circuits, beam forming circuits, and filters in multiplexer configurations. RF coaxial cables support transverse-electromagnetic (TEM) mode of signal transmission. Phase variation occurs due to change in the physical length of the cable that, typically, happens because of the frequent bending and twisting during the normal use or because of the assembly technique of the used cable. Different cable designs and constructions will result in different sensitivity of phase for the same bending radius. Therefore, selecting the suitable test cables with tight tolerance and has the flexibility to withstand the rigors of the test lab environment is critical for providing reliable, repeatable, and accurate test results. Typically, a test cable with $\pm 5°$ of uncertainty in the phase response at approximately 50 GHz, or $\pm 2°$ at 18 GHz, is considered a good test cable. The characteristic impedances of these cables are widely available in two values either 50 Ω or 75 Ω.

There are different measurement setups that can be used to evaluate the phase stability of a test cable. The robustness of these setups relies on their repeatability and reliability. Phase and amplitude stability measurement can be conducted by bending the cable under test around a precision mandrel and then several amplitude and phase measurements over frequency are taken of the cable when it's bent in reference to the cable in its straight condition. Usually, a one-port test method is preferred over a two-port setup where one end of the cable is connected to the VNA and the other end is terminated with a load. This is because the one-port test method provides a linear relationship between phase change and flexure and can be used to measure the cable in its straight state. The measurement accuracy and repeatability can also be influenced by the temperature of the test environment surrounding the cable (typically the room temperature, 25 °C). Thus, monitoring and maintaining a fairly constant test temperature is important to ensure accurate readings.

8.7 Port reduction method

The port reduction method (PRM) is an approach to reconstruct the scattering matrix of a reciprocal n multiport network from the measurement of the network at a reduced order of $(n-1)$-ports using three sets of terminations. This port reduction process can be repeated using two ports of an n-port network [9] or using one port with a precharacterized auxiliary device as demonstrated in [10].

Due to the large number of I/O peripherals of current RF and microwave modules, evaluation of the electrical performance of these modules may be limited by the number of ports of the available VNAs. This approach is beneficial in cases of an incompatible port that cannot be directly connected to the analyzer. Another use of this procedure could be when some interconnecting traces, matching circuits, and embedded LC notch-filters are embedded in the laminate of the module and need to be characterized. During the design iterations, engineers usually try to find causes behind isolation issues and high loss areas in the module by bracketing the embedded designs into small areas. While using field simulators with added ports in the RF module at the suspected areas can provide an answer, more often simulated data of high complex modules are not available or doesn't correlate well to that from the measurements. Therefore, this technique can be used by implementing a number of laminate variants with the required terminations at the location of interest, specifically at port locations that may not be physically accessible by the probing tools in order to calculate S-parameters and thereby troubleshooting the issues.

To explain the underlying principle, consider an example of a two-port VNA with three arbitrarily but known standards for termination and three-port DUT that has a number of ports greater than the VNA. With this assumption, recall from Section 3.4 of Chapter 3 that SFGs can be utilized to obtain both diagonal and off-diagonal elements of an n-port S-matrix while terminating one of its ports with unmatched load. Assuming kth port is terminated with Γ_k, the S-matrix elements of the $(n-1)$-port network can be given by:

$$S_{i,j}^{(k)} = S_{i,j} + \frac{S_{i,k}S_{k,j}\Gamma_k}{1 - S_{k,k}^{k}\Gamma_k}, \quad \forall\, i, \quad j \neq k \tag{8.34}$$

where $S_{i,j}^{(k)}$ is the measured S-parameters of the reduced $(n-1)$-port network with port k terminated by unmatched load, Γ_k, and $S_{i,j}$ is the corresponding S-parameter of the n-port network. To obtain the three-port S-matrix through algebraic manipulation of the two-port measurement data, three sets of two-port measurements are required each with a different termination connected at the kth port of the three-port DUT to solve a set of linear equations. The reflection coefficients of three terminators are given as $\Gamma_{k,A}$, $\Gamma_{k,B}$, and $\Gamma_{k,C}$, respectively. Therefore, (8.34) can be expanded as:

$$\begin{aligned} S_{i,j}^{(k,A)} &= S_{i,j} + \frac{S_{i,k}S_{k,j}\Gamma_{k,A}}{1 - S_{k,k}^{k}\Gamma_{k,A}} \\ S_{i,j}^{(k,B)} &= S_{i,j} + \frac{S_{i,k}S_{k,j}\Gamma_{k,B}}{1 - S_{k,k}^{k}\Gamma_{k,B}} \\ S_{i,j}^{(k,C)} &= S_{i,j} + \frac{S_{i,k}S_{k,j}\Gamma_{k,C}}{1 - S_{k,k}^{k}\Gamma_{k,C}} \end{aligned} \tag{8.35}$$

In the case of a reciprocal network, $S_{i,k} = S_{k,i}$ and (8.35) can be rearranged to have two linear equations in terms $S_{i,j}$ and $S_{k,k}$as follows:

$$\left(\frac{1}{\Gamma_{k,A}} - \frac{1}{\Gamma_{k,B}}\right)S_{i,j} + \left(S_{i,j}^{k,A} - S_{i,j}^{k,B}\right)S_{k,k} = \frac{S_{i,j}^{k,A}}{\Gamma_{k,A}} - \frac{S_{i,j}^{k,B}}{\Gamma_{k,B}} \tag{8.36}$$

$$\left(\frac{1}{\Gamma_{k,A}} - \frac{1}{\Gamma_{k,C}}\right)S_{i,j} + \left(S_{i,j}^{k,A} - S_{i,j}^{k,C}\right)S_{k,k} = \frac{S_{i,j}^{k,A}}{\Gamma_{k,A}} - \frac{S_{i,j}^{k,C}}{\Gamma_{k,C}} \tag{8.37}$$

The other scattering parameters, $S_{i,k}$ and $S_{k,i}$, can then be calculated accordingly by substituting the solved value of $S_{i,j}$ and $S_{k,k}$ into (8.34). However, in the case of a reciprocal network, $S_{i,k}S_{k,i} = S_{k,i}^2$ and solving this quadrature relation ends with two solutions possessing different signs. To solve this sign ambiguity, another $n-1$ measurement is required by terminating a different port other than the kth port. Here, the reflection coefficient of the terminator is not necessarily a known standard.

In summary, the extracted n-port parameters from an $n-1$ measured data set are unique and suitable for both reciprocal and nonreciprocal networks [9].

References

[1] Su D.K., and McFarland W.J. "An IC for linearizing RF power amplifiers using envelope elimination and restoration." *IEEE Journal of Solid-State Circuits*. 1998; **33**(12):2252–58

[2] McKinley M.D., Remley K.A., Myslinski M., Kenney J.S., Schreurs D., and Nauwelaers B. "EVM calculation for broadband modulated signals." *64th ARFTG Conference*; Orlando, FL, Dec 2004. pp. 45–52

[3] Shafik R.A., Rahman M.S., Islam A.R., and Ashraf N.S. "On the error vector magnitude as a performance metric and comparative analysis." *2006 International Conference on Emerging Technologies*; Peshawar, 2006. pp. 27–31

[4] Specifications Groups – 3rd Generation Partnership Project (3GPP). *About 3GPP* [online]. Available from http://www.3gpp.org/about-3gpp/about-3gpp [Accessed Feb 01, 2018]

[5] 3Gpp Technical Specification Group. *User equipment* (*UE*) *radio transmission and reception* (*FDD*) [online]. Available from http://www.3gpp.org/ftp/Specs/archive/25_series/25.101/25101-f10.zip [Accessed Feb 02, 2018]

[6] Friis H.T. "Noise figures of radio receivers." *Proceedings of the IRE*. 1944;**32**(7):419–22

[7] Luzzatto A., and Shirazi G., *Wireless Transceiver Design: Mastering the Design of Modern Wireless Equipment and Systems*. New York, NY: John Wiley & Sons, Inc.; 2007

[8] Bunting C.F. "Shielding effectiveness in a two-dimensional reverberation chamber using finite-element techniques." *IEEE Transactions on Electromagnetic Compatibility*. 2003;**45**(3):548–52

[9] Lu. H.-C., and Chu. T.H. "Multiport scattering matrix measurement using a reduced-port network analyzer." *IEEE Transaction on Microwave Theory and Techniques*. 2003;**51**(5):1525–33

[10] Lin Y.C., and Chu T.H. "Multiport scattering matrix determination from one-port measurements." *IEEE Transactions on Microwave Theory and Techniques*. 2015;**63**(7):2343–52

Index